# ECOLOGICAL THEORY AND INTEGRATED PEST MANAGEMENT PRACTICE

# ECOLOGICAL THEORY AND INTEGRATED PEST MANAGEMENT PRACTICE

EDITED BY

## MARCOS KOGAN

Department of Entomology
University of Illinois

A WILEY-INTERSCIENCE PUBLICATION

**JOHN WILEY & SONS**

NEW YORK • CHICHESTER • BRISBANE • TORONTO • SINGAPORE

*Library of Congress Cataloging in Publication Data:*

Ecological theory and integrated pest management practice.

(Environmental science and technology)
"A Wiley-Interscience publication."
1. Agricultural pests—Integrated control.
2. Agricultural pests—Integrated control—Philosophy.
3. Agricultural pests—Ecology.   4. Agricultural
ecology.   5. Ecology—Philosophy.   I. Kogan, M. (Marcos)
II. Series.

SB950.E36   1986       632'.96       86-9074
ISBN 0-471-82005-9

Printed in the United States of America

10 9 8 7 6 5 4 3 2 1

# CONTRIBUTORS

**Ken A. Bloem**
Department of Entomology
University of California
Davis, California

**Sean S. Duffey**
Department of Entomology
University of California
Davis, California

**Joseph E. Funderburk**
University of Florida
    Agricultural Research
    and Education Center
Quincy, Florida

**Donald C. Herzog**
University of Florida
    Agricultural Research
    and Education Center
Quincy, Florida

**Peter Kareiva**
Department of Zoology
University of Washington
Seattle, Washington

**George G. Kennedy**
Department of Entomology
North Carolina State University
Raleigh, North Carolina

**Marcos Kogan**
Section of Economic Entomology
Illinois Natural History Survey and
    Department of Entomology
University of Illinois
Champaign, Illinois

**Richard Levins**
Department of Population Sciences
School of Public Health
Harvard University
Boston, Massachusetts

**Robert M. May**
Department of Biology
Princeton University
Princeton, New Jersey

**Robert L. Metcalf**
Department of Entomology
University of Illinois
Urbana, Illinois

**David Pimentel**
Department of Entomology
Cornell University
Ithaca, New York

**Paul G. Risser**
Office of the Vice-President
  for Research
University of New Mexico
Albuquerque, New Mexico

**Daniel Simberloff**
Department of Biological Sciences
Florida State University
Tallahassee, Florida

**Donald R. Strong**
Department of Biological Sciences
Florida State University
Tallahassee, Florida

# SERIES PREFACE
## Environmental Science and Technology

The Environmental Science and Technology Series of Monographs, Textbooks, and Advances is devoted to the study of the quality of the environment and to the technology of its conservation. Environmental science therefore relates to the chemical, physical, and biological changes in the environment through contamination or modification, to the physical nature and biological behavior of air, water, soil, food, and waste as they are affected by man's agricultural, industrial, and social activities, and to the application of science and technology to the control and improvement of environmental quality.

The deterioration of environmental quality, which began when man first collected into villages and utilized fire, has existed as a serious problem under the ever-increasing impacts of exponentially increasing population and of industrializing society. Environmental contamination of air, water, soil, and food has become a threat to the continued existence of many plant and animal communities of the ecosystem and may ultimately threaten the very survival of the human race.

It seems clear that if we are to preserve for future generations some semblance of the biological order of the world of the past and hope to improve on the deteriorating standards of urban public health, environmental science and technology must quickly come to play a dominant role in designing our social and industrial structure for tomorrow. Scientifically rigorous criteria of environmental quality must be developed. Based in part on these criteria, realistic standards must be established and our technological progress must be tailored to meet them. It is obvious that civilization will continue to require increasing amounts of fuel, transportation, industrial chemicals, fertilizers, pesticides, and countless other products; and that it will continue to produce waste products of all descriptions. What is urgently needed is a total systems approach to modern civilization through which the pooled talents of scientists and engineers, in cooperation with social scientists and the medical profession, can be focused on the development of order and equilibrium in the presently disparate segments of the human environment. Most of the skills and tools that are needed are already in existence. We surely have a right to hope a technology that has created such manifold environment problems is also capable of solving them. It is our hope that this Series in Environmental Sciences and Technology will not only serve to make this challenge more explicit to the established professionals, but that it also will help to stimulate the student toward the career opportunities in this vital area.

*Robert L. Metcalf*
*Werner Stumm*

# PREFACE

During the 1940s and 1950s pest control, particularly the control of insect pests, consisted primarily of preventive applications of broad spectrum pesticides. The focus of pest control research was heavily chemical/toxicological and pest biology and crop ecology were secondary. The signs of failure of this methodology, however, were already evident by the end of the 1940s, the decade that witnessed the advent of DDT (see Metcalf, Chapter 8, this volume). The agricultural research community of the early 1960s was ready for a change in the direction of pest control technologies. Integrated pest management (IPM) opened an alternative direction. The concept of IPM has quickly captured the interest and imagination of biologists throughout the world.

The 1960s were a period of ecological awakening. Entomologists were eager to search for and adopt an ecologically compatible methodology of pest control. IPM was called applied ecology, and the connection between ecology and IPM was simply taken for granted. In reality, however, only species ecology (life histories and phenologies) and elementary population dynamics were requisite for the design, and even the implementation, of some of the most successful IPM programs. These programs were often detached from contemporary developments in ecological theory. For example, while the controversy around competitive displacement raged in the technical literature, natural enemies were transported around the world with no evidence of advantage or disadvantage to either single or multiple introductions. In fact, interest in the use of parasitoids, predators, and diseases in biological control has increased in recent years, even though the argument focusing on density dependence remains unsettled (see Strong, Chapter 3, this volume). Crops were mixed or rotated simply because the procedure made sense from the point of view of pest natural history and host affinities and not because the diversity–stability hypothesis suggested that approach. Obviously, then, ecological theory does not stand for IPM as particle physics stands for the development of practical uses of nuclear energy. Ecology is not an exact science, and IPM often advances ahead of the establishment and, at times, despite the contradictions, of current ecological theories. Yet, IPM has been hailed as one of the powerful intellectual constructs of recent applied biology. The power of this construct stems precisely from the recognition that ecology is IPM's ideological root. Consequently, the perception that IPM lacks a solid theoretical background is a matter of grave concern to both practitioners and theoreticians of IPM.

We are rapidly approaching the limit of the possibilities of problem-solving strategies that rely solely on empiricism. Modern agricultural systems have created complex environmental, economic, and social problems, the solutions of which may at times not be amenable to direct experimental approaches. Models for those systems are needed, as are better forecasting capabilities. Ecological theory should provide a basis for predicting how specific changes in production practices or inputs might affect pest problems. It may also aid in the design of agricultural systems most appropriate for specific crops under diverse sets of environmental conditions. A theoretical basis is essential for progress in these fronts.

Transition from theory to practice will require considerable adjustment. Much of the theory is not yet robust enough to allow for broad generalizations. Current theory is often reductionist, while IPM requires an expansionist view of nature. Moreover, most ecological theory treats humans as outside observers. Yet in agriculture, the context of IPM, humans are an integral part of the system, often acting as the dominating force. They determine community structure by reducing diversity and inhibiting competition. They control the genetic composition of the dominant plant species and direct inputs of nutrients and water. They enhance crop plant fitness by strengthening plant defenses against herbivores and pathogens. In turn, humans are entirely dependent for their own survival on the success of agricultural production. In Chapter 5, I suggest that humans and crop plants have coevolved into a tight mutualistic relationship in which neither component can survive for long without the other. It is this interdependence that causes agricultural systems to transcend the parameters of most current theories.

The need to explore the connection between IPM and theoretical ecology is therefore evident. Standard ecology texts contain few, if any, examples relevant to IPM, and the few IPM books and the many symposia volumes and review papers restate the need for sound ecological information without demonstrating the absolute dependence of IPM practice on any particular theory, beyond the obvious need to know species ecology (life histories, phenologies, and rudimentary population dynamics).

The review by Levins and Wilson (1980, *Annu. Rev. Entomol.* **25**:287–308) critically appraised this dilemma, and that review provided the seed for the idea for a symposium convened at the annual meeting of the Entomological Society of America, in San Antonio, Texas, December 12, 1984. The idea was to offer a forum in which "basic and applied" ecologists could discuss various aspects of the theoretical foundations of IPM. Within the time constraints of the symposium some relevant areas of theoretical ecology could not be covered; among the omissions were physical ecology, population genetics, and certain aspects of behavioral ecology (most obviously, sexual communication). The areas that were included, however, provided an opportunity for participants to explore approaches in establishing the connections between ecological theory and IPM practice. The papers presented in that symposium were expanded and organized into the present volume.

The volume contains 12 chapters, the scope of which is as diverse as the backgrounds of contributors and their philosophical approach to the topic. In Chapter 1, Richard Levins provides an idealized view of IPM systems that operate with a concern for preserving environmental quality; this opening chapter provides a human dimension to the discussions that follow. Chapters 2, 3, and 4 introduce theory that has been considered critical for the interpretation of population and community level processes. The following five chapters focus on the theoretical background of the major pest control tactics: host-plant resistance, biological control, microbial control, cultural control, and chemical control. Chapter 8 discusses ecological concepts related to arthropod-borne plant diseases and their control. Chapter 11 introduces the economic dimension of agroecology; and finally, Chapter 12 compares and contrasts natural and agricultural ecosystems. At this point readers will recognize that to achieve the level of integration needed for meaningful ecosystems analyses, vastly improved theoretical and experimental bases of IPM will be required.

This book does not claim to represent a beginning or a climactic synthesis of agroecological thought. It is, however, an effort to present an honest assessment of the status of that thought by a group of ecologists. I hope that this effort will stimulate students to devote more attention to the important questions that remain unanswered.

I want to express my sincere appreciation to the contributing authors, to the Entomological Society of America 1984 Program Committee, and to the science editors of John Wiley and Sons for their enthusiastic support. I thank George Kennedy, in particular, for sharing with me some of his ideas that I have used in this preface.

*Marcos Kogan*

*Urbana, Illinois*
*September 1985*

# CONTENTS

12. AGROECOSYSTEMS—STRUCTURE, ANALYSIS,
    AND MODELING                                           321
    *Paul G. Risser*

# ECOLOGICAL THEORY
# AND INTEGRATED
# PEST MANAGEMENT
# PRACTICE

# 1

# PERSPECTIVES IN INTEGRATED PEST MANAGEMENT: FROM AN INDUSTRIAL TO AN ECOLOGICAL MODEL OF PEST MANAGEMENT

*Richard Levins*

## 1.1. INTRODUCTION

Integrated pest management (IPM) has a dual significance in the development of agriculture. On the one hand, it is an alternative to the brute force approach of pesticide saturation. From this perspective, it is an intermediate step along a path from a high-intervention, single-goal, industrial model of agriculture toward a gentle, ecologically and humanly rational production system. It shows the first softening of a stance of hostile confrontation with all of living nature except the crop, and a groping toward a strategy of détente and coexistence with most species. Some aspects of this sequence are shown in Table 1.1.

But IPM is also offered as an alternative to biological and natural control. This side of IPM is a rearguard action by industry in response to the criticism of pesticide use, an effort to preserve and even intensify the commoditization of pest control by absorbing some of the insights of agroecology to invent new products while deflecting the criticism of chemicalization.

Therefore, the development of IPM is not a smooth unfolding of knowledge and technique but rather an uneven, erratic course, the result of powerful encouraging and retarding influences that vary in space and time and that bring people together in complex adversarial and cooperative relations. One symptom is a diversity of definition of IPM. While Corbet (1981) defines IPM as "a method of pest management which decreases (and perhaps even avoids) the use of nonselective methods of suppression...," a USDA (1982) annual report on

1

**Table 1.1 Approaches to Pest Management**

| | Industrial | Present IPM | Ecological Agriculture |
|---|---|---|---|
| Goal | Eliminate or reduce pest species | Maximize profits | Multiple economic, ecological and social goals |
| Target | Single pest | Several pests around a crop and their predators | Fauna and flora of a cultivated area |
| Single for intervention | Calendar date or presence of pest | Economic threshold | Multiple criteria |
| Principal method | Pesticide | Prevention by plant breeding and crop timing, careful monitoring, and multiple interventions | System design to minimize outbreaks and mixed strategies |
| Diversity | Low | Low to medium | High |
| Spatial scale | Single farm | Single farm or small region defined by pest | Agrogeographic regions |
| Time scale | Immediate | Single season | Long-term steady-state or oscillatory dynamics |
| Boundary conditions | Everything as is: crops, cropping system, land tenure, microeconomic, decision rules, social organization | Major crops, land tenure, and decision rules | Societal goals |
| Research goal | Improved pesticides | More kinds of interventions | Minimize need for intervention |

agricultural research includes under the task labeled IPM "...(the combination) of two or more pest suppression methods into practical systems of IPM to reduce pest problems" and another author prefers to avoid the specific content of the methods by retranslating IPM as "intelligent pest management" (Zweig and Aspelin, 1983).

The development of IPM was encouraged by criticisms of the trajectory of our agricultural technology as a whole coming from diverse sources and by advances in the relevant disciplines that made new approaches possible. The critique of the high-tech industrial model of agriculture is not of pesticide alone, but of the package of which it is a part. It makes the following points:

**1.** Damage due to pests has certainly not decreased since the 1940s and has more likely doubled (Pimentel, 1986).

**2.** The high-tech package undermines its own productive base through the evolution of pesticide resistance; the loss of genetic diversity for plant breeding; the loss of soil fertility through erosion, salinization, compaction, and trace element depletion; the mining of water; and the creation of new pest problems.

**3.** Increased vulnerability of agricultural production to natural and human-made uncertainties.

**(a)** Monoculture, especially in the tropics, creates an ideal target both for the invasion by existing pest species and the adaption of natural populations to the introduced crops.

**(b)** Extension of the growing season through irrigation allows year-round increase of pest populations.

**(c)** Plant breeding focused narrowly on yield under optimal conditions makes heroic measures of protection more necessary. This is especially true when crops are introduced outside their regions of adaptation.

**(d)** Particular practices, such as high nitrogen application, make the crops more attractive to herbivores.

**(e)** Increased dependence on imported inputs for agriculture makes production vulnerable to fluctuations in international prices, especially those of energy. Thus, the politics of the Middle East become part of the environment of the roots of wheat in India, and uncertainties in pesticide supplies exacerbate uncertainties about *Heliothis* populations. This vulnerability is increased even beyond market fluctuations when trade becomes a political weapon and economic blockades threaten tomato seedlings. In general, interdependence without equality leads to subordination.

**(f)** World price trends of inputs compared with crops make the balance of payment less favorable to Third World agriculture.

**4.** The energy efficiency of production tends to decrease as inputs increase.

**5.** Chemical pest control has harmful impacts on the health of agricultural labor and on the general population. A general rule seems to be that the more oppressive the government is on behalf of the landed oligarchy or agribusiness, the weaker the protection of the populace from pesticide poisoning, which now affects more than half a million people annually. Central America and other cotton-producing areas show especially high levels of exposure to toxic substances.

**6.** Damage to other species and to the environment through eutrophication of lakes, siltation of reservoirs, pesticides, and deforestation.

**7.** The high-tech package usually increases social inequalities. Differential access to inputs or credits or knowledge or transportation increases the variances among farms, and often results in land concentration, eviction of tenants, and increases in urban unemployed.

**8.** The high-tech industrial model operated from a very narrow intellectual base that leaves it especially unprepared for unusual events. Research specialization often leads to plausible measures in the small that were harmful in the large.

Since I work at a school of public health, I am impressed by the parallels between agriculture and medicine, their patterns of development, and the criticisms they evoke. The entomologists' secondary pest is analogous to iatrogenesis, physician-caused disease. Acquired resistance to pesticides parallels the acquisition of antibiotic resistance by hospital populations of bacteria and by the malarial parasite. In both areas, research has often been guided by the goals of product development rather than the solving of problems, and the adoption of a technique has depended on the sales efforts of the detail men. Therefore, the rising costs of intervention have become increasingly important problems. In both fields a high-tech, magic bullet approach has led to the deprecation of earlier knowledge and to the compartmentalization of research.

Therefore, in both areas critical, alternative movements have arisen seeking more holistic understanding, gentler interventions, and more equitable access to knowledge and services. Despite occasional lapses into sentimentality or mysticism and some errors of detail, these critical movements should be studied seriously. They are an important corrective to existing biases and point toward exciting possibilities.

But there are also powerful forces retarding the development of a gentle agroecological technology. The most important of these is the agricultural chemicals industry. With a world-wide market on the order of $10–12 billion in 1980 (Zweig and Aspelin, 1983), it has a major stake in resisting those changes in technology that would reduce sales. Industry's retarding role is expressed in four ways: First, it is a leading force in pest control research, setting a research agenda aimed at finding ways of turning oil into something to sell farmers rather than exploring the best ways to reduce pest damage. The widespread use of pesticides also sets the research problems for others. The economic analyses of pest control concentrate on pesticide use, comparing it mostly to no pest management, since that is where the data are available (USDA, 1982). Second, the sales efforts by industry representatives swamp the alternative information offered by the extension service, which itself is not uniformly enthusiastic about pushing them. Third, industry actively wages a propaganda campaign in defense of its products and belittling criticism of the pesticide syndrome. Finally, as the chemical industry acquires seed companies, it is likely that complete packages of technology will be marketed within which it will be difficult to substitute alternative practices.

In Third World countries, other obstacles to ecological and social rationality are at work. Often the sales efforts of private industry supplement the advice of international agencies which, accompanied by aid in development packages, have the force of command.

These external influences are able to have so much impact because of an excessive deference toward what seems to be "modern," "advanced," or "scientific." This viewpoint, known in Latin America as developmentalism, imagines progress to take place along a single direction from less advanced to more advanced. The task of the less advanced is to become more advanced, that is, to catch up with the industralized former masters in those practices that are coming under increasing criticism at home. The high cost of research, the sense of urgency to meeting pressing needs, and the weakness of their own research facilities all conspire to encourage an uncritical, passive reliance on the most popular technologies from abroad.

In general, it seems to be the case that those countries that have broken most completely with their colonial past are also most willing to look critically at existing technologies and to undertake new directions in pest control, but the correlation is by no means perfect.

The third major retarding force is the compartmentalization of knowledge and decision making. At the highest level, the minister of health and the minister of agriculture do not talk to each other so that the questions are rarely posed: How will pesticide use against *Heliothis* in cotton affect the incidence of malaria? What will the new irrigation system do to schistosomiasis? Where will the tsetse fly spend its time when gallery forests are destroyed and bananas planted? What will the mosquito control program do to tachinid parasitoids? Within agricultural research, each department takes as its givens the existing practices coming from the department next door. Thus, the engineer designs machinery for use in monoculture because that is the current practice. Plant breeders select for performance in monoculture because we do not have machinery for handling polyculture. Entomologists do not recommend the polyculture because current varieties yield better in monoculture; and farmers plant monoculture because that is what has an already existing technology. The point is, each decision separately makes sense given the rest of the system. In the small, the technological trajectory seems rational to the point of inevitable. But, nobody evaluates the trajectory as a whole, which is determined largely by external forces.

I suspect that most irrationalities are of this kind. They make sense in the small, and their great failing is in accepting the existing boundary conditions as fixed and natural.

## 1.2. DIRECTION OF CHANGE

Some of the major axes of change are shown in Table 1.1:

**1.** From the single goal of eliminating insects, through the goal of profit maximization by a calculus of economic thresholds, toward multiple goals that combine production, environmental protection, health protection, economic independence, and social equity.

Further, these goals must include not only the average consequences but also the variance over time and across farms. Suppose, for example, that we want to insure yields even in worst cases. Then we examine

$$Y_w = \bar{Y} - K\sigma \tag{1}$$

for yield $Y_w$, average yield $\bar{Y}$, variance $\sigma^2$. For $K = 2$ this is the expected 40-year low, for $K = 3$, the 200-year low, and so forth. Now assume that the variance is proportional to the mean. Then the worst-case yields would be

$$Y_w = \bar{Y} - K\sigma_0\sqrt{\bar{Y}} \tag{2}$$

The effect of simple increase in mean yield is seen by taking the derivative with respect to $\bar{Y}$:

$$\frac{\partial Y_w}{\partial \bar{Y}} = 1 - \frac{K\sigma_0}{2\sqrt{\bar{Y}}} \tag{3}$$

Therefore, if $\bar{Y} < K^2\sigma_0^2/4$, yield increases would increase risk. Whether the risk is harmful or not depends on how the unit of production relates to the unit of distribution. If the unit of production is a small plot within a farm, then the consequences can be absorbed within that farm. If it is as large as a farm or several farms, environmental fluctuations may be disasters for local farmers but the national or regional food supply may not be endangered at all. In that case, a crop insurance or other redistributive scheme might be preferable to agronomic measures. Since the worst-case yield over $n$ independent units would be

$$Y_w = \bar{Y} - \frac{K\sigma_0}{\sqrt{n}}\sqrt{\bar{Y}} \tag{4}$$

redistribution lessens the contradiction between average yield and risk.

In the United States, we are mostly at the stage of profit maximization subject to secondary constraints, but in developing countries multiple concerns are more common.

**2.** From concern with a single-target pest, IPM moves on to consider the ensemble of pests around a crop. In developing countries where new areas or seasons are being opened up to agriculture, not only the actual pests must be considered but also the potential pests. This requires more than a taxonomic search. Biogeographic and evolutionary approaches are required. For example, Strong (1979) and others have shown that the number of herbivore species utilizing a plant species increases with the area occupied by the host. Therefore, any scheme for large-scale production must anticipate invasions and adaptations. Further investigation helps identify potential invaders or adapters: Moran (1983) found for South Africa that introduced crops were subject to attack more or less equally by native or introduced insects, whereas native plants faced mostly native herbivores. The inconsistent effects of conservation tillage, which helps both herbivores and their natural enemies, requires systematic analyses. An examination of host ranges for different insect groups would be another step. For instance, see Price's (1983) discussion of herbivore specialization.

Evolutionary and biogeographic ecology are essential for the planning of new agricultural schemes to minimize the big surprises. In the absence of such research, we have been caught time and again by new pest problems that emerge in the wake of new technologies or the cultivation of new areas. The eruption of the brown plant hopper around green revolution rice in the Philippines, stinkbugs on Nigerian soybeans, and stem borers on sugarcane in Ghana (Sampson and Kumar, 1983) are only a few of the more dramatic examples.

**3.** The signal to intervene changes from the mere presence of an insect or even just the calendar date, through economic thresholds derived from microeconomic models and single population projections, to multiple criteria.

**4.** The principal methods change from high-intensity pesticide therapy in the industrial model, through careful monitoring and frequent interventions with multiple tools, and then on toward an emphasis on system design aimed at self-regulating agroecosystems which require minimum intervention with gentle technologies. This strategy will include an evolution from the low diversity of monocultures through the medium diversity of monocultures supported by trap crops or repellent plants, to a system of high plant and animal diversity on different spatial scales. The unit of plant protection will be not the single homogeneous field or a single farm, but rather an agricultural region. And the time horizon must shift from the immediate "name'em and squash'em" approach through within-season population projections to a long-term dynamic view which only makes sense for a larger geography.

**5.** Traditional pest control works within narrow boundary conditions. The crop and cropping system, land tenure, and economic structures are taken as givens, leaving the entomologist few degrees of freedom for innovation. IPM opens up the cropping system for modification at least in terms of planting dates, tillage, varieties, plant spacing, crop rotation, and protective crops. The economy is still treated as given but subject to some intervention through price supports, subsidies, or insurance. In developing countries a broader set of questions can be raised: geographic patterns of land use, choices of crops, land tenure, and criteria for decision making at national and local levels. In the long run, any incompatibility between the ecology and the economy will have to be re-solved in favor of the ecology.

**6.** The goal of research shifts from the improvement of pesticides and other intensive practices through the search for more kinds of interventions, toward design that reduces the need for intervention. This greatly broadens the neces-sary knowledge base. Technologies that override natural processes require less knowledge: If you are going to kill them all, you don't have to know them well. Thus, the industrial pesticide approach depends mostly on taxonomy and toxi-cology. Present-day IPM adds on population dynamics, behavior, some agrocli-matology, plant genetics, and microeconomics. Future developments toward less frequent, gentle interventions will require more understanding of weaker forces, direct pathways, and larger, more complex systems.

## 1.3. RESEARCH STRATEGY

The strategy advocated here will require that we work with more complex sys-tems in which what happens depends not only on the response of individual or-ganisms to natural or human inputs but also on how these initial effects perco-late through the network of interacting species and other variables. Often we will be ignorant of the details of those interactions but can still make plausible infer-ences about the behavior of the whole. To do this, several directions of research are necessary:

First, we have to map out the pattern of interactions at the community level. There has been much work by ecologists around the notion of community, and debate about whether or not we are dealing with integrated wholes. As is usually the case in ecology, the answer seems to be sometimes yes, sometimes no. Most species in a local ensemble probably do not interact directly but may be linked through indirect pathways in such a way as to present unexpected consequences of initial impacts. For example, consider the hypothetical network shown in Fig. 1.1 in which $C_1$ and $C_2$ are the biomasses of plant species with self-limiting growth, $H_1$ and $H_2$ are herbivores, and $P_1$ and $P_2$ are predators. The links show interactions, $\rightarrow$ indicating a positive direct effect and $-\circ$ a negative direct ef-fect. Note that $H_1$ and $P_1$ are specialists on single resources whereas $H_2$ and $P_2$

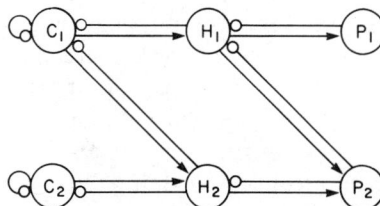

Fig. 1.1 A hypothetical community.

are generalists. The behavior of this system depends on the various parameters that determine growth, reproduction, mortality, and predation. If any of these change, the equilibrium values of the variables will be changed in the direction shown in Table 1.2. (The procedure for these computations is given in Levins, 1975.) Each row shows the direction of change of the equilibrium values of variables when a positive impact enters the system by way of the indicated variable. If there is not equilibrium, the average values will change more or less in those same directions but with some modification.

Note first that almost half the entries are zeroes. Variables that are causally connected may nevertheless not respond to some changes of parameter. The herbivores $H_1$ and $H_2$ are least susceptible while all inputs to the system affect the predators. The predators serve as sinks that absorb impacts passing through the herbivores. The occurrence of sink variables in community networks is an important property that should be searched for since they often thwart our expectations. Note also that $H_1$ responds to changes (e.g., increased reproduction) in its specialized predator $P_1$ but not to changes in $P_2$, whereas $H_2$ is affected both by its own predator $P_2$ and by $P_1$, although $P_1$ does not prey on it. Conversely, changes entering the system at $H_1$ and $H_2$ affect only the predators, the sinks that absorb all the impact. The question marks in entries for the plants are caused by pathways of opposite sign between $C_1$ and $P_1$ ($P_1$—o$H_1$—o$C_1$ is positive but $P_1$—o$H_1 \rightarrow P_2$—o$H_2$—o$C_1$ is negative) and between $P_1$ and $P_2$. Note also that although $H_1$ and $H_2$ are competitors in the sense that they consume the same resource $C_1$, nevertheless conditions that make life easier or harder for one of them will not affect the abundance of the other. The only directly causal correlation between them is generated by factors acting on $P_1$ which is not a shared predator.

The direction of change is shown in Table 1.2. The magnitude of the change depends on the strengths of the various pathways in the graph. In particular, all the effects are divided by the feedback of the system, the determinant of the community matrix. In this case, the feedback is the product of the $(H_1, P_1)$ loop,

Table 1.2 The Response of Equilibrium Levels of Variables to Positive Parameter Changes Entering the System by Way of each Variable

| Positive Input to | Effect on | | | | | |
|---|---|---|---|---|---|---|
| | $C_1$ | $C_2$ | $H_1$ | $H_2$ | $P_1$ | $P_2$ |
| $C_1$ | + | 0 | 0 | 0 | ? | + |
| $C_2$ | 0 | + | 0 | 0 | − | + |
| $H_1$ | 0 | 0 | 0 | 0 | + | 0 |
| $H_2$ | 0 | 0 | 0 | 0 | − | + |
| $P_1$ | ? | − | − | + | + | ? |
| $P_2$ | + | + | 0 | − | ? | + |

the ($H_2$, $P_2$) loop, and the self-dampings of the $C$s. Any strategy for reducing the susceptibility of yield to any external changes would focus both on the introduction of appropriate sinks in the system and on strengthening the feedback of the network. It is not necessarily our goal to maintain the insect populations as constant. Rather, they serve as homeostatic variables, analogous to the fluctuations in insulin level that stabilize blood sugar.

Of course, drawing a graph does not prove nature behaves that way. This hypothetical example under restrictive conditions is intended only to demonstrate that research is necessary to determine the actual patterns of interaction and to explore the behavior of those patterns, to find out what happens when the restrictive assumptions are relaxed, and to identify quickly graspable graph properties such as sinks and feedbacks that will allow us to develop the intuitions suitable to the new problems.

Although the example just presented was a graph based on trophic relations, with species as variables, it is necessary to assert that a community is more than a foodweb, and nontrophic links may be of major importance in the search for détente with the insects.

Some of the nontrophic effects that may be important are:

1. The stimulation of plant growth by moderate herbivory. This is especially the case for root nibblers or the induction of tillering by shootflies and other herbivores.

2. Extension of harvest season by removal of flowers. This would be useful for small farms serving schools, hospitals, and other institutions.

3. Late defoliation to enhance ripening, as in tomatoes and apples.

4. Stimulation of plant defenses. Cross resistance as between *Fusarium* and *Verticillium* in tomato, may be more common than now realized.

5. Alteration of microenvironment, as when consumption of senile leaves increases aeration at the base of tomatoes and reduces late blight.

6. Attraction of predators who then feed on more damaging pests. This is particularly important when the predators are ants. Small, easily captured prey such as *Drosophila* can bring out a sufficient density of foragers to mob the more difficult lepidopteran larvae.

7. Harrassment by unsuccessful predators will often cause the herbivores to keep moving, sometimes to unusual microsites. We have observed roaches, who usually seek out cozy shelters, spend their time up on the vegetation when pursued by ground foraging ants in laboratory populations. Harrassment may disrupt feeding, expose herbivores to more effective predators, cause them to emigrate, and increase the spread of insect-borne diseases.

8. Facilitation of predation. The smooth, round eggs and pupae of many insects prevent ants from grasping them. But any roughening of the surface, such as by the growth of that which do not penetrate the surface, provides ants with a grip.

9. The leftovers from predation provide food for omnivores, building up populations.
10. Nutritional conditions related to plant quality or type of prey affect the size, developmental rate, and feeding behavior of insects.
11. Creation of breeding or overwintering sites, as by weed borers.


Therefore, before we decide on the value of a species in a particular community, we should examine for the less obvious, nonconsumer effects, and include them in our models. Furthermore, these nontrophic effects can be selected for or enhanced, making more of the fauna useful. It is often claimed in the medical context that in nature, host–parasite systems evolve toward more resistant hosts and less virulent parasites. The rationale for this belief is that a parasite that kills its host quickly will have less opportunity to reproduce and find a new host. This is not universally true. The parasite has several ways of increasing its own fitness: The overcoming of most defenses would seem to increase virulence. But the avoidance of the defense system by accumulating in the peripheral tissues would reduce virulence while increasing opportunities for spread. The point is, we could join in the evolution of the parasite, selecting ways that remove the opposition between plant and parasite fitness. This could be done by selecting for time of infection, location within the plant, or specific metabolites.

As we move from single-year optimization to long-term strategies, we have to develop dynamic models that include the farmers' decisions as covariables with natural components. Farmers enter the system in several ways. For instance, by responding to the abundance of one pest they affect the survival of other species and therefore establish indirect interaction linkages between species that may never meet each other. If one pest is an early-season hervibore and the other late, the first species encourages later planting, which favors the second. But if they are both early, the abundance of either will cause measures making life worse for the other. Thus, farmer decisions can create either competitive or mutualistic relations between species.

Farmers' efforts to control pests may give rise to population cycles. When the Hessian fly is abundant, resistant varieties of wheat are sown. But in the absence of the fly, these varieties yield less than the susceptible genotypes. Farmers abandon the resistant varieties once populations decline. Then, of course, the fly comes back. The interests of the individual farmer and of all the farmers alternately coincide and conflict, determining the oscillatory dynamics of the pest.

The following models explore what happens when pest control measures are undertaken based on economic decisions. First, consider a single species of herbivore that reproduces locally within a field and also immigrates from other areas. Its rate of change may be given by

$$\frac{dH}{dt} = H(r - p) + m \qquad (5)$$

or

$$\frac{dH}{dt} = H\left(r - p + \frac{m}{H}\right) \tag{6}$$

where $r$ is the local growth rate (which may be negative), $m$ the immigration, and $p$ the mortality induced by pest control. Under constant conditions, there would be an equilibrium at

$$\hat{H} = \frac{m}{p - r} \tag{7}$$

if $p > r$.
   Since

$$\frac{\partial}{\partial H}\left(\frac{dH}{dt}\right) = \frac{-m}{H} < 0 \tag{8}$$

the equilibrium is stable. Now suppose that the pesticide application or other induced mortality $p$ depends on the price of the crop and, therefore, varies for exogenous reasons. This will induce population variation in $H$, and taking average values, we get

$$mE\left(\frac{1}{H}\right) = \bar{p} - r \tag{9}$$

where $E$ means average or expected value. But

$$E\left(\frac{1}{H}\right) = \frac{1}{\bar{H}} + z > \frac{1}{\bar{H}} \tag{10}$$

where $z$ is a positive increment due to variation. Therefore,

$$\bar{H} = \frac{m}{p - r - mz} \tag{11}$$

so that the average pest population is increased by the variation induced by tracking the economy. Furthermore, the covariance of $H$ with $p$ can be shown to be negative. When prices are low, $H$ is high, so that greater damage conspires with low prices to make bad years worse. Thus, the attempt to maximize profits actually results in higher average herbivore levels, and, therefore, may lower average yields along with bigger variance. This result would hold for any decision rule that determines $p$.
   Now suppose instead that the herbivore reproduces entirely locally but is

preyed upon by a migrating predator or parasite with population size $N$. Then we might represent the dynamics by

$$\frac{dH}{dt} = H(r - p - aN) \tag{12}$$

$$\frac{dN}{dt} = N\left(aH - \theta + \frac{m}{N}\right) \tag{13}$$

where $a$ is a predator rate and $\theta$ the death rate of the predator. Now if $p$ varies due to prices or weather,

$$\bar{H} = \left[\theta - mE\left(\frac{1}{N}\right)\right] \Big/ a. \tag{14}$$

The average value of $H$ is therefore reduced by the fluctuations. Predator and prey will have a positive correlation, so that during outbreaks mortality will be due mostly to predation. But $p$ will be negatively correlated with $H$: Most pest control efforts will coincide with population lows. The two sources of mortality will be out of phase with each other, reducing the amplitude of fluctuation induced by price instability.

One final example: Consider an immigrating herbivore with a predator as before, but with predation showing a Holling type-2 functional response (Holling, 1965). Let

$$\frac{dH}{dt} = H\left(r - p + \frac{m}{H} - \frac{aN}{K + H}\right) \tag{15}$$

and

$$\frac{dN}{dt} = N\left(\frac{aH}{K + H} - \theta\right) \tag{16}$$

This system could have an equilibrium at

$$\hat{H} = \frac{K\theta}{a - \theta} \tag{17}$$

which is independent of $r$ and $m$. However, this will only be stable if $\partial(dH/dt)/\partial H$ is negative or

$$m > \frac{aKN}{(K + H)^2} H^2 \tag{18}$$

Therefore, the setting up of traps or windbreaks to reduce immigration could actually destabilize the dynamics and result in wide fluctuations and an increased $\bar{H}$.

There have been relatively few studies of long-term agricultural production with pest and economic variables. However, fisheries scientists have examined harvesting strategies in which present yield affects future yield. One important innovation in this field is the recognition that our intervention affects not only the average yield but also the stability of the dynamics so that criteria of stability must supplement the goal of maximum sustainable yield. A general observation is that a fixed harvest goal without regard for the state of the population has a destabilizing effect (Beddington and May, 1977). In the agricultural context, the resource that is harvested may be nutrients, water, or even pest tolerance of the system.

Agroclimatological studies have usually concentrated on conditions affecting plant growth or projections of separate insect populations. Average conditions are used for biogeographic purposes, whereas for year-to-year decision making, the actual conditions of that year are used. The statistical distribution of weather conditions has not been explored sufficiently.

An outstanding exception is the work done at ICRISAT (the International Center for Research in the Semi-Arid Tropics). Workers there (Virmani et al., 1980) found that two areas with similar synoptic conditions differed very much in their suitability for particular crops because of differences in the conditional probabilities of a rainy week given a previous rainy week, or of the three consecutive dry days for harvesting at the right time after sowing.

This kind of analysis should also be adapted for insect populations. The end of the dry season is often the signal for the emergence of herbivores as well as for planting. Since rain and temperature control the timing of life-cycle phenomena, their reliability will also determine the correlations between populations and plant growth. A slight acceleration of plant growth relative to herbivore might allow it to escape from serious damage. A slight acceleration of a predator might cause a crash in the prey population. These small shifts can happen easily: Species differ in their temperature threshold for development, the temperature for their microhabitats, and their degree-day requirements.

Suppose that $T_1$ and $T_2$ are the actual temperatures where two species are located, $S_1$ and $S_2$ their thresholds for development, and $R_1$ and $R_2$ the required degree-days for completion of the stage of interest. Then the difference in time of completion of development is

$$D = \frac{R_1}{T_1 - S_1} - \frac{R_2}{T_2 - S_2} \tag{19}$$

The statistical distribution of this bioclimatological measure may be an indicator of the reliability of a control organism in a given location. It would indicate why the same parasite–host pair gives good control in some places and erratic control elsewhere. It would suggest how to choose a set of control organisms to cover environmental uncertainties. To some extent, the actual temperatures are modified by the activity of the herbivores since defoliation affects temperature within a canopy. They are also manipulated by farmers. The thresholds are aspects of temperature adaptation. They may be amenable to selection or conditioning by pretreatment.

The point here is that a gentle technology requires more relevant indices of environment as defined by the biologies of the organisms.

## 1.4. TOWARD DÉTENTE

The expanding knowledge of the workings of agricultural systems as biosocial wholes can lead us in two quite different directions. Knowledge may become increasingly a commodity owned by pest management consultants and data processors. It would be used to create a highly productive but vulnerable agricultural system requiring more external inputs to maintain itself against the problems it creates, and reducing still further the autonomy of the farmers especially in Third World countries.

But it could also develop with a commitment to a gentle ecological technology and evolve from a strategy of confrontation with the insects to one of détente and coexistence based on careful system design and minimum intervention.

If we follow the latter course, we would recognize that we cannot control everything that happens in a region or even a field. We would not attempt to maintain community stability by heroic efforts. Rather, our design would include sinks and feedback loops that divert the effects of the weather or population outbreaks away from harm to the crop and have them absorbed mostly by other variables.

We could reduce the frequency of major outbreaks by herbivores, but not prevent them completely. The frequency of population peaks of some herbivore would be seen as an operating characteristic of the system, a parameter that we could partly reduce and partly compensate for by insurance and redistribution and by depending on a diversity of crops that differ in their vulnerabilities.

Of course, not all pests can be managed in the gentle way advocated here. Robert May (Chapter 7) cited the desert locust as an example of a species whose dynamics is far from a steady state and which may require heroic interventions. However, I quote without advocating the Biblical suggestion for confronting the problems: "And these of them ye may eat: the locust after its kind. . ." (Leviticus 11:22). The Mexican use of corn smut as a delicacy shares that approach. Often, with populations showing tremendous fluctuations, a crash occurs when pathogens build up only at high population densities. It may very well be that the maintenance of moderate locust populations may be a necessity for retaining pathogen levels and damping the oscillation.

We would recognize that not all herbivores are pests *a priori*. Not only can plants tolerate often surprising levels of herbivory, but may also benefit through its impact on growth, ripening, microenvironment, and other insects. We could even attempt to incorporate insects into the productive system. Relatively small manipulations of synchrony could nudge the insects into more benign temporal niches while stronger selective forces guard the niche boundaries.

We would recognize that no production system is final, that evolutionary processes are always at work among the insects, and that social changes alter our needs and goals. Our objective would be to use our understanding of evolution in such a way that selection on wild host plants can retard adaptation to the crops so that extirpation of wild populations is not always a good procedure.

## 1.5. IMPLEMENTATION

The possibilities of adopting a gentle, ecologically and socially rational form of IPM depend on a combination of factors. It will be easier to introduce in the course of developing new agricultural areas than through the reform of an already in place, tightly articulated system of sales efforts, advice networks, customs, and economic interests. An exception occurs when crisis strikes such an area as it did with Texas cotton in the 1970s. Therefore, the change is most likely to start in the periphery of world agriculture in developing countries or the less industrial agricultures of the United States such as Appalachia and New England. However, these areas may first require a strengthening of extension systems.

It depends also on public policy to reduce pesticides. This is most likely in countries where high population density exposes not only farm labor but much of the public to potential problems. Thus, the Netherlands, some states in the United States and some prefectures of Japan have explicit policies in this regard (Corbet, 1981). It is also public policy in countries with strong political commitments to protection of health and environment such as Cuba. Alexander and Anderson (1984) report the Cuban goal to double or triple the area under biological or natural control in the next 5 years, the initial successes of Nicaragua in shifting to IPM for cotton, along with short foreign exchange for pesticide imports and political commitment create a favorable environment for IPM if they are allowed the time to develop in peace.

The gentler technologies are especially site specific. Therefore, ways have to be found to combine the detailed intimate, but local, knowledge that farmers have of their own surroundings with the more general scientific knowledge that comes from research centers. This requires that we reject the arrogant dismissal of nonscientific experience without falling into the sentimentality of claiming that peasants always know best. It requires the active participation of farmers in observation, experimentation, and adaptation of general principles to local conditions. This, in turn, is not possible without mass literacy and a rural society that encourages innovation, makes it safe to take risks and allows farmers to meet scientists on terms of equality. It requires a scientific community closely linked both to agriculture and to world science without excessive deference toward current practices or world centers. And it requires international cooperation between scientists who have the resources and freedom from urgency that permits theoretical exploration but who live in countries with more impediments to ecologically and socially rational agriculture, and countries committed to such a direction but lacking in scientific resources. Out of that collaboration, it will be possible to reap the advantages of backwardness and create and apply new approaches. If they are allowed to continue their development in peace, in a few decades the last shall come first.

## REFERENCES

Alexander, R., and P. Anderson. (1984). Pesticide use, alternatives and workers' health in Cuba, *Int. J. Hlth. Serv.*, **14**, 31–41.

Beddington, J. R., and R. M. May. (1977). Harvesting natural populations in a randomly fluctuating environment, *Science*, **197**, 463–465.

Corbet, P. S. (1981). Non-entomological impediments to the adoption of integrated pest management, *Protec. Ecol.*, **3**, 183–202.

Holling, C. S. (1965). The functional response of predators to prey density and its role in mimicry and population regulation, *Mem. Entomol. Soc. Can.*, **45**, 1–60.

IITA. (1981). *Annual Report for 1980*, International Institute of Tropical Agriculture, Ibadan, Nigeria, 185 pp.

Levins, R. (1975). "Evolution in communities near equilibrium," in M. L. Cody and J. M. Diamond, Eds., *Ecology and Evolution of Communities*, Harvard University Press, Cambridge, Massachusetts, pp. 16–50.

Moran, V. C. (1983). The phytophagous insects and mites of cultivated plants in South Africa: Patterns and pest status, *J. Appl. Ecol.*, **20**, 439–450.

Pimentel, D. (1986). "Status of integrated pest management," in D. Pimentel, Ed., *Some Aspects of Integrated Pest Management*, Cornell University.

Price, P. (1983). "Hypotheses on organization and evolution in herbivorous insect communities," in R. F. Denno and M. S. McClure, Eds., *Variable Plants and Herbivores in Natural and Managed Systems*, Academic, New York, pp. 559–596.

Sampson, M. A., and R. Kuman. (1983). Population dynamics of the stem-borer complex on sugar-cane in southern Ghana, *Insect Sci. Appl.*, **4**(1/2), 25–32.

Strong, D. R., Jr. (1979). Biogeographic dynamics of insect-host plant communities, *Ann. Rev. Entomol.*, **24**, 89–119.

USDA. (1982). *Crop Protection Research—Annual Report*, USDA, Agricultural Research Service, Washington, D.C.

Virmani, S. M., M. V. K. Sivakumar, and S. J. Reddy. (1980). "Climatological features of the semi-arid tropics in relation to the farming systems research program," in *ICRISAT Proc. Intl. Workshop on the Agroclimatological Research Needs of the Semi-arid Tropics*, 22–24 November 1978. Hyderabad, India, 230 pp.

Zweig, G., and A. Aspelin. (1983). "The role of pesticides in developing countries," in United Nations Industrial Development Organization, *Formulation of Pesticides in Developing Countries*.

# 2

## ISLAND BIOGEOGRAPHIC THEORY AND INTEGRATED PEST MANAGEMENT

*Daniel Simberloff*

### 2.1. INTRODUCTION

The dynamic equilibrium theory of island biogeography (MacArthur and Wilson, 1963, 1967) generated enormous interest among ecologists and biogeographers (Simberloff, 1974). It employed straightforward mathematics to characterize a community by a simple statistic (number of species), yet depicted communities as dynamic entities. Furthermore, the conception of nature as divided into more or less discrete microcosms implied that, though nature is large and dynamic, its workings can be understood and represented concisely (Simberloff, 1978).

The theory is simplicity itself. Every island (including a habitat island, such as a field amidst forests) has, at a particular time, a number of species present ($S$), as well as an immigration rate ($I$) of new species onto the island from among those in the vicinity that are not already present, and an extinction rate ($E$) of species that are on the island. The theory states that the immigration and extinction rates are not constants, but rather monotonic functions of the numbers of species present. It says that the immigration rate is highest when the island is empty and declines to zero when all $P$ species in the regional species pool already inhabit the island. The extinction rate, on the other hand, is of necessity zero when there are no species on the island and rises to some maximum value when all $P$ species are present. These two monotonic curves must cross (Fig. 2.1), and it is a tenet of the theory that this intersection constitutes an equilibrium number of species ($\hat{S}$) about which the island's number of species should vary. The equilibrium is dynamic because it is accompanied by a "turnover" of species at rate $\hat{X}$, as species are locally extinguished on the island and are replaced by immigrants from the pool. Thus, it is the number of species, rather than their indentity, that is expected to remain approximately constant.

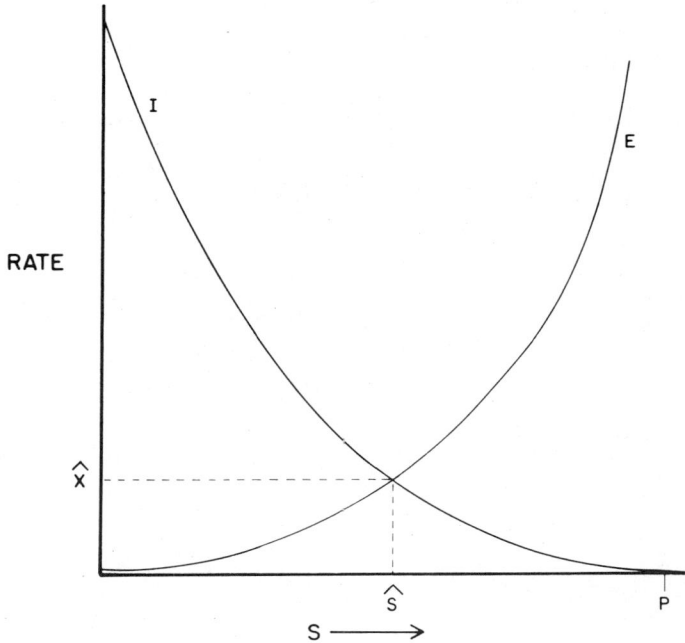

Fig. 2.1 Intersection of immigration ($I$) and extinction ($E$) curves to produce equilibrium number of species ($\hat{S}$) and equilibrium turnover rate ($\hat{X}$). $S$ = number of species present on the island; $P$ = number of species in the regional species pool.

A key attraction of the equilibrium theory is that it provided a ready explanation of the species–area relationship. This empirical relationship, that larger areas tend to have more species than smaller areas do, has been observed in so many systems, beginning as early as 1835 (Connor and McCoy, 1979), that Schoener (1976) refers to it as "one of community ecology's few genuine laws." The equilibrium theory predicts exactly such a relationship by virtue of the assumption that, for any particular number of species present, larger islands will have lower extinction rates (Fig. 2.2). If larger islands also have higher immigration rates, then the species–area relationship would be even more pronounced.

The nexus between equilibrium theory and the species–area relationship is crucial to the continuing fascination with the theory. Area is easily measured, and, though it is not trivial to determine the number of species of some taxa, like insects, species richness for many groups is not too difficult to find; one can count tree species, trap mammals, and so on. For some taxa and archipelagoes, there are already species lists in the literature, and it is a straightforward matter to regress richness on area. The result is that almost all "support" for the validity of the theory as a model consists simply of showing that species richness does, in fact, increase with area (Simberloff, 1974; Gilbert, 1980; McGuinness, 1984). The problem with this approach is that several other models, resting on

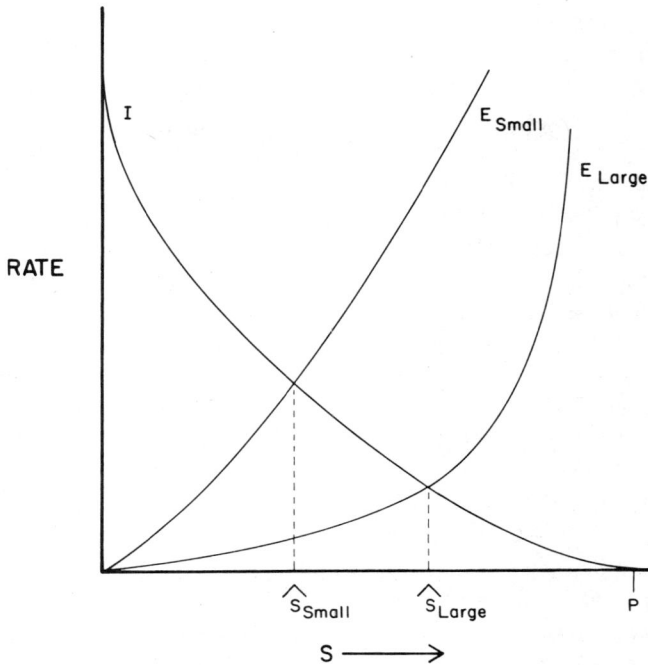

Fig. 2.2 Species–area relationship as explained by the equilibrium theory. Smaller islands have higher extinction rates ($E_{small}$) than do large islands ($E_{large}$) for any number of species present. The equilibrium number of species ($\hat{S}$) is therefore larger for large islands.

completely different causal mechanisms, also predict a species–area relationship (Simberloff, 1974; Connor and McCoy, 1979; McGuinness, 1984). Consequently, just demonstrating a species–area relationship cannot provide strong support for the equilibrium model. Attempting to find the best statistical fit for a particular species–area curve has not been a productive way to determine which model faithfully depicts the underlying causal mechanisms (McGuinness, 1984; Simberloff, 1976b).

The equilibrium theory has been directly tested very rarely, and then with mixed results (Gilbert, 1980; McGuinness, 1984). One reason that the theory cannot be easily substantiated is that, as originally stated, it was unfalsifiable: There was no objective definition of "equilibrium," so any observed change in number of species through time can be construed as within acceptable limits (Simberloff, 1983a). Even if one can define "equilibrium" objectively and exactly, it is difficult to determine whether changes in the species list for an island really constitute turnover of the sort envisioned by the theory (Smith, 1975; Simberloff, 1976a; Gilbert, 1980). As will be discussed below with respect to pest control, one can conceive of wide-ranging populations whose ranges encompass several islands and possibly the mainland. If it should happen at a particular time that no individuals of such a population occupy one of the islands in its range, this would not constitute extinction in the equilibrium theory sense. The extent to which tallied turnover is actually "pseudo-turnover" of this sort is difficult to measure, but for several systems this effect appears to be important (Simberloff, 1983b). Finally, there are several examples known in nature where either there is no turnover whatever (pseudo- or otherwise) and so the equilibrium is not dynamic, or where number of species appears to be changing monotonically so that an equilibrium (however defined) would not appear to obtain (Simberloff, 1983b).

Despite the limited empirical support for the equilibrium theory, it has been hailed as one of the great successes of community ecology. The more perceptive admirers of the theory are at pains to explain and qualify their admiration. For example, Brown (1981) and Haila and Järvinen (1982) admit that such tests as have been conducted seem much more often than not to falsify the theory, but believe that it has been heuristically valuable, inspiring people to ask the right sorts of questions, to view old issues (like species–area relations) from new, more fruitful perspectives, and generally providing a framework relating various data to a cohesive picture of a community or an island biota.

It is all well and good to take the ecumenical view that a widely cited theory, even if wrong, is useful in inspiring good research. This view certainly avoids controversy and fosters the generally progressive view of science that we find comforting, or at least that we feel presents the best picture of science to the lay public—the "Whig interpretation of history" (Kearney, 1971). In this interpretation,

> there was a simple tale to be told . . . . The history of science appears as the story of the emergence from clouds, irrationalism, or superstition of a rational method of interpreting nature. The early scientists are seen as groping for the discovery of a scientific method which was to be finally revealed in all its clarity by later generations.

There is a hierarchial progression, from worse to better, and each generation builds on the work of its predecessors—the *right* predecessors, of course. The relativism that the Kuhnian paradigm seems to imply is absent, and there is certainly no sense that a large group of intelligent scientists could be led for long into a wrong or unproductive line of research.

Without producing an essay on the history, philosophy, and sociology of science, I imagine the recent ferment of activity in these fields will have convinced most of my scientific colleagues that science, at least sometimes, does not work this way. The physical data the universe can provide to us should ultimately prevent interminable adherence to an incorrect theory, but there is no institutional safeguard in the operation of science, as opposed to other intellectual endeavors, that prevents an unproductive or even demonstrably incorrect theory from persisting for quite a while. I do not think that science is pervaded by such problems, but I think that such an aberration occurs occasionally. Why such things happen is, of course, the subject of intense debate among scientists and those who study them. My views on the reasons are not critical to my argument, so I will not indulge myself in this direction.

Whether the dynamic equilibrium theory of island biogeography has, in fact, been a great heuristic success, and has illuminated community ecology, or whether it is one of these false leads that seduces a substantial number of scientists for a good while into a rather unproductive line of research, or whether it is somewhere in between, is probably not possible to say at this time. Brown (1981) and Haila and Järvinen (1982) would favor the first view; Gilbert (1980) and McGuinness (1984) the second. Since all informed parties on all sides agree that, whatever success the theory has had, it must have been heuristic rather than as a direct elucidation of mechanisms in particular natural systems, it is especially difficult to measure "success." We would have to know what ideas and readings inspire researchers to do the kind of work they do, and there has been no such study for the equilibrium theory by adherents of either verdict.

However, whether or not a theory turns out to be correct or even heuristically valuable, if it is bruited about for long enough, it eventually achieves credibility in the eyes of the public, or even in the eyes of scientists in other disciplines who have neither the time nor the inclination to examine the evidence. Just as there is usually a lag between the advent of a theory and its spread to visibility outside its discipline, there is a lag between time when it is questioned or even discredited in the cold light of evidence and the time when nonexperts realize there is a dispute. Most laymen probably still believe that LSD uniquely breaks chromosomes and that extensive English twin studies establish the high heritability of IQ.

## 2.2. EQUILIBRIUM THEORY AND CONSERVATION

Interest in a theory beyond the narrow discipline in which it arose is especially high in two situations: (1) when the subject is of great public interest, as are drugs and intelligence tests, and (2) when the theory has potential application in a problem of economic or perceived societal importance. Because the dynamic equilibrium theory aroused such great interest among ecologists and biogeographers, it is not surprising that speculation quickly arose on extensions beyond concern with the diversity of island biotas. Among these were two similar potential applications. I will briefly discuss the idea that the theory has implications for nature reserve design and more fully address attempts to apply the theory to insect pests of crop plants.

The two ideas are motivated by the same analogy, that of the "habitat island." In each instance, islands of some habitat are embedded in a matrix of different habitats that presents a barrier to dispersal into the island. For nature reserves, the habitat island generally consists of some sort of native community, surrounded by a sea of agricultural, silvicultural, or other modified and managed land. For crop plants, the island consists of the field of some crop, and the matrix of other crops or uncultivated land. Indeed, agricultural and natural archipelagoes may be the inverse of one another. It has often been noted (e.g.,

Burgess and Sharpe, 1981) that natural habitats, such as forests and prairies, are becoming increasingly insular as more and more land is devoted to agriculture, and thus the equilibrium theory becomes increasingly applicable. Crop fields may concurrently become less insular, unless the interspersion of different crops renders each field an island. One sees immediately that there are *degrees* of insularity to be considered, for both natural areas and agricultural fields. If several crops have similar insect species, then a patchwork quilt may not constitute an insular situation. For the terrestrial inhabitants of ocean islands that were the original focus of the theory, insularity was usually a qualitative, not quantitative, condition, but in these applications insularity is relative.

Willis first applied the dynamic equilibrium theory to conservation in 1971, with suggestions on refuge design. He believes (Willis, 1984) that his ideas have not been properly cited, especially by Diamond (1973), but his subsequent publication (Wilson and Willis, 1975) enunciated two principles that have since been frequently repeated (e.g., Terborgh, 1974, 1975; Diamond, 1975; Diamond and May, 1976). First is the claim that the theory shows that single large refuges are preferable to groups of small ones with equal total area. Second is the contention that the theory dictates circular refuges rather than long, thin ones.

The brouhaha surrounding the first idea was immediate and did not abate in spite of the demonstration (Simberloff and Abele, 1976) that (1) the equilibrium theory did not lead to this prediction, and (2) actual data might show exactly the opposite pattern: more species in the network of small refuges. Indeed, subsequent study has shown the opposite pattern to be far more common (see references in Simberloff, 1982; Simberloff and Abele, 1982), and nowadays even early admirers of the application concede that (1) the equilibrium theory is not applicable and (2) empirical data on present diversity in single large sites and groups of several small sites do not usually show higher diversity in the large sites (e.g., Soulé and Simberloff, 1986). This reassessment results partly from a more tempered consideration of support for the equilibrium theory itself (e.g., Gilbert, 1980), partly from consideration of the logic of the application (e.g., Kitchener et al., 1980; Higgs, 1981), and partly from the simple accumulation of counterexamples (e.g., Simberloff and Abele, 1982). Exactly the same three factors have led to rejection of the contention that refuges should be circular instead of long and thin (e.g., Blouin and Connor, 1985).

What is interesting here is that, in the semipopular literature, the idea that the equilibrium theory dictates single large instead of several small refuges and, to a lesser extent, the claim that the theory mandates circular refuges have taken on lives of their own. They have become policy in certain arenas and entered the public consciousness as proven scientific fact at the very time that they have been shown to be unfounded in either the equilibrium theory or empirical assessments of diversity. In fact, they are still viewed as fact, sometimes to the real detriment of conservation. The reason is that popular writers (e.g., Iker, 1982) or conservation administrators (e.g., IUCN, 1980) often fail to keep up with details of scientific debates. If the science is not beyond them (and in this instance it assuredly is not), the other demands on their energies may preclude their giving sufficiently thorough attention to an idea, especially if the idea promises clear guidelines about what to do in a situation that had seemed terribly complicated.

The key publication in this saga was *World Conservation Strategy* (IUCN, 1980), published jointly in 1980 by the International Union for Conservation of Nature and Natural Resources, the United Nations Environmental Program, and the World Wildlife Fund. This manifesto for dealing with the crisis of threatened extinction stated that the dynamic equilibrium theory provides a sound basis for designing refuges and repeated Wilson and Willis's design recommendations, including the supposed superiority of single large refuges and of circular refuges. With this imprimatur, it is unsurprising that policymakers and the public would be impressed.

It is nonetheless a pity and has impeded conservation more than once. For example, I have a draft (marked "not for quotation," so I will not cite it) produced in June, 1983, by the Office of Environmental Affairs of one of the world's largest lending and development organizations entitled "Wildlife management in [. . .] projects: A policy proposal." This proposal is already in informal use and is intended to govern when the organization will grant loans for projects with environmental or conservation components. We find (p. 86) that, "Several simple but important guidelines for the optimal shape of WMUs [Wildlife management units] have been derived from studies in 'island biogeography'." Thereafter are given the principles of Wilson and Willis (1975), including the need for single large rather than several small refuges and the importance of circular refuges. The citation is of the IUCN–UNEP–WWF (1980) document. As another example of the tenacity of a scientific idea, even one that is discredited, I can cite the refuge system of Israel, consisting of some 200 sites, of which many are very small. These are protected and managed to various degrees by the Nature Conservation Authority, and the Authority is under increasing pressure to abandon some of the smaller refuges, not on the basis of research into the specific characteristics of the sites, but on the grounds that island biogeography theory has shown such sites to be impractical (R. Ortal, personal communication, 1984). Of course, in Israel there are very few large sites that are not desert, and many of the small refuges harbor populations or species that are unique in Israel.

## 2.3. EQUILIBRIUM THEORY AND PEST CONTROL

No doubt the fascination with the dynamic equilibrium theory among conservationists will wither, one hopes before too much damage is done. In the field of agricultural pest control, it has not gotten as far, but neither is it exactly a nonstarter. It is often cited as potentially useful (see references in Levins and Wilson, 1980), but has not, to my knowledge, actually been applied. I think the reason is that, in agriculture, the economic impact of a particular course of action is both apparent and quick, so that no one is likely to take a chance with a questionable scheme. In conservation, on the other hand, the exact effect of designing a refuge one way or another may not be manifested for decades, and no one suffers a short-term economic loss from an extinction. Also, agriculture must be one of the most conservative of academic disciplines whereas conservation biology, often driven by desperation, is frequently innovative.

The first question if one is to apply the equilibrium theory of island biogeography to agriculture is, "What are the islands?" The islands in such applications potentially range all the way from single plants through the entire range of a plant species. Janzen (1968, 1973) has even suggested that all individuals of a plant species can be construed as an island evolving through time. To determine the levels at which the theory can be applied, one must assess how insular an island really is. Since insularity is a relative concept, no island is completely immune to intercourse with other sites. Otherwise there would be no biota whatever, much less ongoing immigration and extinction. On the other hand, the concept of insularity can be trivialized by the assumption that any site is an island. Smith (1975) made this point with his "birds-in-a-tree" model. Any tree can be construed as an island, and one can simply record, at intervals, which species are present. One can then call any change in the list an immigration and/or an extinction and, if the number of species in the tree does not change much, say it is an island in equilibrium.

The equilibrium model, at least in terms of the causal mechanisms described by its founders (MacArthur and Wilson, 1963, 1967), would not be an appropriate framework with which to study birds in a tree, though changes in the numbers of species found in a tree might be an interesting topic. Rather, the equilibrium model envisions recruitment to the insular populations as resulting largely from *in situ* reproduction, not from the arrival of exogenous propagules (Simberloff, 1976a).

### 2.3.1. Individual Plants as Islands

By this criterion, I doubt if individual plants would usually qualify as islands. Brown and Kodric-Brown (1977) applied their "rescue effect" to insects of individual thistles, but most recruitment must have been exogenous. That is exactly what the rescue effect is—prevention of extinction by immigration into the population. What I am saying, in essence, is that as rescue frequency approaches that of reproduction, an island ceases to be an island. Faeth and Simberloff (1981) moved individual oak trees in an attempt to make them islands, but were unable to achieve sufficient isolation to suggest an equilibrium theory analysis. Probably insects on individual plants are usually more fruitfully treated by diffusion, Markov, and other population models (e.g., Kareiva, 1982; see also Kareiva, Chapter 4).

### 2.3.2. Small Stands as Islands

Small stands of plants have also been treated as islands. For example, Davis (1975) attempted to apply the model to insects colonizing clumps of nettles. His data seem to indicate that, for a few species, the clumps are isolated sufficiently that recruitment is mostly within the clump, but for most species, they are not. Seifert (1975) applied the equilibrium theory to insects of clumps of *Heliconia* inflorescences. However, his rationale for viewing the clumps as islands is their species–area curve, rather than direct observation of immigration, extinction, and transient movement. Since even samples from a homogeneous field would show a species–area relationship (Connor and McCoy, 1979), the existence of the relationship can hardly be an indication of insularity. Strong (1977) doubts that larger regions of *Heliconia*, or even more local patches, are islands in the equilibrium theory sense because he feels turnover is very rare.

### 2.3.3. Fields as Islands

Crop plants probably seem most analogous to islands at the level of the field. Price and Waldbauer (1982) see many potential advantages in viewing crop fields as islands and applying equilibrium theory. For example, they feel predictions may be possible about colonization rate of a field, given its distance from sources of invading insects, about the equilibrium number of species in a field, and about differences in this equilibrium between clumped and dispersed archipelagoes. Certain aspects of the analogy have empirical support. For example, the weevil *Phyllobius pomaceus* Gillenhal took 3 years to reach an isolated clump of nettles (Davis, 1975). It is not a great leap from such observations (see also Wolfenbarger, 1946) to the equilibrium theory prediction (Simberloff, 1969) that distant islands will approach their equilibrium number of species more slowly than near islands will.

Exactly how to use such an analogy in effective pest control is far from clear. One might propose that simply locating agricultural fields far from sources would lower the pest load (see references in Altieri and Letourneau, 1984). Consider the problems.

**1.** Crop loss to pests does not rest on the number of pest species in the field. Yet this is the statistic that equilibrium theory treats. It is, instead, numbers of individuals of certain species that are key determinants of reduced yield.

**2.** Even if it should prove beneficial to locate fields of a particular crop far from one another, is this the best economic strategy? What is to be done with the intervening land? Is it feasible for it to be fallow or planted in another crop?

**3.** Suppose the particularly damaging pests overwinter *in situ*? Then it would help, if they are host-specific, to shift a particular crop from site to site, but not simply to have fields widely separated. This is part of the general problem of defining a source pool for a crop field (Price, 1976). An oceanic island often has an obvious primary source; even when no single source is apparent, it may be possible to construct a composite source pool (Graves and Gotelli, 1983). For one field of many, embedded in a complicated matrix, such a construction would be very difficult.

**4.** Results on soybean insects (Price, 1976; Mayse and Price, 1978; Shepard et al., 1977) and *Spartina* insects (Rey and McCoy, 1979) suggest that barriers slow down colonization by parasitoids and predators more than by herbivores. If distance acted as a barrier in this way, whatever decrease in herbivore colonization one achieved by isolating fields might be more than compensated for by increased survival of the herbivores because their enemy populations were lower.

5. Virus incidence in some plants is increased when plants are isolated, apparently because some insect vectors colonize isolated plants at a higher rate (see Kennedy, Chapter 8). So a particular spacing pattern may have different effects on different pests.

Price (1976), recognizing problems (3) and (4) above, as well as the desire in pest control to stabilize populations, especially those of beneficial insects, suggested yet another strategy based on the equilibrium theory. Viewing stabilization as attainment of equilibrium, he proposed that one might want to increase colonization rates so that the equilibrium is quickly achieved in each year's growing season. Reducing field size is one of his suggestions for increasing colonization rates. For each species $\alpha$ with immigration rate $i_\alpha$ and extinction rate $e_\alpha$, the expected time $t_x$ for a fraction $x$ of a group of replicate defaunated islands to be colonized by $\alpha$ is $t_x = -[\text{ln}(1 - x)]/(i_\alpha + e_\alpha)$ (Simberloff, 1969). Since, in the simplest equilibrium model, $i_\alpha$ is unaffected by area and $e_\alpha$ is larger for smaller islands, $t_x$ decreases with island size. Price (1976) also proposes retaining uncultivated land between rows or including refugia for predators and parasites, thus raising their immigration rates, again lowering $t_x$.

One recognizes that such habitat management proposals for facilitating immigration by beneficial insects are well known to entomologists, completely independently of the equilibrium theory (e.g., Southwood, 1972; Root, 1973; see also Herzog and Funderburk, Chapter 10). Intercropping and leaving uncultivated areas can also affect viral transmission (see Kennedy, Chapter 8). It is not clear what advantages derive from viewing this problem in the context of the theory, when what is really needed is careful study of the ecology, including habitat requirements, of the beneficial insects (Rey and McCoy, 1979). Root (1973) found that collard fields were not acting as isolated islands colonized each year by chance arrival of propagules from afar, largely because the insect pests were able to disperse from one seasonal crucifer host to another; collards were not isolated. In a more general vein, Haila (1983) argued that the equilibrium model is not appropriate for a system in which populations are eliminated each winter and recolonize each spring (exactly as in many fields of crops). Even more generally, McGuinness (1985) suggested that the equilibrium model is not very useful for treating any demonstrably nonequilibrium system, such as one in which relatively frequent disturbances mold the community.

It is interesting and heartening that the specific recommendations conservationists proposed—single large sites and circular sites—have not cropped up in the literature on applying equilibrium theory to pest control. At least they have not cropped up in the same way. Price (1976) has, in fact, offered the inverse of both recommendations, but for different reasons than the direct effects on species richness that motivated the conservationists. For Price, the group of small fields might equilibrate faster than the single large one (because the equilibrium would be lower), while the narrow field would allow better access for predators and parasitoids.

I believe his motivation for the second recommendation suggests why workers in pest control have not paid much attention to the conservation recommendations. It could be that agriculturalists recognize the lack of empirical or theoretical support for the recommendations, but I suspect that the major hindrance for these recommendations in agriculture is confusion over possible effects on species richness at different trophic levels (Altieri and Letourneau, 1984). If one were to argue, for example, that a large field of a crop would attract more pest species than would several small fields of the same total area (not unlike Root's "resource concentration" hypothesis), the immediate rejoinder would be, "What about beneficial insects?" Similarly, if a long, thin field would house fewer pest species than a round one, would it also house fewer predators and parasitoids? It is not surprising that Stenseth (1981) has found little acceptance for his applications of island biogeography to pest control, since he prefaces his models with the statement that "the environment is assumed constant; thus parasitism and predation cannot be treated."

### 2.3.4. All Fields in a Region as an Island

The next level at which the island metaphor may be applied to insect pests is all fields of a particular plant species in some geographic entity, such as a state or nation. To my knowledge, the equilibrium model has not been explicitly applied at this level, though the species–area relationship, which is frequently erroneously viewed as synonymous with the equilibrium model (see above), has often been used in this context. Species–area curves are so wedded to equilibrium theory that Strong, who has published most extensively on them for phytophagous insects, was falsely accused of having applied equilibrium theory (Gilbert, 1980; Kuris et al., 1980) in spite of his explicit statement to the contrary (Strong, 1979). For Strong (1979), the main reason why the equilibrium theory is not useful is that "normal" turnover seems very rare, with extinction, when documented at all, usually associated with anthropogenous habitat change.

Nevertheless, simply as a phenomenological description, a species–area curve may be a convenient way to express information about numbers of insect species on a plant in some regions (Strong, 1979), though differing sampling intensities can confound results (Karban and Ricklefs, 1983). Southwood's (1961) study of insects on British tree species was an early attempt to relate a host insect fauna to area. Southwood (1961) viewed both area and density through time as key to faunal size, but Strong (1974a) showed that, whether equilibrial or not, the number of insect species on British trees can be explained largely by area. Whatever "deficit" a species has by virtue of being newly introduced is made up very quickly, and an introduced tree species quickly acquires an entomofauna appropriate to its geographic range in the region of introduction. Subsequent studies on just the leafhoppers and leafminers of British trees show significant species–area relationships, but area explains rather little of the variance in species richness (Claridge and Wilson, 1982a,b).

Capinera et al. (1984a,b) have examined pests of many crop species in Colorado. When they examined all pest species of 11 vegetables, plant height and crop value were significant predictors, but area was not. However, when the results were restricted to major pests only, area was the only significant predictor. For 11 field and forage crops, there was no significant regression for all pests, but when the analysis was restricted to major pests only, several variables (including area) yielded significant regressions. This result suggests that some of the phytophages in any comprehensive list are found frequently or primarily on other plants and are more or less stochastic confounding elements. The extent of this problem, of course, depends on such things as the physical locations of fields of the crop of interest relative to other fields and the phytochemistry and phenology of the various crops.

Studies of species–area relationships for phytophagous insects on cacao in many nations of the world (Strong, 1974b) and pests of sugarcane in 51 regions (Strong et al., 1977) take a different tack, by looking at one species in several regions, rather than several species in one region. The key results for both cacao and sugarcane are that a significant species–area relationship explains much of the variance in number of species, and time does not appear to be a major factor, as initial date of introduction into a region does not add to the variation explained by area alone.

Tentative explanations for species–area relationships at this level abound in the literature (Cornell and Washburn, 1979; Rey et al., 1981; Lawton et al., 1981; Lawton, 1982). None of these invoke the equilibrium theory, and most explicitly discount its importance in this context. The picture that emerges seems to be a version of the "habitat explanation" for species–area relationships generally: Larger areas have, on average, more habitats, and this is why they have more species (Connor and McCoy, 1979; McGuinness, 1984). Lawton (1982) summarized this view for plant phytophage faunas: "Widespread host plants experience different climatic regimes and grow in different habitats, in different places, with the inevitable consequence that a more widespread plant is more likely to be colonized by different species of insects in different parts of its range."

This said, it is difficult to determine how the existence of the relationship can be helpful in pest management. One can envisage that a continuous range expansion of a crop could cause the crop as a whole to accrue more pests, but unless the pests themselves expand from the new range into the original range, the damage in the original range would not increase by this means. The more normal course of addition of a pest to a host fauna is probably by "jump dispersal" (Pielou, 1979) of the pest: a discontinuous range expansion, such as the arrival of an exotic phytophage (e.g., Sailer, 1983). Just increasing the acreage devoted to a crop will, on average, increase the number of its pest species. By whatever means, an increase in area ultimately leads to an increase in number of phytophagous species. For example, Strong et al. (1977) showed a tight species-area relationship between sugarcane acreage and number of pest species, with the exponent $z$ about 0.46 in the typical species–area curve $S = cA^z$. Thus, if one were to double acreage in sugarcane, one might expect an increase in number of pest species by about 1.38. Would such a prediction lead planters, even rational ones, *not* to double acreage if other ecological and economic considerations suggested this course? I doubt it! As above, it is not the number of phytophagous species that determines damage, but numbers of individuals of particular species.

## 2.3.5. A Plant Species as an Island

The most imaginative version of the island metaphor for plant hosts of insects is at the largest scale: an entire plant species as an island for the insects that feed on it (Janzen, 1968, 1973). The key argument is that all insects associated with a host plant constitute an equilibrium assemblage, mediated by host plant resources and chemistry. When a host species acquires a new insect, the drain on the plant's resources forces competitive exclusion of a preexisting pest, at least within the geographic region in which the new immigrant is found. There should, however, be some long-term coevolution allowing a slow but limited increase in the total number of insects on a plant species, analogous to the evolutionary increase in the oceanic island equilibrium envisaged by Wilson (1969).

For Janzen, in addition to the straightforward analogies of island size to total biomass of a plant species and island isolation to geographic distance of a plant species' geographic ranges from other ranges, isolation has a taxonomic component. The more isolated a plant species is taxonomically from other species, the lower will be the immigration rate of insects from other plants, because the chemistry of the prospective host will be too unusual for many immigrants to survive. There is some slight evidence that taxonomic isolation does indeed restrict immigration to a host (e.g., Claridge and Wilson, 1982a; Connor et al., 1980), but the processes by which a host species acquires and loses pests are still virtually unknown. In particular, there is little evidence of equilibrium extinction; most deletions from crop plant pest lists probably result from changing cultivation methods, rather than competitive pressure from the resident phytophage community for a limited plant resource (Rey et al., 1981).

Janzen (1968, 1973) meant his metaphor to call attention to the processes by which a plant's host fauna changes, rather than as a detailed resemblance between an oceanic island and a plant species. One can imagine the proliferation of schemes for pest control based on the analogy. For example, one might suggest replacement of insects that damage a particularly valuable plant part (e.g., leaves) by deliberate introduction of species that feed on other, less valuable parts (e.g., flowers), assuming that competition through the plant nutrient budget will eventually have the desired effect. As more flowers are eaten, the plant is selected to produce more flowers. It will then produce fewer leaves so there would be a resource shortage for folivores. Such a tactic would be ill-founded. Certainly Janzen never intended such a literal reading of his metaphor, and with the basic equilibrium model apparently having limited application (Gilbert, 1980), it would be particularly inappropriate to propose such a scheme. Fortunately, even were it proposed, I doubt that the USDA would allow importation of known plant pests based on such a rationale. Once again, the economic importance and basic conservatism of agriculture as opposed to conservation insulate pest control from idle speculation and from applications based on such speculation. We often bemoan the conservatism, but here is an instance where it serves us well.

## 2.4. CONCLUSION

As with conservation, the possibility that the dynamic equilibrium theory could inform pest control efforts was an interesting one, and the exploration of the possibility has perhaps, as Price and Waldbauer (1982) predicted, helped us to focus a bit more sharply on the colonization process. However, it appears that we have learned what lessons were there to be gleaned, and it is important now to move on to other approaches. In particular, we should view the application of biogeographic theory to conservation as an object lesson and not be seduced away from the careful study of natural history that is a prerequisite for any sound pest control method (Rey and McCoy, 1979).

## REFERENCES

Altieri, M. A., and D. K. Letourneau. (1984). Vegetation diversity and insect pest outbreaks, *CRC Crit. Rev. Plant Sci.*, **2**, 131–169.

Blouin, M., and E. F. Connor. (1985). Is there a best shape for nature reserves? *Biol. Conserv.*, **32**, 277–288.

Brown, J. H. (1981). Two decades of homage to Santa Rosalia: Toward a general theory of diversity, *Am. Zool.*, **21**, 877–888.

Brown, J. H., and A. Kodric-Brown. (1977). Turnover rates in insular biogeography: Effects of immigration on extinction, *Ecology*, **58**, 445–449.

Burgess, R. L., and D. M. Sharpe, Eds. (1981). *Forest Island Dynamics in Man-Dominated Landscapes*, Springer-Verlag, New York, 310 pp.

Capinera, J. L., C. S. Hollingsworth, D. C. Thompson, and S. L. Blue. (1984a). Ecological correlates of pest species richness on crop plants, *Protec. Ecol.*, **6**, 287–297.

Capinera, J. L., D. C. Thompson, and C. S. Hollingsworth. (1984b). Taxonomic and ecological characteristics of crop pests in Colorado, *Environ. Entomol.*, **13**, 202–206.

Claridge, M. F., and M. R. Wilson. (1982a). Insect herbivore guilds and species-area relationships: Leafminers on British trees, *Ecol. Entomol.*, **7**, 19–39.

Claridge, M. F., and M. R. Wilson. (1982b). Species–area effects for leafhoppers on British trees: Comments on the paper by Rey *et al.*, *Am. Nat.*, **119**, 573–575.

Connor, E. F., S. H. Faeth, D. Simberloff, and P. A. Opler. (1980). Taxonomic isolation and the accumulation of herbivorous insects: A comparison of introduced and native trees, *Ecol. Entomol.*, **5**, 205–211.

Connor, E. F., and E. D. McCoy. (1979). The statistics and biology of the species–area relationship, *Am. Nat.*, **113**, 791–833.

Cornell, H. V., and J. O. Washburn. (1979). Evolution of the richness-area relation for cynipid gall wasps on oak trees: A comparison of two geographic areas, *Evolution*, **33**, 257–274.

Davis, B. N. K. (1975). The colonization of isolated patches of nettles (*Urtica dioica* L.) by insects, *J. Appl. Ecol.*, **12**, 1–14.

Diamond, J. M. (1973). Distributional ecology of New Guinea birds, *Science*, **179**, 759–769.

Diamond, J. M. (1975). The island dilemma: Lessons of modern biogeographic studies for the design of natural reserves, *Biol. Conserv.*, **7**, 129–146.

Diamond, J. M., and R. M. May. (1976). "Island biogeography and the design of natural reserves," in R. M. May, Ed., *Theoretical Ecology*, Saunders, Philadelphia, pp. 163–186.

Faeth, S. H., and D. Simberloff. (1981). Experimental isolation of oak host plants: effects on mortality, survivorship and abundances of leafmining insects, *Ecology*, **62**, 625–635.

Gilbert, F. S. (1980). The equilibrium theory of island biogeography: Fact or fiction? *J. Biogeog.*, **7**, 209–235.

Graves, G. R., and N. J. Gotelli. (1983). Neotropical land-bridge avifaunas: New approaches to null hypotheses in biogeography, *Oikos*, **41**, 322–333.

Haila, Y. (1983). Land birds on northern islands: A sampling metaphor for insular colonization, *Oikos*, **41**, 334–351.

Haila, Y., O. Järvinen. (1982). "The role of theoretical concepts in understanding the ecological theatre: A case study on island biogeography," in E. Saarinen, Ed., *Conceptual Issues in Ecology*, D. Reidel, Dordrecht, Holland, pp. 261–278.

Higgs, A. J. (1981). Island biogeography theory and nature reserve design, *J. Biogeog.*, **8**, 117–124.

Iker, S. (1982). Islands of life in a forest sea, *Mosaic*, **13(5)**, 24–30.

IUCN. (1980). *World Conservation Strategy*, International Union for the Conservation of Nature–United Nations Environmental Program–World Wildlife Fund, Morges, Switzerland, 450 pp.

Janzen, D. H. (1968). Host plants as islands in evolutionary and contemporary time, *Am. Nat.*, **102**, 592–595.

Janzen, D. H. (1973). Host plants as islands. II. Competition in evolutionary and contemporary time, *Am. Nat.*, **107**, 786–790.

Karban, R., and R. E. Ricklefs. (1983). Host characteristics, sampling intensity, and species richness of Lepidoptera larvae on broad-leaved trees in southern Ontario, *Ecology*, **64**, 636–641.

Kareiva, P. (1982). Experimental and mathematical analyses of herbivore movement: Quantifying the influence of plant species and quality on foraging discrimination, *Ecol. Monogr.*, **52**, 261–282.

Kearney, H. (1971). *Science and Change. 1500–1700*, McGraw-Hill, New York, 255 pp.

Kitchener, D. J., A. Chapman, J. Dell, B. G. Muir, and M. Palmer. (1980). Lizard assemblage and reserve size and structure in the Western Australian wheatbelt—Some implications for conservation, *Biol. Conserv.*, **17**, 25–62.

Kuris, A. M., A. R. Blaustein, and J. J. Alio. (1980). Hosts as islands, *Am. Nat.*, **116**, 570–586.

Lawton, J. H. (1982). Vacant niches and unsaturated communities: A comparison of bracken herbivores at sites on two continents, *J. An. Ecol.*, **51**, 573–595.

Lawton, J. H., H. Cornell, W. Dritschilo, and S. D. Hendrix. (1981). Species as islands: Comments on a paper by Kuris *et al.*, *Am. Nat.*, **117**, 623–627.

Levins, R., and M. Wilson. (1980). Ecological theory and pest management, *Ann. Rev. Entomol.*, **25**, 287–308.

MacArthur, R. H., and E. O. Wilson. (1963). An equilibrium theory of insular zoogeography, *Evolution*, **17**, 373–387.

MacArthur, R. H., and E. O. Wilson. (1967). *The Theory of Island Biogeography*, Princeton University Press, Princeton, NJ, 203 pp.

Mayse, M. A., and P. W. Price. (1978). Seasonal development of soybean arthropod communities in east central Illinois, *Agro-Ecosystems*, **4**, 387-405.

McGuinness, K. A. (1984). Equations and explanations in the study of species–area curves, *Biol. Rev.*, **59**, 423-440.

McGuinness, K. A. (1985). "Communities of organisms on intertidal boulders: The effects of disturbance and other factors," Ph.D. dissertation, University of Sydney, Sydney, Australia, 310 pp.

Pielou, E. C. (1979). *Biogeography*, Wiley, New York, 351 pp.

Price, P. W. (1976). Colonization of crops by arthropods: Non-equilibrium communities in soybean fields, *Environ. Entomol.*, **5**, 605-611.

Price, P. W., and G. P. Waldbauer. (1982). "Ecological aspects of pest management," in R. L. Metcalf and W. H. Luckman, Eds., *Introduction to Insect Pest Management*, 2nd ed., Wiley-Interscience, New York, pp. 33-68.

Rey, J. R., and E. D. McCoy. (1979). Application of island biogeographic theory to pests of cultivated crops, *Environ. Entomol.*, **8**, 577-582.

Rey, J. R., E. D. McCoy, and D. R. Strong. (1981). Herbivore pests, habitat islands and the species–area relationship, *Am. Nat.*, **117**, 611-622.

Root, R. B. (1973). Organization of a plant–arthropod association in simple and diverse habitats: The fauna of collards (*Brassica oleracea*), *Ecol. Monogr.*, **43**, 95-124.

Sailer, R. I. (1983). "History of insect introductions," in C. Graham and C. Wilson, Eds., *Exotic Plant Pests and North American Agriculture*, Academic, New York, pp. 15-38.

Schoener, T. W. (1976). "The species–area relation within archipelagoes: Models and evidence from island land birds," in H. J. Firth and J. H. Calaby, Eds., *Proceedings of the 16th International Ornithological Congress*, Australian Academy of Science, Canberra, pp. 629-642.

Seifert, R. P. (1975). Clumps of *Heliconia* inflorescences as ecological islands, *Ecology*, **56**, 1416-1422.

Shepard, M., G. R. Carner, and S. G. Turnipseed. (1977). Colonization and resurgence of insect pests of soybeans in response to insecticides and field isolation, *Environ. Entomol.*, **6**, 501-506.

Simberloff, D. (1969). Experimental zoogeography of islands: A model for insular colonization, *Ecology*, **50**, 296-314.

Simberloff, D. S. (1974). Equilibrium theory of island biogeography and ecology, *Ann. Rev. Ecol. Syst.*, **5**, 161-182.

Simberloff, D. (1976a). Species turnover and equilibrium island biogeography, *Science*, **194**, 572-578.

Simberloff, D. (1976b). Experimental zoogeography of islands: Effects of island size, *Ecology*, **57**, 629-648.

Simberloff, D. S. (1978). "Colonization of islands by insects: Immigration, extinction, and diversity," in L. A. Mound and N. Waloff, Eds., *Diversity of Insect Faunas*, Blackwell, Oxford, pp. 139-153.

Simberloff, D. (1982). Island biogeographic theory and the design of wildlife refuges, *Ékologiya* (Moscow), No. 4, 3.

Simberloff, D. (1983a). When is an island community in equilibrium? *Science*, **220**, 1275-1277.

Simberloff, D. (1983b). "Biogeographic models, species distributions, and community organization," in R. W. Sims, J. H. Price, and P. E. S. Whalley, Eds., *Evolution, Time and Space: The Emergence of the Biosphere*, Academic, London, pp. 57-83.

Simberloff, D., and L. G. Abele. (1976). Island biogeography theory and conservation practice, *Science*, **191**, 285-286.

Simberloff, D., and L. G. Abele. (1982). Refuge design and island biogeographic theory: Effects of fragmentation, *Am. Nat.*, **120**, 41–50.

Smith, F. E. (1975). Ecosystems and evolution, *Bull. Ecol. Soc. Amer.*, **56**, 2–6.

Soulé, M. E., and D. Simberloff. (1986). What do genetics and ecology tell us about the design of nature reserves? *Biol. Conserv.*, **35**, 19–40.

Southwood, T. R. E. (1961). The number of species of insects associated with various trees, *J. An. Ecol.*, **30**, 1–8.

Southwood, T. R. E. (1972). Farm management in Britain and its effect on animal populations, *Tall Timbers Conference on Ecological Animal Control by Habitat Management*, **3**, 29–51.

Stenseth, N. C. (1981). How to control pest species: Applications of models from the theory of island biogeography in formulating pest control strategies, *J. Appl. Ecol.*, **18**, 773–794.

Strong, D. R. (1974a). Nonasymptotic species richness models and the insects of British trees. *Proc. Natl. Acad. Sci. USA*, **73**, 2766–2769.

Strong, D. R. (1974b). Rapid asymptotic species accumulation in phytophagous insect communities: The pests of cacao, *Science*, **185**, 1064–1066.

Strong, D. R. (1977). Insect species richness: Hispine beetles of *Heliconia latispatha, Ecology*, **58**, 573–582.

Strong, D. R. (1979). Biogeographical dynamics of insect–host plant communities, *Ann. Rev. Entomol.*, **24**, 89–119.

Strong, D. R., E. D. McCoy, and J. R. Rey. (1977). Time and number of herbivore species: The pests of sugarcane, *Ecology*, **58**, 167–175.

Terborgh, J. (1974). Preservation of natural diversity: The problem of extinction-prone species, *BioScience*, **24**, 715–722.

Terborgh, J. (1975). "Faunal equilibria and the design of wildlife preserves," in F. Golley and E. Medina, Eds., *Tropical Ecological Systems: Trends in Terrestrial and Aquatic Research*, Springer, New York, pp. 369–380.

Willis, E. O. (1984). Conservation, subdivision of reserves, and the antidismemberment hypothesis, *Oikos*, **42**, 396–398.

Wilson, E. O. (1969). The species equilibrium, *Brookhaven Symp. Biol.*, **22**, 38–47.

Wilson, E. O., and E. O. Willis. (1975). "Applied biogeography," in M. L. Cody and J. M. Diamond, Eds., *Ecology and Evolution of Communities*, Harvard University Press, Cambridge, MA, pp. 523–534.

Wolfenbarger, D. O. (1946). Dispersion of small organisms. *Am. Midl. Nat.*, **35**, 1–152.

# 3

## POPULATION THEORY AND UNDERSTANDING PEST OUTBREAKS

*Donald R. Strong*

### 3.1. INTRODUCTION: ECOLOGY AND ITS SIAMESE TWIN, MATHEORY

Outbreaks are unusually high and/or harmful populations of a pest. Their perception and definition, though roughly useful, is arbitrary ecologically because outbreaks are only phases in population dynamics and dispersion, phases with several different sorts of antecedents for most species. For pest arthropods as a group, the "causes" of outbreaks are many; they are the sum total of different conditions that can lead to population buildup. Therefore, unless the mechanics and aliatory nature of population dynamics and dispersion are understood, population ecology really has little to contribute to integrated pest management (IPM). There is probably no simpler set of reasons for pest outbreaks in general than for population buildup in general.

This chapter explores a paradox about population dynamics, the potentially symbiotic but often insidious association between mathematics and theory, "matheory." Sophisticated mathematics is only the end of a gradient of theory that begins qualitatively, with verbal and graphical images that are vague and extremely general. In ecology, the vital theories, which have driven the discipline, inspired empiricism, steered experimentation, and focused critical measurements have rarely been sophisticated mathematically and are far from the mathematical end of the gradient. The vital ecological theories have been sets of key words and simple graphs, more like allegories, general and sketchy, than like specific and detailed algorithms. We see how little sophisticated mathematics has contributed to vital ecological theories in such classic examples as the density controversies, where semantics and definition were the key issues (Strong, 1984), in the competition controversies, where the major questions concerned the nature of evidence (Hairston, 1985; Connell, 1983; Schoener, 1983), and in the area of predator/parasite studies, where, to this day, the quantitative effects of natural enemies on their prey/host populations are not well resolved (Dempster, 1983; Murdoch et al., 1985; M. P. Hassell, in prep.). This is not to say that each of these areas and others in ecology have not had their matheory (long has it gushed!), but that the matheory has mostly led to dead ends that were unable to redirect the major ideas of the discipline; the matheory in ecology has been largely derivative and mannered rather than germinal and inspirational, as it has in physics.

In exploring why vital theory in ecology has so often remained so qualitative and inchoate, I assume without question the immense potential value of sophisticated mathematics for our discipline; theory and math in ecology can be tremendously symbiotic, as they are in the physical sciences. Obviously, ecology would be greatly served by fertile and vital matheory that correctly predicted wonderful, undiscovered features of ecology.

I shall argue that what many regard as the mainstream of mathematical theory in ecology is actually applied mathematics that, though excellent and thoroughly worthwhile in its own right, has been guided by the inherently nonecological esthetics of order ("stability") and balance ("equilibrium") in homogeneous systems. Because of its overriding concern with mathematical elegance, this matheory might be called prim, and its results have been pretty much peripheral to the ecology of real organisms in nature. Optimistically, we might view prim matheory as an early stage leading to realistic matheory, which is tuned to the substantial disorder, fluctuation, and heterogeneity of real ecological systems, agricultural and natural. In this view, prim matheory is technical background to realistic theory.

Before addressing the relationship between math and theory in population dynamics, I shall outline the ecological nature of relations among insects and plants, because my thesis emerges from a verbal sketch of the mechanics of these systems.

## 3.2. TROPHIC WEBS AS COMMUNITY MODELS

Communities are groups of populations occurring sufficiently close together in space and time for the *potential* of *some* interaction. This is an operational definition. To the degree that genetic differences within species affect community processes, a finer focus is necessary and evolutionists have something to teach us. To the degree that proximate population processes are affected by grand abiotic and biotic happenings, such as continental weather, energy flow, and nutrient cycling, autecology or ecosystem approaches are necessary.

In discussing intraspecific interactions within populations and interspecific actions among them, I emphasize "potential" effects because co-occurrence need not beget effective interaction; I emphasize "some" because interactions that do occur need not be intense, pervasive, consistent, or even important to the ecology of all populations involved.

### 3.2.1. "Tall and Narrow-Waisted" Trophic Webs for Insects on Plants

Trophic webs (Pimm, 1982; Paine, 1979) are a useful way to understand communities of insects on plants; feeding activities determine much of the influence among species, the dimensions and directions of interspecific forces that affect the growth and spatial extent of populations (Fig. 3.1). Fortunately, insects on plants form fairly discrete trophic levels and most species have distinct trophic levels, from the basal plants, through the intermediate herbivores, to the carnivores at the top end of the web, which eat herbiovores and each other. [Some planktonic communities do not sort out into discrete trophic levels because both animals and plants in the plankton leak nutrients that are reabsorbed by other organisms. Thus, the plankton can lack the order of a trophic web in the sense of insects on plants because its species have indistinct and changing trophic roles in the community (Peters, 1977)].

Among insects on plants, there are rare exceptions to distinct vertical layering and consistent, simple trophic roles for species in communities. For example, few plants are hemi-parasites, facultative heterotrophs, which graft to other plants via interconnecting root haustoria (Atsatt, 1973), and some herbivorous

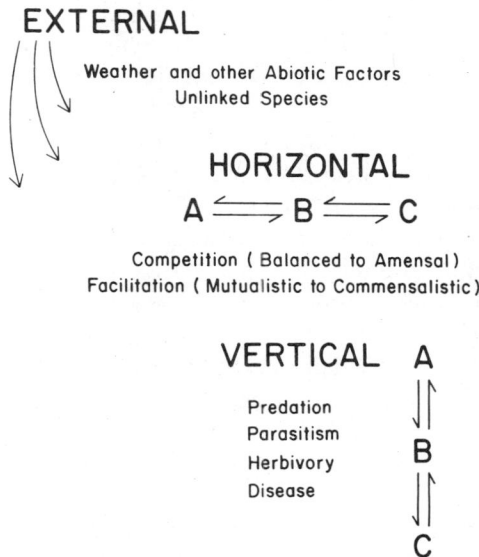

Fig. 3.1 Trophic webs for insects on plants are composed of vertical influences that move up and down between trophic levels, horizontal influences that act within trophic levels, and unilateral external influences.

insects are obligately carnivorous during some periods of development [e.g., mirid bugs ([Southwood, 1972)] or facultatively so [e.g., cannibalistic caterpillars (Fox, 1975)].

Vertical influences are predation, parasitism, herbivory, and disease, which occur between different trophic levels, between plants, herbivores, and carnivores. Horizontal influences are interspecific competition and mutualism, which occur within trophic levels. Both vertical and horizontal influences have the potential to involve sets of populations in reciprocal coactions with influences travelling from one to another species then back again. However, as I emphasized above, only a subset of actions actually occur, and reciprocity need not develop from either vertical or horizontal interspecific activities. For insects, interspecific competition that does occur is normally unilateral. One species is typically affected negatively with little if any returned influence (Lawton and Hassell, 1981).

For the vertical actions of trophic webs, reciprocity needs to be viewed in a slightly different light than for horizontal actions. Though most if not all vertical actions in food webs are reciprocal in the sense that individuals of both species are affected, the reciprocity need not extend to the population level, for which the trophic web has meaning. Feeding by a particular species may be too infrequent, isolated spatially from the majority of the particular host or prey population, or too late in ontogeny or phenology for much if any influence. Even a monophagous herbivore, predator, parasitoid, or disease can cause harm that is too little or sporadic for any consistent effect on a host or prey population, given the variety of influences in the life cycle and the high variance of all influences. In such "one-way" interactions, the natural enemy feels strong positive influence of nutrition from the prey or host population, but only a statistically insignificant effect is reflected back to the host population. It is my belief from experience that a substantial fraction of interspecific activities in trophic webs of insects on plants are usually weak and/or nonreciprocal, with some potential for intermittent reciprocity; a natural enemy might be effective at reducing the density of an herbivore under some circumstances but often would not do so to any statistically discernible extent.

External influences come from outside the group of populations that we define as a community. External influences are always unilateral and flow into the community with little if any potential to be changed by the populations that they affect. For example, the tide is an external influence upon littoral marine organisms; this physical force is not influenced much, if any, by the plants and invertebrates of the seashore. Weather is the archetypical external influence upon ecological communities, and justifiably so. Weather and its larger-scale climatic influences are a tremendously important aspect of community ecology; they set the stage for the community drama by affecting the roles, picking the particular species of the cast, and timing the acts. This is a fact that has been well appreciated by entomologists and botanists, but less so by community ecologists working with vertebrates, during the past 20 years or so. External influences are not only abiotic; the definition can be applied to other species that affect communities without any returned influence. Some polyphagous or migratory predators are likely candidates for biotic external influences. It is possible, but poorly investigated, that external factors like the weather affect internal factors like interspecific competition (DeBach, 1958). Such an influence could be viewed as stochasticity (Chesson, 1985) or contingency (Strong, 1986) in a density function.

The "shapes" of trophic webs are determined by the relative importance of the different influences operating within them. For insects on plants, much information suggests that trophic webs are "tall and narrow-waisted," with great external influences (Strong et al., 1984). "Tallness" means that great vertical influences, up from the host plants and down from the natural enemies, are the predominant biotic forces among the trophic levels. The "narrow waistedness" means that interspecific competition is relatively less frequent and intense among the herbivores at the "waist" of the food web than among the plants at its base or, perhaps, among the natural enemies at its shoulders (whether these food webs are "broad shouldered," with intense interaction among carnivores, is unresolved in any general sense).

The strength, variety, and volatility of ascending influences from plants to herbivorous insects is one of the truly distinctive facets of these most common trophic webs. Host plants are Janus-faced to even the most adapted herbivores. Sometimes sustenance and othertimes bane, plants can be simultaneously food, home, poison, and a deadly trap (Rosenthal and Janzen, 1979), nutritious but vacuous (Rodriguez, 1972; Brodbeck and Strong, 1986), attractive but insipid (Whitham, 1981), and hospitable but the lurking place for natural enemies (Hassell and Southwood, 1978) (see also Kogan, Chapter 5 and Duffey, Chapter 6).

The balance between benevolence and malignance of influences from host plants to herbivorous insects is quite sensitive to external influences and is much of the reason for the volatile nature of population dynamics of many insects on plants. I believe that one major source of this volatility is the usual condition of nutritional insufficiency of plant tissue for herbivorous insects; concentration of nitrogen is commonly an order of magnitude lower in plant tissue than in insect tissue (Southwood, 1972). Crude nitrogen concentration of host plants is normally [but not universally (Mopper et al., 1984)] well correlated with performance (growth rate and survivorship) of immature insect herbivores, provided plant water concentrations are sufficient (Scriber and Slansky, 1981). Volatility in population dynamics can be caused by windfalls of nitrogen made available to herbivores during times of stress to host plants. Drought, root flooding, and other forms of abiotic stress often prompt plants to mobilize amino acids from vulnerable photosynthetic proteins and transport them to safer tissues, often in the roots, yielding a nutritional bonanza to both chewing and "sucking" herbivores that might cause "outbreaks" of herbivores (White, 1978; Brodbeck and Strong, 1986).

### 3.2.2. The Stochastic Complexity of Trophic Webs

A major point of this chapter is the richness of population dynamics of insects on plants. The preceding ideas of how insect outbreaks may be caused by the physiological reactions of host plants to abiotic stresses give one source of richness: External abiotic forces eccentrically drive biotic actions of the trophic web. The external forces of extreme weather are inherently stochastic with no negative feedback, and their influence upon the biotic forces that may regulate interspecific associations, show how "stochastically complex" and "open" these community dynamics can be.

Stochastic complexity in partially open systems is different from the "deterministic complexity" in closed machines like computers, space craft, and nuclear power plants, which have many parts coacting with intricately precise negative feedback. With deterministic complexity, components with very consistent behavior are highly integrated to produce precise system control. In contrast, the stochastic complexity of ecological communities involves only imprecise negative feedback among components with wide latitude of behavior; many of the influences within and among populations are at best vaguely reciprocal, with ill-defined returned influences. The character of influences varies with time, season, phenology, and other external conditions, and migration into and out of a community opens it to biotic influences that are only loosely linked to local conditions. The negative feedback that does occur in trophic webs of insects on plants is often weak and is likely to occur only at extremes of population level. A likely outcome of stochastic complexity in open communities is weak and imprecise biotic control, density-vagueness, and "regulation" that moves populations toward broad medial bands of density but does not cause point equilibrium or stability in the simple mathematical sense.

## 3.3. PRIM AND REALISTIC THEORY IN ECOLOGY

A great deal of orthodox matheory in ecology is more appropriate for modeling deterministic than stochastic complexity; I call this "prim" matheory (Table 3.1). Its esthetic is mathematical elegance of a sort prized in physics, and its object is rationalizing "equilibrium" and "stability" for ecological systems. Prim matheory is deterministic and models tight linkages in imaginary food webs, with interactions that lock consumer and resource species into inexorable codynamics that are constantly effective and so intense as virtually to exclude effects of other factors. Prim matheory in ecology has population dynamics caused by constants that are single-number composites of actual demographic functions. Intense negative feedback loops among populations and resources are often an assumption of prim theory. Finally, prim theory concerns closed and internally homogeneous systems. Of course, no particular theory has all of the features in Table 3.1, and none is so extreme as to be strictly prim. Not all deterministic matheory is strictly prim, and intriguing developments have recently occurred in deterministic theory that are leading it toward ecological reality (Table 3.2 and below).

While little theory in ecology has been purely prim, virtually no matheory can be totally realistic. Even shallow ecological inspiration will nudge the most naive

**Table 3.1 Comparisons of Features of Ecological Theories between the Prim Extreme and Various Realistic Approaches[a]**

|  | Prim | Realistic |
|---|---|---|
| Mathematics | The object | One of several subjects |
| Trophic linkage | Tight | Loose |
| Population influences | Constant, absolute | Variable, contingent |
| Negative feedback | Assumed: consistent, strong | Searched for: inconsistent, variable; vague |
| Limits to population | $k$: constant point | Variable, hazy region |
| Rates | Instantaneous: $r$, composite | Finite: $R_0$, components |
| Factors | Single to few | Several to many |
| Determinism | Absolute | Moderate |
| Stochasticity | Ignored or trivial | Moderate to great |
| Key variables | $k$ | Environmental mosaic |
|  | Density and kin | Weather Phenology Migration |
| System | Closed, single cell | Variously open, multicelled |
| Environment | Homogeneous, constant, cyclical | Heterogeneous, changing, fluctuating |

[a]The realistic approaches are less extreme in terms of mathematical simplicity and elegance but more representative and tractable for real organisms in nature.

and oblivious matheorist away from sterile mimicry of electronic circuits, and, certainly, profound understanding of the natural variability of ecological systems does not quench our desire for theory that finds order, predictability, and firm relationships at some scales and levels. The difference between prim and realistic theory often boils down to the scale or level on which ecological relationships are sought. Prim theory often has its kernel at very fine scales (individuals), with populations and communities viewed as the average behavior of homogeneous sets of similar individuals, whereas realistic theories address the factors that would foul up simple homogeneity, such as external factors, heterogeneous individuals, different parts of the environment, and phenology (Table 3.1). Because of the great diversity of organisms and environmental influences among real communities, there is quite a variety of form, approach, and objectives among realistic theories; some concern stochastic variation in homogeneous closed systems, others deterministic change among partially isolated subsystems, and so forth. While some realistic theories are distinctly mathematical and hence fairly specific, others are purely verbal, without the details of quantitative models, and hence much more general. Verbal models have the desirable feature of not being bound to a mathematical technique, like continuous or difference or stochastic

**Table 3.2  Some Realistic Perspectives on Transgenerational Population Theory**[a]

Andrewartha and Birch (1954). V, E(L&F). General theory of multiple factors for fluctuations in natural populations.

Milne (1957). * R, E(F), V, G. "Imperfect" versus "perfect" density effects.

Ito (1961). * R, E(F), V, G. Graphical model of population irruptions.

Reddingius and den Boer (1970). * R, V, S. "Spreading the risk," "Gambling for existence."

Roff (1974b,c). * S. Spatial heterogeneity and population persistence.

Chesson (1978, 1985). * S. "Stochastic boundedness" and the "Storage effect."

Levins (1979). D. Competitively structured communities with no "equilibrium" may persist under some kinds of nonlinearities and fluctuating environments.

Armstrong and McGehee. (1980). D. More species than resources may persist in competitive communities if resource uses are nonlinear and carrying capacities are not fixed.

Caswell (1982). D. Nonequilibrium life histories.

Oster (1981). E(L), D & S. A "statistical" model for nonlinear birth and death rates in a lab population with age structure.

Paine and Levin (1981). E(F), D. Patch dynamics. Birth (= recruitment episode in open area of rock), death, and standing crop of mussel patches on a rigorously wave-swept shore. Patches could be analogs of arthropod colonies like those of aphids or mites.

Hastings (1984). D. Simplicity lost? The mathematical form of modeling demography in seasonal environments, using composite ($r$) versus component (birth and death) rates, and the kind of limits to population all greatly affect mathematical result.

Gutierrez et al. (1984). R, E(F), D. Complex realistic models of concrete sets of plant, herbivores, and natural enemies. Emphasizes actual seasons, phenologies, and energetics. Careful with match between model and actual data. Getz and Gutierrez (1982) is a broad review of complex realistic modelling.

Steele and Henderson (1984). E(F), S. "Red noise" and rapid shifts in fish populations. Red noise is environmental stochasticity with variance that increases over intervals of several generations for the fish. The stochasticity enters this model via the predation term. See May (1981) for comment and overview of this model, and Beddington and May (1977) for deterministic antecedents.

Royama (1984). E(F), S & D. Second-order autoregression: Fluctuations about a longer-termed basic oscillation. Quite complex model, several abiotic and biotic components. Coordinated regional population dynamics, as opposed to independent subpopulations. Specifically for spruce budworm in Canada, contrasts with orthodox model of epidemic spread from local foci. Density-dependent factors, which would account for the basic oscillation, are poorly understood, believed to be due to parasites and a mysterious disease-like factor called the "fifth agent."

Cappuccino and Kareiva (1985). E(F), V. "Capricious," fleeting, and hostile environments for a butterfly and its caterpillars. Phenology of the plant that is only briefly appropriate and viral diseases in the soil combine to keep this insect rare and prevent it from coordinating its abundance with that of its host.

---

| | |
|---|---|
| [a]R = literature review | G = graphical theory, |
| E = empirical, concerned with concrete | D = deterministic math, |
| example in laboratory (L) or field (F), | S = stochastic math, |
| V = verbal theory, | * = extended comments in text. |

equations, techniques with math assumptions that can affect a theory as much or more than the ecological features. As Wang and Gutierrez (1980) have pointed out, "The results of stability analyses appear to reflect the formulation, assumptions, and parameter values of the models rather than testable field biology..." (p. 435).

A list of some of my favorite realistic theories in ecology appears in Table 3.2. Certainly, not all are very mathematical. I will discuss at some length the ideas that, because of their age or location in an exotic literature, are likely to be missed by young students of populations and communities. In Table 3.2, I also refer to, but do not discuss at any length, some current realistic approaches to community ecology that appear in mainstream journals that we see as a matter of course.

Ito's (1961) review of the literature on the ideas about population dynamics is among the best but least known of older perspectives on this subject. Of course, some of the material in the paper is dated, but the number and diversity of early views that the author has drawn together about how trophic webs affect population dynamics leads to an appreciation of how rich the possibilities of population dynamics are.

Another insightful old classic that is still quite useful, but largely unfamiliar to young ecologists, is Milne's (1957) review and verbal/graphical theory of insect population dynamics. It contrasts some early simple mathematical theories by Smith and Nicholson with the early realistic theories, by Thompson (1928) and Andrewartha and Birch (1954). In a sort of Hegelian synthesis to the thesis–antithesis of the density dependent versus density independent combat, Milne points out that any density influences are usually "imperfect," are conditional on other factors, and have high variance. He believed that "perfect" density dependence, with inexorable and overwhelming effect, would be virtually restricted to very high densities and would often be associated with intraspecific competition. Thus, rather than maintaining an equilibrium or stable density in populations, "density dependence" should restrain populations below some hazy broad band of densities. Milnes's ideas were important in leading me to notions of "density vagueness" and "liberal" population dynamics (Strong 1984, 1985).

Reddingius and den Boer (1970) offer the most unusual and, to my mind, one of the most beautiful approaches to population and community theory. It is certainly one of the most realistic and is essential for an appreciation of the multifaceted problem of modeling population behavior of insects on plants. There are two antecedents to this work. One is Reddingius's "Gambling for Existence" (1971), which is a cogent and extensive challenge to prim theory's most basic assumption, fine scale, explicit density dependence that operates continuously over the full range of densities traversed by a population. He effectively refutes the old saw that explicit, continuous density dependence is logically necessary for population persistence, to prevent frequent excursions to very high and zero densities. Reddingius (1971) presents several theoretical counterexamples to explicit, continuous density dependence; his populations change according to eco-

logically realistic Markovian rules that are not unlike the external forces acting on real populations.

The second antecedent to the paper referred to in Table 3.2 is den Boer's (1968) idea of "spreading the risk" among subpopulations of a species, which might dampen the amplitude of population fluctuations of the total population. This idea is basic to deterministic subpopulation models of persistence or coexistence (Cohen, 1971; Levins and Culver, 1971; Slatkin, 1974).

Reddingius and den Boer (1970) shape their theories to mesh with the sorts of variables measured by people who deal with real organisms in nature rather than for the elegance of particular kinds of equations. One of the findings is that an increased number of different factors affecting net change in population causes increasing stabilization of population fluctuations. This result is controversial (May, 1971; Roff, 1974a; Levandowsky, 1974). To me, an independent finding is more interesting; explicit density dependence is not needed for populations to persist over a large number of generations. In one model, without explicit density dependence and with average growth rates ($\ln R_0$) slightly greater than zero, populations persisted for large numbers of generations even with externally imposed random "crashes" of the sort that might be caused by the weather or a catastrophic disease. In general, they found that density dependence is not conspicuously more important in population persistence than migration among subpopulations with dissimilar environments (Fig. 3.2*a*) or than age structure (Fig. 3.2*b*).

Roff's (1974b,c) "spatial heterogeneity" is a detailed and analytical treatment with few parameters of movement among subpopulations that is similar in spirit to that of Reddingius and den Boer (1970), who used computer simulations with a rather large number of parameters. Among many interesting results, Roff found that dispersal among subpopulations generally increases persistence time by several orders of magnitude and that dispersal in a heterogeneous environment reduces the variance in the total population that is caused by environmental fluctuations. Both the sort of dispersal (to all versus just some of other subpopulations) and the total number of subpopulations can greatly affect persistence time without affecting mean subpopulation size.

Chesson (1978) gives an inclusive review of stochastic approaches to population and community ecology. "Stochastic boundedness" is his mathematical view of species persistence within hazy limits that involves altered average population growth rates as a function of population size. As the lower boundary region is approached from above, a stochastically bounded population will have an increasing trend of densities that indicates a tendency of positive net growth rates. The opposite occurs as the the upper boundary region is approached from below; net population change is negative on average over a stretch of time or series of generations. One must emphasize boundary "regions" because the stochastic nature of the idea is distinct from the absolute boundaries such as $k$ in determistic Lotka–Volterra equations. Stochastic boundaries may vary, and need be established empirically. Stochastic boundedness can be seen as vague density relationships with "soft" floors and ceilings (Fig. 3.3).

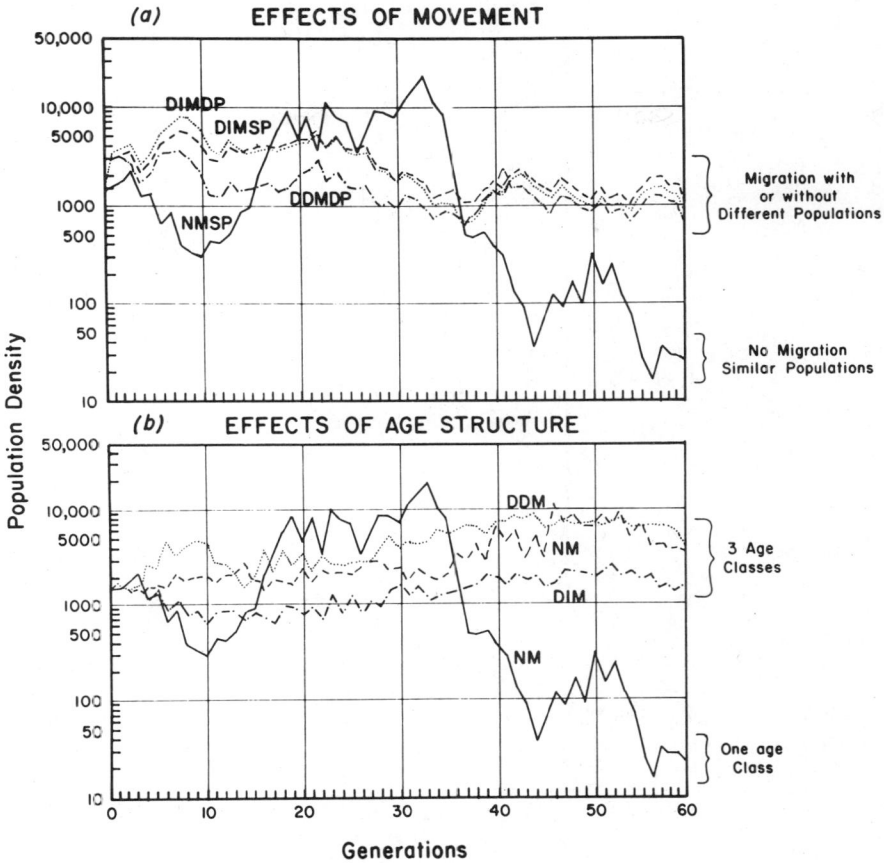

Fig. 3.2 Effects of (*a*) dispersal and (*b*) age structure among subpopulations on population dynamics in the stochastic population models of Reddingius and den Boer (1970), redrawn from their Figs. 6 and 7.

The "storage effect" (Chesson, 1985; Warner and Chesson, 1985) is a way that stochastic environments can reverse a deterministic process of local competitive exclusion and lead to local competitive coexistence; all species have positive growth rates at low density even though one is sufficiently superior at higher densities to cause the other to have a negative population growth rate.

## 3.4. QUASI-EQUILIBRIUM AND BROAD BANDS FOR DENSITY "REGULATION"

Virtually all ecologists, including myself, believe in processes at least reminiscent of stability, in a vague sense. Our intuitions are shaped by geographical distributions that are fairly constant for many species and by relative abun-

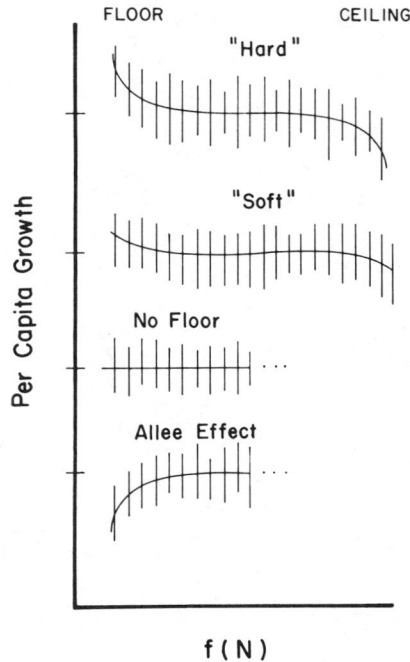

Fig. 3.3 Density vagueness is the lack of trend between per capita population change and density at medial and low densities. Standard deviations are shown by vertical lines and means by horizontal lines. Ticks on the vertical axis indicate no net change in $N$.

dances of populations that remain within rough broad bands of density. In the absence of man's more pernicious influences, immigration and irruption of locally novel species and extinctions of established ones seem rare, at least within periods that our lifespans allow us to see. Features of distribution and abundance are certainly reminiscent of equilibria, in a certain sense. Balancing processes of some sort that prevent, or at least retard, the extremes of outbreak and extinction must work, somehow, in ecological nature.

"Equilibrium" has played a role in ecological mythology much like that of the perpetual motion machine in ancient and medieval physics, a platonic ideal of how things would work sans mundane imperfections. The second law of thermodynamics is perpetual motion's downfall, and in ecology the bugbears of ideal equilibrium are logically equivalent to the second law in physics. The bugbears are climatic fluctuations, seasons, and internal but loosely linked features of real trophic webs, phenologies of host species, spatial heterogeneity, and the highly aliatory character of dispersal and migration.

Prim matheory derives from an intellectual tradition with a distinct view of ecological nature, of the sort given by Slobodkin (1961, p. 19): "if a sufficiently long time period is examined it is found that the mean number of animals in any particular species or in any particular region is constant. We can therefore assume that the ecological world is in a steady state." Here, I am less concerned with the logical curiousness and statistical credulousness of this assertion than in its ideological basis. I dare say that few, if any, ecologists now consider such a "steady state" to be a fruitful view of organismal nature. But we are tentative about alternatives and only recently have begun to address them in any comprehensive way. Some new terms have cropped up to describe alternatives to simple steady states and equilibria (Connell and Sousa, 1983), and mathematics have found that even constant values of parameters in deterministic models can produce an array of unsteady dynamics, from constant population size, through stable cycles, diverging or converging cycles, to mathematically deterministic "chaos" (May, 1981) and "strange attractors" (Schaffer, 1985).

The empirical front in ecology traditionally has been less sanguine about "equilibrium" and "stability" than has the matheory front. This is shown powerfully by Connell and Sousa (1983) in a review of the empirical literature. They conclude that there is no clear demarcation between assemblages that may exist in an equilibrium state and those that do not. They found only a few real examples of what might be stable limit cycles, and no evidence of multiple steady states in unexploited natural populations or communities. "Rather than the physicist's classical ideas of stability, the concept of persistence within stochastically defined bounds is, in our opinion, more applicable to real ecological systems" (p. 808). Schoener (1986) also reviewed the literature and comes to what I appraise as a conclusion functionally similar to that of Connell and Sousa.

### 3.4.1. Density Vagueness

Density relationships occur in many, if not all, populations (Table 3.3), but density relationships need not be simple. The logically simplest form of regulation is to a point, given by the elementary version of the logistic equation, in which population is forced to "carrying capacity" $k$, by density dependence (DD) that operates deterministically, forcefully, and continuously over the full range of densities (Fig. 3.3). Other deterministic density relationships are less simple and give stability (cycles and "strange attractors") that is less obvious or, paradoxically, in some cases no stability in any sense (chaos of cascading bifurcations). But deterministic density relationships are only one possible source of quasi-equilibria, fading equilibria, and equilibria-lost.

Just as convergence to a constant $k$ is only one simple extreme of deterministic population behavior, explicit, deterministic, and continuous DD is only one extreme of a rich spectrum of real density effects in natural populations, all of which diffuse any tendencies for equilibria and stability into broad indeterministic bands. The stochastic, discontinuous, and irregular sorts of density relationships that one finds in many real populations might be called "density vagueness" (Strong, 1984, 1985).

Density vagueness is natural variability in density relationships, rich and multifaceted variability in slope, shape, continuity, and the range of population

**Table 3.3    Kinds of Density Effects and Sorts of Regulation in Some Population Models[a]**

| Density Effect | Sort of Regulation | Model |
|---|---|---|
| Explicit, low slope | Fixed point | Logistic and Lotka–Volterra (LV) with no time lags, no variance |
| Explicit, moderate slope | Moving point | Logistic and LV with time lags, no variance |
| Explicit, high slope | Deterministic "chaos" | Logistic and LV with cascading bifurcations (May, 1981) |
| Explicit and fuzzy, moderate slope | Narrow band | Logistic and LV with continuous density dependence and a bit of irregularity |
| Explicit and changing slope; nonlinear | Deterministic "chaos" | Strange attractors (Schaffer, 1985) |
| Vague | Broad band or quasi-stability | Bounded stochasticity (Chesson, 1985; Connell and Sousa, 1983) Liberal regulation (Strong, 1984) Heterogeneity/dispersal models (Reddingius and den Boer, 1970; Paine and Levin, 1981; Roff, 1974b,c) |
|  |  | Envelopes (Ehrlich et al., 1972) |
|  |  | Pest outbreaks (Southwood and Comins, 1976; Ito, 1961) |
|  |  | Imperfect regulation (Milne, 1957) |
|  |  | "Counter model" (Horn, 1968) |

[a] The "slope" of the density effect is a reference to the sort of slope in the equation $L = (B)N_{eq}$ used by Krebs (1985) in Fig. 12.2 that gives rise to the different kinds of population growth in his Fig. 12.3.

sizes over which negative feedback occurs in density. At least some density vagueness must occur even in the most deterministic real populations, for example, even in quickly responding linear birth and death processes (Fig. 3.4). Even slight density vagueness in birth and death rates causes (Fig. 3.3) population change that is largely independent of density over medial densities, which some sort of DD at high densities and either inverse DD (an Allee effect), strict density independence (DI), or some kind of DD at the lowest densities. The key element for density relationships is the variation around any sort of measure of central tendency, such as a regression line. Variation can be resolved into some degree of "inherent" or residual variance and contingent variance that is caused by known variables other than density. One shortcoming of this approach (which is inspired by analogy to analysis of covariance) is that it may artifactually impose a density relationship on medial densities, where the population spends most time, when the real responses to density occur only at extreme densities or only at

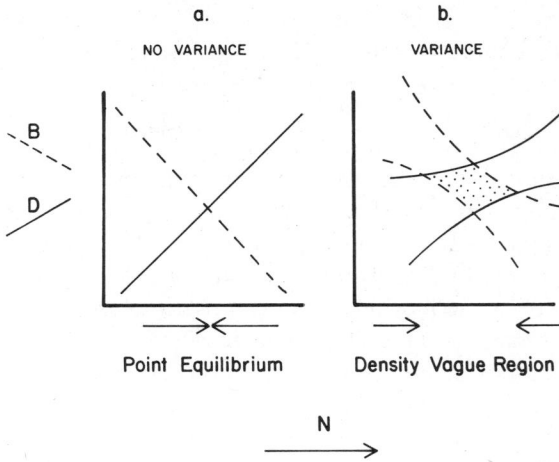

Fig. 3.4 Per capita birth and death rates as a function of population density. (*a*) Completely deterministic rates give a point equilibrium. (*b*) Variance in birth and death rates yield an intermediate region where population change is unrelated to density.

high densities; inferences about density functions depend upon experimental design and techniques of curve-fitting, and interpolation to medial densities from extremes is a serious technical issue for ecologists; density relationships in ecology are at least as much an empirical and statistical subject as a mathematical one.

### 3.4.2. Implications for IPM

IPM shows promise for permitting appreciation of the rich and density-vague population dynamics of insects by differentiating among some broad categories of population behaviors that theoretical ecology has usually been oblivious to or, at least, uninterested in. A good example is the clear distinction between "classical" biological control (BC) and "augmentative" or "inundative" BC. In classical BC, the explicit assumption is that the persistence of a natural enemy and its reduction of the target pest species are part and parcel of its "density-dependent" interactions with the host and with its host specificity, which prevents other, potentially beneficial hosts from being attacked by introduced species.

> The concept of classical biocontrol "serves to direct scientists efficiently to sources of the most host-specific, and therefore most density-dependent . . . and effective, agents for the reduction of populations of introduced pests" (Batra, 1982, p. 136).

Inundative or augmentative BC, where the natural enemy does not persist until the next growing season or until the next outbreak, is believed to differ fundamentally from classical BC in lacking the "density-dependent" relationship with its host (Table 1 in Batra, 1982). Even though the basis of this distinction is now known to be incorrect, as I shall explain below, the mere acknowledgment of population behavior based upon something other than explicit density dependence is an essential step for any conceptual progress in the population biology of plant protection. The error is in basing distinction upon differences in density relationships. Introduced natural enemies that persist and suppress their host populations for fairly long runs of generations do not necessarily have any clear density-dependent relationship with their host population.

A good example is the pair of introduced parasitoids of the olive scale that have persisted for many years in California olive groves and have caused populations of the scale to decrease tremendously in some areas. The authors of the original report on the details of this success in "classical" biocontrol fretted over the apparent lack of density dependence for these parasitoids (Huffaker and Kennett, 1966, p. 366), as I have noted elsewhere (Strong, 1984, 1985). Murdoch et al. (1984) have examined in great detail and with much statistical rigor the local patterns of scale mortality caused by the parasitoid and conclude that there is not any evidence for the assumed density dependence, at least at the level of individual branches and trees; there are no critical data on any higher levels, such as entire groves or different geographical regions.

In a further challenge to the dogma about the population behavior that causes "classical" biocontrol to work, Murdoch et al. (1985) point out the problems with assuming that "equilibrium" and "stability" are characteristics of successful natural enemies; local extinction and reinvasion in the spirit of the realistic models of Tables 3.1 and 3.2 above are much more likely. This conclusion leads one to question any inference of IPM based upon long-term equilibrium among herbivorous pests and their natural enemies. I have argued above that any such equilibrium or stability among plants and their herbivore populations is also quite unlikely.

### 3.4.3. Murdoch Experiments

The implication of my message, of the inexorable looseness of trophic connection, high contingency, great external influence, stochastic complexity, and density vagueness for insects on plants is that population dynamics very much needs critical empiricism, which illuminates variability as much as central tendencies of population mechanisms and density relationships. It has been long recognized that one sort of evidence is central to our understanding of these phenomena; this evidence comes from perturbation experiments. For population ecology, I call these "Murdoch experiments," after Murdoch's (1970) clear statement of what is required for knowing population dynamics in any critical way: Displace population both above and below natural densities and study subsequent population behavior. Bender et al. (1984) have elaborated upon Murdoch's original suggestions. "Equilibrial" tendencies are shown by convergent behavior in net population change, and density relationships are shown by patterns and variances in the magnitudes of change at displaced densities. Of course, birth and death rates, as well as the net rate itself, should be studied in this way. As simple as is the basic idea, results are likely to be very complex, and proper implementation in field situations will be quite involved, statistically sophisticated, and expensive. This is by no means an argument against these experiments (there is little alternative!), but it is an argument for both ecologists and granting agencies to view experimental population dynamics as a worthwhile undertaking.

Implementing Murdoch experiments will always involve replication, because real populations never have a density, they have a range of densities among the places and times that the species occupies and among samples within places and times. How this range of densities is affected by other variables (habitat heterogeneity, host plant phenology, vigor, age, or natural enemies, etc.) will greatly influence interpretations of the experimental outcome and may lead to additional experiments for resolution. Every factor or "contingency" (Strong, 1985) that one considers as having a potential influence upon population behavior would involve other independent variables. Some of the most interesting variables, such as season and plant phenology, might require the basic experiment to be repeated through time. Finally, the experimental technique must be good enough for the controls (experimental densities set within the range of natural densities) to behave during the experiment in the manner that the natural densities behave.

Temporally, the simplest kind of Murdoch experiment is done during a distinct ecological period like spring. In this schema, standard time series studies, which search net signal for cycles or trends, are often impractical because they require strings of observations that are very long (Schaffer, 1985). [When changes in behavior of individuals, within generations, rather than changes in density among generations are the object of study, standard time series analyses sometimes are used because long strings of behavioral data are much easier and cheaper to obtain than accurate long strings of intergeneration density data (Roubik, 1983).] The experiment can address seasonal or phenological phenomena by being repeated during several characteristic periods or slices of time like spring, summer, fall, and winter, or preleafing, leafing out, and expanded-leaf stages of a host plant, for example.

Within time slices, the range of possible outcomes for Murdoch experiments is pretty broad. The simple point equilibrium that is stationary during the time slice is a logical reference point, but a highly unlikely one. Other more likely possibilities are temporary density independence, in a statistical sense, and vague density signals at the extremes of the density range.

If the elevated densities of the experiment were not high enough to detect any upper density range that caused density dependence during one period or season, most of us would be inclined to expect some density dependence during a subsequent period or season [but recall Reddingius's (1971) results, Fig. 2!]. Stationary density independence in the context of a Murdoch experiment would be no net trend in densities and no trend in magnitude of change as a function of density. Directional density independence would have a net trend in density up or down, but no trend in magnitude of change as a function of density. Directional density independence can be positive, for example, in the spring with a multivoltine population that is sparse relative to its host plant or natural enemies, that grows through several generations without feeling much ill effect of density. In the fall the same population could experience negative growth and directional DI again as conditions deteriorated and populations fell. If conditions remained salutary for sufficiently long during the growing season there could be some vague density dependence in the most crowded colonies, patches, or aggregations, with none in the sparser aggregations.

Of course, cycles with a period longer than the series of experiments would be resolved only by much larger runs of data through the phases of a cycle. I have been several times challenged by people who argue that the *possibility* of long cycles or fluctuations is some sort of mark against experimental population dynamics. I find this argument illogical simply because, without experimentation, one always knows much less about anything; in systems where experiments are impossible, one automatically knows less than if they were possible. If cycles are real, then intense DD might occur intermittently, and one would repeat a Murdoch design during nadirs and zeniths of the cycle to learn how the density relationship worked, about its central tendencies and variance.

## 3.5. SUMMARY

Trophic webs for insects on plants are relatively "tall and narrow-waisted." Vertical influences between the basal plants and herbivores, and between herbivores and their natural enemies, are the most generally forceful in these webs, with horizontal, competitive, and facilitative forces between herbivores less often important.

Linkages in foodwebs for insects on plants are usually "loose," weak, or variable in intensity in time and space. The variety of linkages suggests that these webs are complex, and the looseness of linkages suggests that they are stochastically, as opposed to deterministically, complex.

Stochastic complexity suggests that realistic ecological theory, which is sensitive to real heterogeneity of environments, instability of species associations, migration, and dispersal, thus fluctuations of population, is more appropriate to ecology than "prim" theory, which can be the antithesis of realistic ecological theory.

Several lines of reasoning suggest that real populations are not "equilibrial" or "stable," with central set points or deterministic cycles stochastically bounded to extremes of population excursion. Density relationships are part and parcel of any kind of population model, and even the smallest amount of variance in density relationships will generate an indeterminate medial range of population within which "regulation" or "control" does not operate. The variance of density relationships is often great for natural populations; their dynamics are density-vague, making quasi-equilibria with broadly separated bounds the only possible sort of "regulation."

One of the most productive approaches to the study of population variability is the perturbation experiment, which I term the Murdoch experiment, after designs by Murdoch (1970). This sort of experiment is simple in basic concept but often difficult and complex in practice because the null expectation, given seasonality, phenology, and interannual change in host plants and other factors, is not constancy but change that is not tightly related to population density.

## ACKNOWLEDGMENTS

I thank Peter Kareiva and Richard Green for vigorous discussion on these topics and Mike Antolin, Nick Gotelli, Mick Keough, Richard Levins, Robert May, Daniel Simberloff, Anne Thistle, and Joe Travis for comments on the manuscript. Work partially supported by NSF grant BSR 8206856 to author.

## REFERENCES

Andrewartha, H. G., and L. C. Birch. (1954). *The Distribution and Abundance of Animals*, University of Chicago Press, Chicago.

Armstrong, R. A., and R. McGehee. (1980). Competitive exclusion, *Am. Nat.* **115**, 151–170.

Atsatt, P. R. 1973. Parasitic flowering plants: How did they evolve? *Am. Nat.* **107**, 502–510.

Batra, S. W. T. (1982). Biological control in agroecosystems, *Science* **215**, 134–139.

Beddington, J. R., and R. M. May. (1977). Harvesting natural populations in a randomly fluctuating environment, *Science,* **197**, 463–465.

Bender, E. A., T. J. Case, and M. E. Gilpin. (1984). Purturbation experiments in community ecology: Theory and practice, *Ecology,* **65**, 1–13.

Brodbeck, B. V., and D. R. Strong. (1986). "Amino acid nutrition of herbivorous insects and stress to host plants," in P. Barbosa and J. Schultz, Eds., *Insect Outbreaks,* Academic, New York.

Capuccino, N. and P. Kareiva. (1985). Coping with a capricious environment: A population study of the rare woodland butterfly *Pieris virginiensis, Ecology,* **66**, 152–162.

Caswell, H. (1982). Life history and the equilibrium status of populations, *Am. Nat.,* **120**, 317–339.

Chesson, P. (1978). Predator–prey theory and variability, *Ann. Rev. Ecol. Syst.,* **9**, 323–347.

Chesson, P. L. (1985). "Variation and species interactions," in J. Diamond and T. J. Case, Eds., *Community Ecology,* Harper & Row, New York, pp. 229–239.

Cohen, J. E. (1971). Mathematics as metaphor, *Science,* **172**, 674–675.

Connell, J. H. (1983). On the prevalence and relative importance of interspecific competition; evidence from field experiments, *Am. Nat.,* **122**, 661–696.

Connell, J. H., and W. P. Sousa. (1983). On the evidence needed to judge ecological stability or persistence, *Am. Nat.,* **121**, 789–824.

DeBach, P. (1958). The role of weather and entomophagous species in the natural control of insect populations, *J. Econ. Entomol.,* **51**, 474–484.

den Boer, P. J. (1968). Spreading of risk and stabilization of animal numbers, *Acta Biotheor.,* **18**, 165–194.

Dempster, J. P. (1983). The natural control of populations of butterflies and moths, *Biol. Rev.*, **58**, 461–481.

Ehrlich, P. P., D. E. Breedlove, P. F. Brussard, and M. A. Sharp. (1972). Weather and the "regulation" of subalpine populations, *Ecology*, **53**, 243–242.

Fox, L. R. (1975). Cannabalism in natural populations, *Ann. Rev. Ecol. Syst.*, **6**, 87–106.

Getz, W. M., and A. P. Gutierrez. (1982). A perspective on systems analysis in crop production and insect pest management, *Ann. Rev. Entomol.*, **27**, 411–446.

Gutierrez, A. P., J. U. Baumgaertner, and C. G. Summers. (1984). Multitrophic models of predator–prey energetis. *Can. Entomol.*, **116**, 923–963.

Hairston, N. G. (1985). The interpretation of experiments on interspecific competition. *Am. Nat.* **125**, 321–325.

Hassell, M. P. Manuscript. Insect natural enemies as regulating factors.

Hassell, M. P., and T. R. E. Southwood. (1978). Foraging strategies of insects, *Annu. Rev. Ecol. Syst.*, **9**, 75–98.

Hastings, A. (1984). Evolution in a seasonal environment: Simplicity lost? *Evolution* **38**, 350–358.

Horn, H. S. (1968). Regulation of animal numbers: A model counter-example, *Ecology* **49**, 776–778.

Huffaker, C. B., and C. E. Kennett. (1966). The biological control of *Parlatoria oleae* (Colvee) through the compensatory action of two introducd parasites, *Hilgardia*, **37**, 283–334.

Ito, Y. (1961). Factors that affect the fluctuations of animal numbers, with special reference to insect outbreaks, *Bull. Natl. Inst. Agric. Sci. Tokyo*, **13**, 57–89.

Krebs, C. J. (1985). *Ecology: The Experimental Analysis of Distribution and Abundance*, 3rd ed., Harper & Row, New York, 678 pp.

Lawton, J. H., and M. P. Hassell. (1981). Asymmetrical competition in insects, *Nature*, **289**, 793–795.

Levandowsky, M. (1974). A further comment on the model of Reddingius and den Boer, *Am. Nat.*, **108**, 396–397.

Levins, R. (1979). Coexistence in a variable environment, *Am. Nat.*, **114**, 765–783.

Levins, R., and D. Culver. (1971). Regional coexistence of species and competition between rare species, *Proc. Natl. Acad. Sci. USA*, **68**, 1246–1248.

May, R. M. (1971). Stability in model ecosystems, *Proc. Ecol. Soc. Australia*, **6**, 18–56.

May, R. M. (1981). The role of theory in ecology, *Am. Zool.*, **21**, 903–910.

Milne, A. (1957). The natural control of insect populations, *Can. Entomol.*, **89**, 193–213.

Mopper, S., S. H. Faeth, W. J. Boecklen, and D. S. Simberloff. (1984). Host-specific variation in leaf miner population dynamics: effects on density, natural enemies and behaviour of *Stilbosis quadricustatella* (Lep. Cosmopterigidae), *Ecol. Entomol.*, **9**, 169–177.

Murdoch, W. W. (1970). Population regulation and population inertia, *Ecology*, **51**, 497–502.

Murdoch, W. W., J. Chesson, and P. Chesson. (1985). Biological control in theory and practice, *Am. Nat.*, **125**, 344–366.

Murdoch, W. W., J. D. Reeve, C. B. Huffaker, and C. E. Kennett. (1984). Biological control of olive scale and its relevance to ecological theory, *Am. Nat.*, **123**, 371–392.

Oster, G. (1981). Predicting populations, *Am. Zool.*, **21**, 831–844.

Paine, R. T. (1979). The third Tansley Lecture. Food webs: Linkage, interaction strength and community structure, *J. Anim. Ecol.*, **49**, 667–686.

Paine, R. T., and S. A. Levin. (1981). Intertidal landscapes: Disturbance and the dynamics of pattern, *Ecol. Monogr.*, **51**, 145–178.

Peters, R. H. (1977). The unpredictable problems of tropho-dynamics, *Env. Biol. Fish.*, **2**, 97–101.

Pimm, S. L. (1982). *Food Webs,* Chapman Hall, 219 pp.

Reddingius, J. (1971). Gambling for existence. A discussion of some theoretical problems in animal population ecology, *Acta Biotheor.,* (Suppl.), **XX,** 208 pp.

Reddingius, J., and P. J. den Boer. (1970). Simulation experiments illustrating stabilization of animal numbers by spreading of risk, *Oecologia,* **5,** 240-284.

Rodriguez, J. G., ed. (1972). *Insect and Mite Nutrition,* North-Holland, Amsterdam, 702 pp.

Roff, D. A. (1974a). A comment on the number-of-factors model of Reddingius and den Boer, *Am. Nat.,* **108,** 391-393.

Roff, D. A. (1974b). Spatial heterogeneity and the persistence of populations, *Oecologia,* **15,** 245-258.

Roff, D. A. (1974c). The analysis of a population model demonstrating the importance of dispersal in a heterogeneous environment, *Oecologia,* **15,** 259-275.

Rosenthal, G. A., and D. H. Janzen, Eds. (1979). *Herbivores: Their Interaction with Secondary Plant Metabolites,* Academic, New York, 718 pp.

Roubik, D. W. (1983). Experimental community studies: Time series tests of competition between African and neotropical bees, *Ecology,* **64,** 971-978.

Royama, T. (1984). Population dynamics of the spruce budworm *Choristoneura fumiferana, Ecol. Monogr.,* **54,** 429-462.

Schaffer, W. M. (1985). Order and chaos in ecological systems, *Ecology,* **66,** 93-106.

Schoener, T. W. (1983). Field experiments on interspecific competition, *Am. Nat.,* **122,** 240-285.

Schoener, T. W. (1986). "Patterns in terrestrial vertebrate vs. arthropod communities: Do systematic differences regularly exist?" in, T. Case and J. Diamond, Eds., *Community Ecology,* Harper & Row, New York.

Scriber, J. M., and F. Slansky, Jr. (1981). The nutritional ecology of immature insects, *Annu. Rev. Entomol.,* **26,** 183-211.

Slatkin, M. (1974). Competition and regional coexistence, *Ecology,* **55,** 128-134.

Slobodkin, L. B. (1961). *Growthand Regulation in Animal Populations,* Holt, Rinehart, Winston, New York, 184 pp.

Southwood, T. R. E. (1972). "The insect/plant relationship—An evolutionary perspective," in H. G. van Emden, Ed., *Insect/Plant Relationships,* Symposia of the Royal Entomological Society of London, Number 6, Blackwell Scientific, Oxford, pp. 3-30.

Southwood, T. R. E., and H. N. Comins. (1976). A synoptic population model, *J. Anim. Ecol.,* **45,** 949-965.

Steele, J. H., and E. W. Henderson. (1984). Modeling long-term fluctuations in fish stocks, *Science,* **224,** 985-987.

Strong, D. R. (1984). "Density vague ecology and liberal population regulation in insects," in P. W. Price and C. N. Slobodchikoff, Eds., *A New Ecology: Novel Approaches to Interactive Systems,* Wiley, New York, pp. 313-327.

Strong, D. R. (1986). "Density vagueness: Abiding the variance in the demography of real population," in J. Diamond and T. J. Case, Eds., *Community Ecology,* Harper & Row, New York, pp. 257-268.

Strong, D. R., J. H. Lawton, and T. R. E. Southwood. (1984). *Insects on Plants: Community Patterns and Mechanisms,* Blackwell, Oxford, 313 pp.

Thompson, W. R. (1928). A contribution to the study of biological control and parasitic introduction in continentl areas, *Parasitology,* **20,** 90-112.

Wang, Y. H., and A. P. Gutierrez. (1980). An assessment of the use of stability analyses in population ecology, *J. Anim. Ecol.,* **49,** 435-452.

Warner, R. R., and P. L. Chesson. (1985). Coexistence mediated by recruitment fluctuations: A field guide to the storage effect, *Am. Nat.*, **125**, 767–787.

White, T. C. R. (1978). The importance of a relative shortage of food in animal ecology, *Oecologia*, **33**, 71–86.

Whitham, T. G. (1981). "Individual trees as heterogeneous environment: adaptation to herbivory or epigenetic noise," in R. F. Denno and H. Dingle, Eds., *Insect Life History Patterns: Habitat and Geographic Variation*, Springer-Verlag, New York, pp. 9–28.

# 4

---

# TRIVIAL MOVEMENT
# AND FORAGING
# BY CROP COLONIZERS

*Peter Kareiva*

## 4.1. INTRODUCTION

In crop systems, most of which are short-lived in comparison to natural communities, damage is typically caused by herbivores that migrate into fields from the outside (Joyce, 1976; Price, 1976). Consequently, much attention has been paid to the dispersal behavior of insects, with the hope of understanding how dispersal influences patterns of pest attack (Cameron et al., 1979; Baltensweiler and Fischlin, 1979; Stinner et al., 1983). Despite the accumulation of data on the movements of numerous pest species (Lamb et al., 1971; Hawkes, 1972; Howell and Clift, 1974; Gentry et al., 1979; Fletcher and Kapatos, 1981) and despite the formulation of analytic and simulation models of dispersing pest populations (Ludwig et al., 1979; Jones 1979), we have made little progress toward this goal (Stinner et al., 1983). A meaningful incorporation of dispersal into pest management theory has been hampered by a series of problems: the mathematical complexity of all but the simplest dispersal models (Shoemaker, 1981), the technical difficulties of detecting sources of migrants, and the improbability of observing long-range movements. But even if we were to solve these problems, it is unlikely that we could manage this dispersal in a way that would reduce crop losses. Instead, I argue in this chapter that we should become more concerned with how pest insects move after they have arrived at a crop system, with so-called "trivial movement" (*sensu* Southwood, 1962). Although trivial movement lacks the drama of locust swarms or massive flights of rice planthoppers, its ecological consequences are potentially just as profound. And because all insects engage in trivial movement in the course of feeding and oviposition, a mechanistic understanding of these movements could be of very general importance. Although its study has been neglected in the past, there is no technical reason for this neglect;

trivial movement is well suited to direct observation, manipulative field experiments, and testable mathematical models.

Since trivial movement has been so little studied, much of the discussion that follows describes broad areas in need of investigation. There are no definitive experiments to report. Models are suggested, but these are models whose behavior has been only scantily explored. I rely heavily on my students' and my own work because that is what I am most familiar with. First, I examine how the trivial movement of herbivores is affected by changes in the diversity of the food plant "community." I follow this section with a cautionary note about the interpretation of standard agricultural "randomized block" experiments; by creating spatial patterns these experiments influence the movements of associated pest insects, and thus complicate attempts at extrapolation to large-scale uniform plantings. Second, I discuss how the foraging movements of predators and parasitoids are affected by spatial patterns in the distribution of their prey or their prey's habitat. These foraging movements are of practical concern because they may influence the success with which natural enemies regulate pest populations. Finally, I consider how the nonuniform distribution of pest insects within a field (the outcome of trivial movement) might affect the reduction in crop yield caused by pests; in particular, I point out those circumstances for which simple averaging of the numbers of pests per plant is unlikely to produce accurate predictions of damage. I close with some suggestions for promising theoretical directions, and with a justification for the use of partial differential equations in the development of models for pest management.

## 4.2. THE EFFECT OF SYSTEM DIVERSITY ON MOVEMENT OF PEST SPECIES

When crops are grown in polycultures, the numbers of pest herbivores per plant are often reduced relative to crop monocultures (Dempster and Coaker, 1974; Perrin and Phillips, 1978; Cromartie, 1981; Andow, 1983a; Risch et al., 1983). This effect, however, is not consistently obtained and its magnitude varies enormously among plant–pest combinations (Kareiva, 1983). As yet, we have no theory to predict whether a particular pest will increase or decrease in response to a particular cropping manipulation. Thus, despite an ever-growing list of experimental results, we cannot identify the ecological situations or life history traits that make some crop pests sensitive and others insensitive to cropping patterns. To advance beyond merely cataloging the results of crop-patterning experiments, the mechanisms underlying pest responses must be identified.

Cropping pattern may influence pest populations primarily through its effect on the movements of pests. In the few cases that mechanisms have been explored, reduced infestations in polycultures seem to have been due to higher (relative to monoculture) herbivore emigration rates, rather than to higher herbivore mortality rates (Risch, 1981; Risch et al., 1983; Bach, 1984; Kareiva 1985a,b). These systems all involve specialist herbivores for which the enhanced emigration observed in polycultures can be easily explained by changes in patterns of trivial movement. In particular, such an "emigration effect" should be evident in polycultures when the pest's trivial movement involves: (1) mistakenly alighting on nonhost plants, (2) moving off nonhost plants more frequently than moving off host plants, and (3) being at risk of leaving the crop area during movement. It is easy to imagine herbivores satisfying these three criteria. Specialist herbivores often land on unsuitable plants, and by definition (i.e., they are specialists) these errant insects will leave a nonhost much more readily than a host. It has also been shown that, in the course of flight, herbivores occasionally wander away from patches or fields of their food plants (Kareiva, 1983, 1985a). I suspect that quantitative studies relating the frequency of nonhost encounters to subsequent accidental emigration will explain much of the variation in success of polycultures at reducing pest attack. To examine this hypothesis we need models that predict net emigration losses from observations of trivial movement among plants.

Emigration models can be mathematically formalized as continuous time, finite-state Markov processes [see Kareiva (1982) for mathematical background in a biological setting] in which insects move among three states (on host plant, on nonhost plant, outside plot) in polycultures, but only between two states (on host plant, outside plot) in monocultures. Because computations and parameter estimation are simple for these models, this approach is ideal for exploring the connection between herbivory and trivial movement. Instantaneous transition rates between the above states can be easily obtained by releasing and recapturing marked insects. Equilibrium herbivore densities for polycultures versus monocultures can then be calculated from the model; with this approach it might be possible explicitly to attribute reduced pest pressure in polycultures to high rates of movement, either from host to nonhost plants or from nonhost plants to areas outside the crop. The influences of pest mobility and pest host-selection behavior are summarized in the transition matrix [or "intensity matrix," see Kareiva (1982) or Karlin and Taylor (1975)] that is the heart of the model. Markov models such as this should help us both interpret experimental results and predict the response of pests to novel cropping manipulations (e.g., using larger plots of crop mixtures). It is worth noting that Markov models of movement have proven successful for a variety of insect species [two species of flea beetles, Kareiva (1982); the Mexican bean beetle, Turchin (1986); ovipositing cabbage butterflies, Root and Kareiva (1984)].

Emigration by accident is not the only hazard faced by pest insects in managed systems. Many insects are surprisingly inept at searching for food plants and may consequently die before they locate even seemingly abundant food resources (Dethier, 1959; Kareiva, 1985a). In addition, while searching for food, herbivorous insects may be especially vulnerable to predation (e.g., Dethier, 1959), which places a further premium on finding a host as soon as possible. From the viewpoint of pest management, it may be useful to consider features of cropping systems that will increase the expected time an herbivore spends searching before locating a host. Patchiness of food plants is one such feature—

recent models and experiments indicate that for a fixed density of food plants, clumping of the plants reduces the success of randomly searching herbivores (Cain, 1985; Cain et al., 1985). Although plant clumping is probably not amenable to manipulation in most agricultural ventures, it could become a feature of practical concern in forestry. In particular, unlike homogenous tree plantations, natural forests are usually a patchwork of different species of trees, with individuals of the same species growing together in clusters (e.g., Hubbell, 1979). This difference in the degree of patchiness may help explain why pest problems are less severe in "natural" or unmanaged forests than in plantations (Dahlsten and Dreistadt, 1984). For example, spruce budworm tends to outbreak most severely in pure stands of host trees, and less severely where their hosts are only patchily available (Crook et al., 1979; Kemp and Simmons, 1979). In fact, Kemp and Simmons (1979) report that the intermingling of hosts and nonhost trees in a patchwork significantly increases the mortality rate of dispersing budworms. Thus, there may be an advantage to forestry practices that maintain some degree of clumping or patchiness in the dispersion of host trees. While this necessarily implies diverse as opposed to single-species forests, diversity need not prohibit profitable harvest in forest systems.

## 4.3. PEST MOVEMENT AND PROBLEMS WITH THE INTERPRETATION OF RANDOMIZED BLOCK CROPPING EXPERIMENTS

Agriculturalists have long used randomized block field experiments to determine the effects of crop variety and crop spatial patterning on pest populations. A standard approach is to divide common fields or gardens into subplots that are assigned contrasting treatments such as crop monocultures versus crop polycultures. Pest populations are then censused in all replicates of these plots, and after applying an analysis of variance, inferences are made about the influence of treatments (agricultural practices) on pest infestation. This approach has yielded much valuable information. But the information may have limited applicability in the less contrived arena of real agricultural systems. For example, even though a particular cropping practice might achieve a reduced pest population in a randomized block experiment (relative to other treatments included in the design), it may be ineffective when adopted on a widespread or continuous basis. Therefore, an understanding of the mechanisms behind "significant effects" in such experiments is necessary before the results can be translated to policy recommendations.

   I suggest that nonrandom pest movement is the process often responsible for the different damage levels observed in randomized block field experiments. This argument best applies to mobile pest species. Individual insects of these mobile species will move frequently among the various cropping treatments included in a randomized block design; typically they will spend disproportionate amounts of time in the treatments that represent a preferred food or preferred habitat (e.g., Kareiva, 1982). Similarly, ovipositing females will move among cropping treatments, committing their eggs disporportionately to preferred plants or habitats. The key idea in these examples is that *"preferences" are usually relative, with nonrandom movement distributing insects among habitats according to these preferences.* The result will be that some plots (less preferred treatments) end up with fewer insects per plant than other plots. But those treatments receiving the fewest insects cannot necessarily be thought of as protected—they may well have received fewer insects *only* because the pests were presented with more appealing alternative choices. Without a menu of more favored surrounding plots nearby to attract away the pests, the so-called protected treatments in these experiments would not fare as well.

My attention was first drawn to the practical consequences of nonrandom pest foraging when I undertook field studies of movement behavior in phytophagous insects. I found that many specialized herbivores move even after they have located adequate food plant resources (Kareiva, 1981, 1982, 1983, 1985a). Experiments with flea beetles (*Phyllotreta* spp.) indicated that these specialized herbivores concentrated their numbers on lush, fertilized *Brassica oleracea* plants (collards) as a result of nonrandom foraging, but that their ability to do so required that the plant patches of differing qualities were in close proximity. When favorable and unfavorable *Brassica* patches were widely separated, beetle movement was reduced and infestation levels became virtually independent of food plant quality (Kareiva, 1982). Recently, we have extended our analysis of pest movement to the effects of crop diversity and crop density on *Phyllotreta cruciferae* Goeze (J. Bergelson and P. M. Kareiva, in preparation). We picked this system because, as was mentioned above, *Phyllotreta* forages nonrandomly. In addition, *Phyllotreta* has been shown to exhibit lower populations in collard–potato polycultures than in collard monocultures (Root and Kareiva 1984; R. Root, Cornell University, unpublished data). Our preliminary results reveal an important complication that probably arises in much agricultural research (a complication in which trivial movement plays a crucial role).

To evaluate the dependence of apparent crop protection on local pest movement, we established one set of treatment plots in close proximity to one another with free access between them, and a second set of treatment plots (also in close proximity) separated from one another by tall (1.5 m) curtain barriers. These curtain barriers interfere greatly with flea beetle movement (personal observation). One block of this experimental design is illustrated in Fig. 4.1. The cropping treatments that we contrasted were: (1) pure stands of collards versus collards intercropped with potatoes, and (2) collards planted at a high density of $6.7/m^2$ versus collards planted at a lower density of $3.3/m^2$. Where there were no barriers between intercropped and monoculture treatments, flea beetles were less abundant in the intercropped plots than in adjacent monocultures; where there were barriers between the treatments, intercropping yielded no reduction in beetle infestation (Fig. 4.2*a*). Barriers also affected the influence that collard density had on beetle populations: Collard density had the greatest effect where intervening barriers were absent (Fig. 4.2*b*).

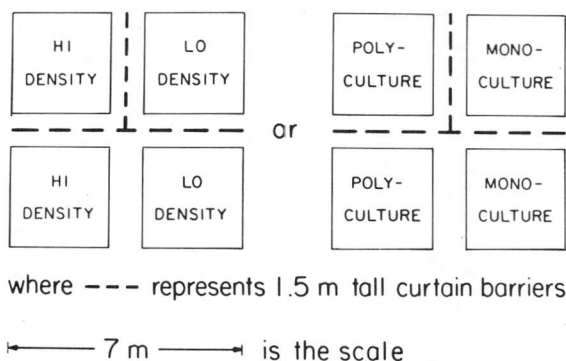

where − − − represents 1.5 m tall curtain barriers

⊢——— 7 m ———⊣ is the scale

Fig. 4.1 Experimental design for evaluating the effect of between-treatment movements on the response of flea beetles to cropping systems (polyculture versus monoculture, or low density versus high density). Movement between the top two subplots in each block was restricted by a curtain barrier. These cultivated blocks were surrounded by old field vegetation comprised mainly of goldenrod (*Solidago* spp.)

I suspect that if movement among treatments were restricted in other polyculture/monoculture experimental designs, the effect of crop diversity on pest attack would often be reduced, or even absent. This does not imply that the results of previous crop diversity experiments should be neglected, but rather that more attention needs to be directed towards underlying mechanisms. Polyculture advantage that is due to the presence of neighboring monoculture "sinks" will not hold up if polycultures are the only available cropping system. When mobile pests are selective about where they spend their time (or place their offspring), extrapolation from randomized block "menu experiments" is a risky proposition.

## 4.4. FORAGING MOVEMENT AND THE EFFECTIVENESS OF PREDATORS AS BIOCONTROL AGENTS

By foraging nonrandomly, insect predators and parasitoids often concentrate their attacks in regions of high prey density (Hassell and Southwood, 1978; Waage, 1979; but also see Murdoch et al., 1985). The details of these nonrandom foraging movements may be important in designing agroecosystems that make effective use of natural enemies. For example, since the likelihood of reliable biocontrol depends substantially on how well predators find their prey, habitat modifications that influence the foraging movements of predators may either interfere with or enhance the control of pest populations. To illustrate this possibility I will briefly describe some of my own research on coccinellid–aphid interactions (Kareiva, 1984, 1985b; Kareiva and Odell, 1986).

The ladybird beetle *Coccinella septempunctata* L. contributes to the control of several aphid pests in Europe and has recently been imported into the United

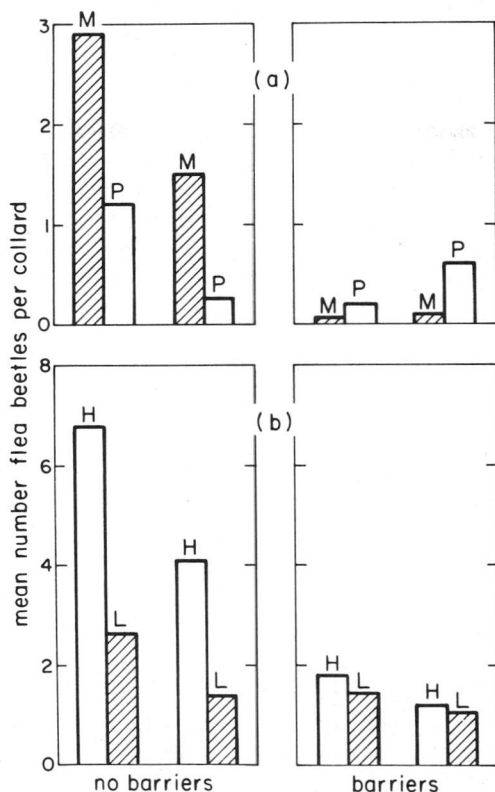

Fig. 4.2 (*a*) Monocultures (M) contain higher densities of beetles than do polycultures (P) when there are no barriers (*left graph*); but this trend disappears where there are barriers between treatments (*right graph*). (*b*) High-density collard plantings (H) contain higher densities of beetles than do low-density plantings (L) when there are no barriers; but this trend disappears where there are barriers between treatments.

States by the USDA (Obrycki et al., 1982). For four years (1981–1984) I studied *Coccinella*'s impact on the population dynamics of the goldenrod aphid, *Uroleucon nigrotuberculatus* Olive. Although not a commercial cropping system, goldenrod grows in large pure stands (much like a crop) and *Uroleucon* occasionally outbreaks to such levels that plants are conspicuously damaged. Conveniently, the spatial pattern of goldenrod plants can be easily manipulated (by mowing and weeding), and goldenrod is a self-maintaining monoculture that does not require energy-intensive cultivation. In the Rhode Island fields of *Solidago canadensis* L. where I have been studying this predator–prey interaction, *Uroleucon* is usually the dominant herbivore and *Coccinella* is consistently the dominant predator.

My research began as a study of the effect of habitat patchiness on *Coccinella–Uroleucon* dynamics, and evolved into a study of how nonrandom foraging influences a predator's capacity for regulating prey populations. The connection between these two inquiries is the observation that habitat patchiness alters *Coccinella–Uroleucon* dynamics primarily by altering the foraging movements of *Coccinella* (Kareiva, 1985b).

Applied entomologists have been interested in patchiness ever since Huffaker (1958) published his classic laboratory study of predator mites and prey mites in heterogeneous environments. These mite experiments have had an enormous impact on general ecological theory, as well as on our ideas about biocontrol (Murdoch et al., 1985). Mathematical models inspired by Huffaker's experiments have even been used to suggest the characteristics of ecosystems and of predators that should facilitate effective biological control of pests (Beddington et al., 1978; Heads and Lawton, 1983; Gould and Stinner, 1984). But I know of no field experiments that have actually determined the effect of patchiness on the ability of predators to control prey populations. To perform such experiments, I have been mowing goldenrod fields into $20 \times 1$-m continuous strips (hereafter called the "continuous" treatment) and into 20-m rows of 1-m$^2$ goldenrod patches (the "patchy" treatment) as shown in Fig. 4.3. These manipulations dramatically influence aphid densities, with aphid populations in the patchy treatment often averaging 10 times the aphid density recorded in paired continuous treatments (Fig. 4.4). Scattered local outbreaks of thousands of aphids per stem occur in the patchy treatments, but are an extremely rare event within continuous stretches of goldenrod. Subsequent experiments have implicated *Coccinella* as the mediator of this effect; aphids sporadically outbreak in the patchy treatments because they have escaped from *Coccinella* control (Kareiva, 1985b).

Observations of aphid demography and *Coccinella* foraging indicate that the key to aphid escape in the patchy treatment is a reduction in the rate at which ladybugs search the habitat, which in turn reduces the rate at which ladybugs aggregate at clumps of high aphid density (Kareiva, 1984, 1985b; P. M. Kareiva, in preparation). In patchy goldenrod, clusters of aphid colonies are more likely to be overlooked by searching ladybird beetles; this increases the likeli-

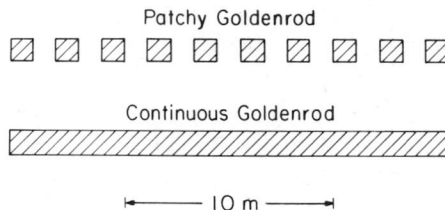

Patchy Goldenrod

Continuous Goldenrod

⊢———— 10 m ————⊣

Fig. 4.3 Geometry of goldenrod in the "patchy" versus "continuous" treatments used in ladybird–aphid experiments. Surrounding vegetation is mown grass. Although there is some movement by ladybirds between arrays, most ladybird foraging movement proceeds along the linear axis of each strip or archipelago. For aphids, after spring colonization by alates, virtually all movement (there is not much) is along the linear axis of each array.

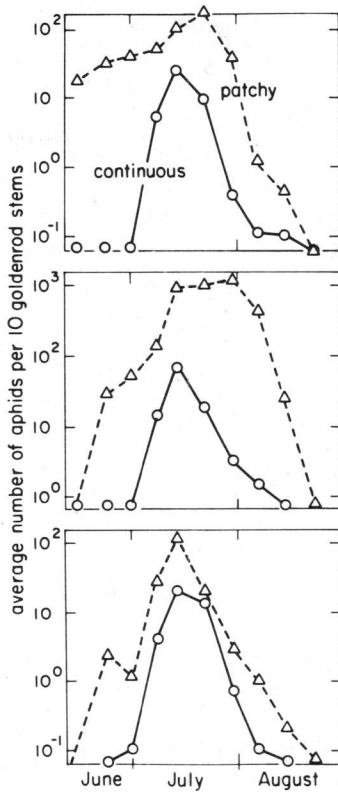

Fig. 4.4 Aphid density in paired arrays of patchy versus continuous goldenrod vegetation as pictured in Fig. 4.3. Each point represents a mean from censusing 10 stems each in ten m² blocks of goldenrod. The three separate figures correspond to paired arrays in three distinct and separate fields—thus, there are three replicates of the experiment. These data are from 1982.

hood that aphids attain an "escape density," at which, even if ladybugs arrive, the total production of new aphids swamps any losses imposed by predators. This effect is exaggerated because ladybug search is not a simple random process, but involves behavior that can yield density-dependent aggregation at regions of high prey density. Patchiness thus not only interferes with random search, but by reducing ladybug mobility, also interferes with the aggregating response of ladybugs at clumps of aphids.

It cannot be concluded, however, that patchiness will always reduce a predator's ability to regulate prey populations. The effect of patchiness will depend on the details of how it influences movement in both the predator and the prey species. For example, if patchiness impedes the ability of prey to "run away from enemies," it may well facilitate the control of prey populations by a predator. The important point is that patchiness will commonly alter the movements of predators and their prey, which in turn has the potential to affect the likelihood of successful biocontrol.

In agricultural systems, the type of patchiness that is relevant will rarely be anything as simple as continuous strips versus rows of patches. But in a broad sense, the texture or patchiness of a habitat will vary with crop diversity, the morphology of crop plants, and the presence of hedgerows or windbreaks between fields; these are all features of agroecosystems that can be managed (see Herzog and Funderburk, Chapter 10). They are also features that may influence predator foraging movement, and as a result the success of biocontrol programs. Only recently have researchers begun to examine the dependence of predator foraging patterns on cropping systems and its implications for the success of biocontrol programs (Wetzler and Risch, 1984; Risch, 1985). The potential for useful work along these lines is great. But once again, an understanding of the movement behavior of predators and prey will be necessary before management policy can be developed.

### 4.4.1. Modeling of Predator–Prey Interactions in Patchy Environments

In addition to their immediate practical implications, my experiments on ladybird–aphid interactions have caused me to reevaluate models of predator–prey interactions in patchy environments. Most such models do not contain explicit representations of habitat patches or movement among patches (Hassell and May, 1985). Nonrandom foraging that leads predators to spend a disproportionate amount of their time in regions of high prey density is also neglected by existing predator–prey theory. One way of making predator–prey theory more realistic is to use partial differential equation models to describe the interaction and redistribution of predators as a function of spatially heterogeneous prey density. With Gary Odell (Department of Mathematics, Rensselaer Polytechnic Institute), I have been developing such models for the *Uroleucon–Coccinella* interaction. Our initial ventures assume a homogenous habitat (i.e., continuous goldenrod) and focus on how the foraging movements of ladybirds govern their ability to contain or annihilate patches of high aphid density that may arise in continuous strips of goldenrod (Kareiva and Odell, 1986). It has been possible to relate the potential for prey control to the microscale details of predator foraging movements (i.e., velocity and turning frequency). The key behavior is *Coccinella*'s tendency to restrict its foraging attention to the vicinity of a recent aphid capture before continuing a wider ranging exploration ("area restricted search"). Simply by assuming that *Coccinella* increases its turning frequency as its stomach fills up from eating aphids, one can derive a partial differential equation that entails aggregation of ladybugs in regions of relatively high aphid densities. In its most general form, the equation that results is

$$\frac{\partial L}{\partial t} = \frac{\partial}{\partial x}\left[ D(A)\frac{\partial L}{\partial x} - L\Phi(A)\frac{\partial A}{\partial x}\right] \tag{1}$$

where $x$ is position in a one-dimensional environment (for convenience of presentation), $A$ is the density of aphids, $L$ is the density of ladybugs, and $\partial A / \partial x$ is the spatial gradient of aphid density. The first term on the right hand side of Eq. (1) portrays changes in ladybug density due to diffusive movement, with the complication that the diffusion coefficient, denoted $D(A)$, is a function of aphid density; the second term portrays ladybug aggregation at a rate proportional to $\Phi(A)$, which also is a function of aphid density. This second term represents aggregation in the sense that it describes ladybugs tending to move up gradients of aphid density (i.e., in the direction of increasing aphid abundance). In particular, if $\partial A / \partial x = 0$ there is no aggregating movement; if $\partial A / \partial x < 0$, ladybugs move towards decreasing values of $x$ (since that will correspond to regions of higher aphid density); and if $\partial A / \partial x > 0$, ladybugs move towards increasing $x$. This model predicts that foraging ladybugs move most rapidly toward regions of relative aphid superabundance (i.e., "aggregate") where they encounter the sharpest changes in aphid density along a spatial transect (i.e., where $\partial A / \partial x$ is large). Thus, Eq. (1) provides a complete description of how ladybugs will redistribute themselves in space by "diffusing" (random motion) and "aggregating" (oriented or nonrandom foraging) at rates and in directions that are responsive to the heterogeneous dispersion of their aphid prey. Moreover, specific functions and parameters can be substituted for $D(\cdot)$ and $\Phi(\cdot)$ on the basis of straightforward observations of foraging ladybugs, so that Eq. (1) can be applied to specific cases (Kareiva and Odell, 1986).

This is an approach that seeks explanations of ecological phenomena such as predator–prey dynamics by considering in detail the behavior of organisms. Its power is that short-term (30–60 min) measurements of individual behavior yield an equation for global population redistribution in a spatially heterogeneous environment. Eq. (1) can then be coupled to traditional mathematical models of predator functional responses and prey reproduction to create a *mechanistic model of predator–prey interaction and redistribution systems.* Such models are especially useful for explicitly quantifying the role of predator mobility or predator aggregation in prey population dynamics (Kareiva and Odell, 1986). Analyses of these interaction and redistribution equations may help relate the effectiveness of different biocontrol systems to species-specific movement behaviors or to cropping patterns that might influence predator movements (e.g., polycultures influencing rates of predator movement). For example, we have found that slight modifications of area restricted search behavior sometimes correspond to dramatic changes in the diffusion coefficients or aggregation coefficients appearing in Eq. (1), which in theory greatly influence a predator's promise as a biocontrol agent (G. M. Odell and P. M. Kareiva, in preparation). It might thus be possible to breed predators, using models such as Eq. (1) as guidelines, for behavior patterns that are most effective at containing prey outbreaks.

Partial differential equation models of aggregation and random searching do have some important limitations. If either predators or prey are scarce, or if predator mobility is so low that areas are likely to escape predator scrutiny by chance alone, then stochastic models will be more appropriate (Chesson and

Murdoch, 1986; Murdoch et al., 1985). But because computations and parameter estimations are usually more difficult in stochastic than in deterministic models, it may remain advisable to resort to models such as Eq. (1), even when the systems under study are quite stochastic. The interpretation of aggregation models in a stochastic framework, however, needs to be modified. For example, we have used Eq. (1) to study a predator–prey interaction involving scarce ladybugs; but we interpret the predictions of the model to represent probability distributions for ladybug position rather than absolute, deterministic densities (Kareiva and Odell, 1986).

## 4.5. PEST DISPERSION, PEST MOVEMENT, AND YIELD REDUCTION

Even in what appear to be homogeneous fields of crop plants, insects feeding on those crops are almost always patchily distributed. Consequently, a great variety of statistical measures of patchiness are routinely applied to pest census data [e.g., Iwao's patchiness index, $K$ of the negative binomial, Taylor's power law; see Southwood (1978) for a thorough discussion of these dispersion indices]. In pest monitoring programs, dispersion indices are typically used to calculate the sample sizes needed to estimate population sizes within some desired confidence limits (e.g., Lampert et al., 1984). Ironically, once a mean pest density has been estimated, pest populations and their interactions with crop plants are almost always treated as homogeneous in space. But in some instances, the spatial heterogeneity of pest populations could significantly affect the amount of yield reduction caused by the pest insects. In such cases, models for forecasting crop damage need to include both pest density and pest dispersion as input variables. Below I discuss this dispersion complication. I begin by identifying the factors that are likely to make crop yields most sensitive to pest dispersion. I then show how trivial movements of pests might influence levels of damage by modifying the dispersion of pests among crop plants.

Nonlinearities in damage functions or nonlinearities in pest demographic rates will make the dispersion of pests an important factor. For example, whenever the effect of herbivore density on yield reduction (per plant) is nonlinear, then the details of herbivore dispersion will govern field-wide losses. To visualize this in concrete terms, consider Fig. 4.5, in which fields A and B contain pest populations that average identical numbers of insects per plant (in this case 20) but that differ greatly in how those individuals are distributed among plants. Next, assume that Fig. 4.6 represents the damage–yield function for this hypothetical crop–pest interaction. Then using Fig. 4.6 to calculate the yield obtained from each plant in the two fields of Fig. 4.5, we find that field A will suffer a 40% reduction in total yield, whereas field B will suffer only a 27% reduction in total yield. Although the magnitude of this "dispersion effect" depends on the precise shape of the damage–yield curve, the potential for pest dispersion to have such an effect requires only that its associated damage–yield curve is nonlinear. This requirement is often met since nonlinearity in damage–yield curves is com-

**Field A**

| 20 | 20 | 20 | 20 | 20 |
|----|----|----|----|----|
| 20 | 20 | 20 | 20 | 20 |
| 20 | 20 | 20 | 20 | 20 |

mean = 20, variance = 0

**Field B**

| 5  | 5  | 5  | 5  | 50 |
|----|----|----|----|----|
| 5  | 5  | 50 | 50 | 5  |
| 50 | 50 | 5  | 5  | 5  |

mean=20, variance=450

Fig. 4.5 Two hypothetical dispersions of a pest insect in fields containing 15 plants. Field A represents an even distribution of pest attack, whereas in field B the insects are highly clumped in their dispersion.

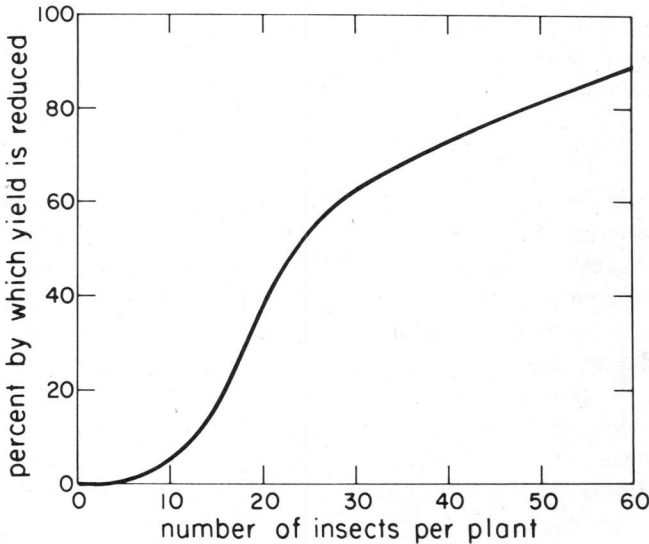

Fig. 4.6 An "attack–yield reduction" curve in which at low insect densities, an increase in pest load has little effect, and at high insect densities there is a tapering off in the increment of yield reduction per additional insect. The average yield reduction for field A of Fig. 4.5 is obtained by simply noting that according to the above curve a density of 20 insects per plants causes a 40% reduction in yield. For field B of Fig. 4.5, the average yield reduction is 27% (no reduction for 10 plants and 80% reduction for the remaining five plants).

mon (for examples, see Hallman et al., 1984; Huffaker and Rabb, 1984; Semtner, 1984; Waddill et al., 1984; Welter et al., 1984).

The relationship between pest demography and pest density (per plant) is another function whose possible nonlinearity is of interest. When the per capita rates at which pests die or reproduce is a nonlinear function of their plant abundance, two pest populations of identical mean densities but different dispersion patterns can grow at markedly different rates. Again, the literature is replete with examples of such nonlinear density-dependent demography (Way and Banks, 1967; Way and Cammell, 1970; Kidd, 1982; Tsubaki and Shiotsu, 1982; Kareiva, 1985b). There are probably several other ways in which nonlinear relationships would make pest dispersion an important variable. The general idea is simply that the average result of a nonlinear function applied to several different variables (plants bearing pest insects) *does not equal* the function applied to the average of those variables. Because nonlinearities of one form or another seem to be the rule in pest–crop interactions, pest dispersion has the potential to influence crop yields in many agricultural systems.

Trivial movement enters the picture by its ability to modify patterns of pest dispersion. If a pest's movements are random, movement will homogenize dispersions at a rate proportional to the rate of movement. Nonrandom movement can do the opposite; if insects are congregating to mate or to feed on some mutually recognized "tasty" or "vulnerable" plant, their movement can increase the degree of patchiness.

### 4.5.1. Modeling the Influence of Herbivore Trivial Movement on Patterns of Crop Damage

There is no obvious way to analyze crop–pest interactions and trivial movement simultaneously. Joint experimental manipulations of the dispersion of pests and their opportunities for movement are difficult to perform and include an overwhelming variety of permutations. Furthermore, without some sort of underlying model, the results of such experiments would be difficult to interpret. Consequently, mathematical models that include movement and local plant–insect interactions are worth exploring for suggestions as to which manipulations are likely to reveal the most interesting effects. But even a modeling approach to the issue of movement and pest–plant dynamics is daunting: The mathematics can be difficult, the computations are lengthy, and there are virtually no existing models (even simulations) of spatially distributed crop-pest systems on which to build (Shoemaker, 1981). Recent advances in the power and availability of personal computers and in numerical methods for partial differential equations (Banks et al., 1985), however, make such models feasible and timely. To illustrate what might be gained from spatially distributed population models, let us assume that the following coupled differential equations describe the plant–herbivore (pest) interaction at a point in space (or on a single plant):

$$\frac{dP}{dt} = \lambda P(1 - P) - (aHP)^k \tag{2a}$$

$$\frac{dH}{dt} = rH \left( 1 - \frac{H}{cP} \right) \tag{2b}$$

Here $P$ represents plant biomass, scaled so that in the absence of herbivory ($H = 0$), it will grow logistically at rate $\lambda$ to an equilibrium of $P = 1$; the term $(aHP)^k$ represents a nonlinear effect of herbivores on plant growth; and Eq. (2b) describes logistic herbivore population growth up to the carrying capacity, $cP$ (i.e., to some number of insects per unit of plant biomass). I have selected Eqs. (2a) and (2b) as simple representatives of the sorts of nonlinearities that might make pest dispersion important; similar nonlinearities appear in a wide variety of equations that have been proposed as biologically plausible models for plant–herbivore dynamics [see Chapter 4 of Crawley (1983) for an excellent review and analysis of these models].

To add the effect of pest movement on population dynamics, we need partial differential equations that include transport in space. Such equations will permit us to keep track of changes in numbers of pests that result from movement as well as from birth and death (so-called "local dynamics" or "local interactions"). These partial differential equation models are complicated because at a particular point in space, the dynamics of the system are not just governed by local densities and plant characteristics, but also by the surrounding pest densities and plant characteristics. Exactly how neighboring positions influence local dynamics depends on the details of pest movement behaviors, a wide variety of which can be explicitly modeled by carefully chosen transport terms. To illustrate how this can be done, consider the following partial differential equation model describing changes in plant biomass (denoted $\partial P/\partial t$) and changes in herbivore or pest densities (denoted $\partial H/\partial t$):

$$\frac{\partial P(x, t)}{t} = P\lambda(1 - P) - (aHP)^k \tag{3a}$$

$$\frac{\partial H(x, t)}{\partial t} = rH \left( 1 - \frac{H}{cP} \right) + \frac{\partial}{\partial x} \left[ D(P) \frac{\partial H}{\partial x} - \Phi(P) H \frac{\partial P}{\partial x} \right] \tag{3b}$$

Although it is not traditional to write it out in full, wherever $H$ or $P$ appear in partial differential equations such as Eqs. (3a) or (3b), we really mean $H(x, t)$ and $P(x, t)$ — that is, the values of $H$ and $P$ at position $x$ and time $t$. These equations thus model a plant–herbivore "interaction and redistribution system" in which there is local interaction described by Eqs. (2a) and (2b), and additional dynamics attributable to pest movement ("redistribution"). Although Eqs. (3a) and (3b) are written to reflect a one-dimensional field (where position in that field is indicated by $x$, they could be easily generalized to two dimensions (but not so easily solved on a computer). Since we are implicitly considering an annual crop, Eq. (3a) contains no provision for crop "movement"; the distribution of crop biomass ($P$) may, however, be spatially heterogeneous (i.e., vary with $x$). Even if a crop started as spatially homogenous, clumped feeding patterns by pests would probably generate heterogeneities in $P$. Pest insects, on the other hand, are mobile even within a growing season, and Eq. (3b) includes the last two terms on the right-hand side so that a variety of pest movement behaviors might be modeled. There is an enormous range of movement behaviors that can be represented by special choices of the $D(\cdot)$ or $\Phi(\cdot)$ terms in Eq. (3b):

**1.** *Simple random movement* corresponds to holding $D$ (which reflects the rate of movement) constant and fixing $\Phi = 0$.

**2.** *Random movement at a rate that depends on plant biomass* is represented by allowing $D$ to vary with $P$ [i.e., $D(P)$], with the precise functional form to be determined experimentally. For example, it might be reasonable to assume that the pest herbivores are least mobile when in the midst of high density or lush foliage ($P$ large), and most mobile when the available food is scarce (i.e., as $P$ tends to 0). Although the rate of movement in this formulation varies with food availability, the herbivores are still assumed to move in random directions.

**3.** *Movement that includes both a random aspect and directional taxis towards regions of relatively higher plant biomass* (or greater food availability) can be represented by having both $D$ and $\Phi$ appear as functions of $P$. The $\Phi$ term here is a taxis term portraying the tendency of herbivores to move from regions of low $P$ to high $P$ (up-gradients of $P$—hence $\partial P/\partial x$ is multiplied by $\Phi$; if there is no gradient in $P$, $\partial P/\partial x = 0$ and there will be no oriented movement). The magnitudes of $\Phi$ and $D$ reflect the relative strengths of oriented versus random motion. The functional relationship between $\Phi$ (or $D$) and $P$ can be estimated from empirical data (Kareiva and Odell, 1986).

**4.** *Density-dependent pest dispersal* in random directions but at a rate that increases as pests become crowded can be described by writing $D$ as an increasing function of $H$ and fixing $\Phi$ to be zero.

In general, partial differential equations can be used to portray quite complicated and realistic movement behavior [see Okubo (1980) for the authoritative review of animal dispersal models]. When population growth or population interaction terms are added to these transport models [as is the case in Eq. (3b)], the behavior of the models is staggeringly rich. In fact, these "reaction-diffusion equations" as they are often called in the mathematical literature, represent an area of vigorous mathematical research (e.g., Conway, 1983; Arcuri and Murray, 1986; Levin and Segel, 1986). But for specific pest management applications, models such as Eqs. (3a) and (3b) can be useful for suggesting experiments or interpreting data without requiring a full understanding of reaction-diffusion theory.

To indicate how Eqs. (3a) and (3b) might stimulate experiments, I report here the results of two sets of numerical solutions. My point is simply to show that by varying the rate of pest movement, total yields for a field can be markedly influenced. Figs. 4.7a and 4.7b present an unfolding of both pest ($H$) and plant biomass ($P$) spatial patterns through time, when $\Phi = 0$ and the diffusion rates

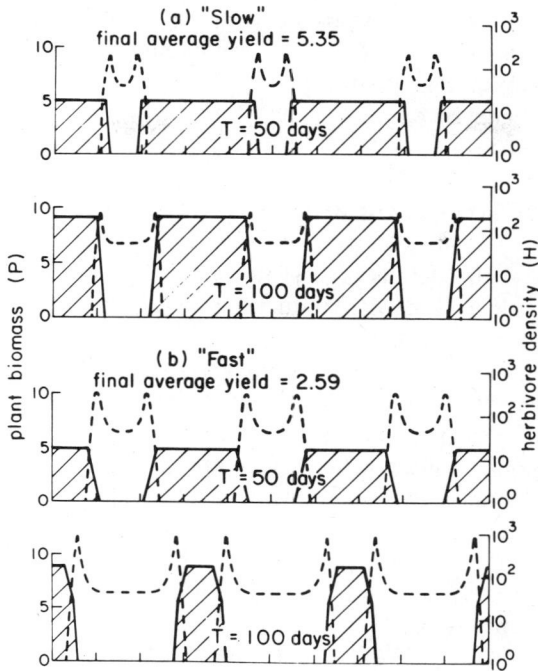

Fig. 4.7 Snapshots of the spatiotemporal dynamics of plant biomass (solid lines) and insect density (dashed lines) described by Eqs. (3a) and (3b) (except I used $(10 - P)/10$ in place of $1 - P$). The only difference between the "slow" and "fast" systems is that the herbivore's movement rate is 10 times faster in the fast system [that is $D$ in Eq. (3b) is 10 times larger]. Otherwise, initial conditions were identical ($P = 1$ everywhere, and three initial peaks of herbivores at three points along the horizontal axis), and the other parameters for the system were identical ($\Phi = 0$, $k = 1$, $a = 1$, $c = 5$, $\lambda = 0.044/\text{day}$, $r = 5\lambda$). If there were no herbivores to attack the plants the average crop biomass or yield would have grown to 9.0 by day 100.

differ by a factor of 10 but all other parameters are the same (values given in figure legend). Since I am interested in within-season crop growth, I have followed the dynamics up to only 100 days; the idea of exploring equilibrium solutions in a crop–pest system is meaningless because we purposely maintain agroecosystems as nonequilibrium systems.

Fig. 4.7 shows that simply by altering pest movement within a field (no pests were allowed to escape the field), crop yield (which is the area under the solid line) can be substantially affected—in this example by 50%. Although no general conclusions can be drawn from limited numerical "experiments" such as Fig. 4.7, the possibility of significant changes in crop yield as a result of differing rates of trivial movement is evident. Furthermore, the interaction equations used to produce this result are not unduly restrictive—they embody no more than the following sensible biological phenomena: (1) resource-limited growth of plants, and (2) nonlinear effects of herbivore density on plant growth. Other models that share these features produce a similar dependence of yield on movement (in preparation). It is not difficult to imagine plausible models in which pest movements dramatically affect crop yields; experimental field tests of this possibility should now be done.

## 4.6. NEW DIRECTIONS FOR FUTURE THEORETICAL WORK

I have argued that the details of pest movement are important whenever the dispersion of pests influences yield reductions and, in particular, whenever there is a nonlinear relationship between the number of insects on a plant and the insect's impact on the plant. To learn more about such nonlinear interactions we need to explore the physiological intricacies of how plant growth and quality (as a resource for herbivores) are altered by and in turn mediate pest attack.

Whereas economic considerations have prompted applied entomologists to emphasize crop biomass as the key plant variable, predictions about pest outbreaks and yield reductions may require that we also keep track of temporal changes in plant quality. For example, modest defoliation can have no effect on plant growth, yet reduce the quality of plants as food for insects so much so that pest growth and reproduction is greatly depressed [see Crawley (1983) for examples of this "asymmetry"]. We can profitably think of plants as recording a history of herbivore damage through changes in tannins, phenols, fiber and nitrogen content, and so forth (see Rhoades, 1985 for a review). When crop–pest systems are modeled without reference to plant quality, this history of damage is ignored. For plants that possess an inducible defense, as cotton does against mites (Karban and Carey, 1984), it is a substantial misrepresentation of the plant–pest interaction to neglect plant quality (see Kogan, Chapter 5). Movement enters the scenario because herbivores generally respond to variations in plant quality by altering their movement behavior; pests may be attracted to and linger on plants of high quality or they may avoid and quickly move away from low quality plants (Kareiva, 1982, 1983; Parker, 1984). Intuition is a poor guide to the consequences of such nonrandom movement in conjunction with induced defenses—in such situations models can offer help and insight as initial exploratory devices.

Although I know of no models that explicitly consider induced defenses and pest movement, a few theoreticians have begun to treat plants in agroecosystems as something more complicated than a passive and homogeneous carpet of digestible biomass. For example, Gutierrez and colleagues have modeled the bollworm–cotton interaction by using mechanistic balance equations that describe the interplay of herbivory and carbon allocation flows within individual cotton plants (Gutierrez et al., 1977, 1979; Wang et al., 1977). Edelstein (1986) has developed models in which crops are represented as a population of plants of varying quality. As a result of herbivory, the frequency distribution of plants with respect to quality changes and in turn determines the overall performance of the consuming insects. My own approach to this problem (in collaboration with Michael Cain of Cornell University) has been to analyze the effects of plant quality by explicitly introducing an equation that describes changes in plant quality as a result of herbivory and plant development. This extra equation keeps track of the accumulated responses of plants to pest attack that are not necessarily a simple matter of biomass reduction. Herbivore demography is then coupled to plant quality so that the model includes the possibility of herbivore (pest) population dynamics that are driven by induced defenses (M. Cain and P. M. Kareiva, in preparation). An ultimate goal is to link these plant quality models with transport equations such as Eqs. (3a) and (3b). Then it might be possible to make predictions about the effect of induced plant defenses on mobile pest populations.

Another goal for pest management theory should be the application of optimal control algorithms to interaction and redistribution systems. Current integrated pest management (IPM) theory has ignored both the spatial component of pest population dynamics and the contribution of pest movement to patterns of crop damage (Shoemaker, 1981). The most advanced approaches to spatially distributed crop–insect systems involve large simulation models (for example, Jones, 1979). But these massive simulations must often assume uncomfortably simplistic rules for population growth and dispersal, or include so many parameters that their interpretation is ambiguous. Recent developments in control theory indicate that powerful algorithms are on the horizon for computing optimal controls in partial differential equation systems (Lions, 1971; Curtain and Pritchard, 1978; Ray and Laioniotis, 1978; Banks and Kunisch, 1983). In this theory, the "amount of control" (insecticide) that should optimally be applied to a particular system at time $t$ and location $x$ depends on the current spatially heterogeneous "state" (population densities) of the system. Because partial differential equations can be used to represent pest movement and crop–pest interactions, these advances in control theory for what are technically called spatially distributed systems may be of practical value. Numerical experiments on the computer could be used to test whether the extra sophistication of spatially distributed control theory offers notable improvement over current simulation or dynamic programming approaches. For example, one might generate hypothetical crop–pest dynamics using Eqs. (3a) and (3b), and apply spatially distributed control theory to these systems. The effectiveness of control measures determined from the full spatial model could then be compared to the effectiveness of control measures obtained by more traditional dynamic programming techniques that neglect the spatial dimension. Some management problems, such as the containment of spreading gypsy moths or Mediterranean fruit flies, are unavoidably spatial. Furthermore, although the mathematics are nontrivial, the necessary spatial data may be already available, since pests are often sampled via a regular gridwork of traps.

## 4.7. THE USE AND USEFULNESS OF MATHEMATICAL MODELS

The pest management literature is replete with equations and mathematical models. The usefulness of these models is, I think, arguable. Still, I have spent much of this chapter introducing yet another battery of models. To make matters worse I have emphasized partial differential equations, which probably are more unpopular with biologists than are difference or differential equations. Why have I pursued such an uninviting and questionable approach?

Trivial movement is a biological process important to the outcome of crop-pest interactions. Its consequences, however, are likely to depend on quantitative rates (e.g., how many meters per day on average an insect moves), and this dependence is not something whose effects can be anticipated by biological intuition. For example, even if we knew intercropping would reduce a pest's movement by 30% and we were well aquainted with the pest's biology and the crop's biology, it would be difficult to predict whether such a change in movement would influence average yield over a whole field. Experiments would help; but they would simply tell us whether or not an effect existed. We would still need to know more if we wanted to generalize or extrapolate to a large-scale agroecosystem. Here then is the reason for quantitative mechanistic models—they are tools for exploring the influences of pest movement on crop–pest interactions. Partial differential equations provide a natural framework for analyzing such interaction and redistribution dynamics. Although the "vocabulary" of partial differential equations may be unfamiliar to many entomologists, these equations actually offer a much more compact and precise summary of dynamics than do many highly touted simulation models. For example, the effect of doubling a pest's mobility rate is parsimoniously captured by doubling the diffusion coefficient. Diffusion coefficients or taxis terms can be concretely estimated by releasing marked insects and following their displacements in space. Although the solution of partial differential equations is a matter of advanced mathematics, their connection to the real world is explicit and mechanistic. Their parameters are few, and the role of these parameters in dynamics is readily exposed. In contrast to system-specific simulation models, partial differential equations of interaction and redistribution facilitate comparisons of pest movement behavior (and its consequences) among experimental manipulations or among different crop-pest systems. I advocate partial differential equations, therefore, as one approach to exploring the consequences of trivial movement in agroecosystems. At this stage they represent heuristic models, which will be most useful if they are wedded to field data. Too often we judge models on their convenience or familiarity. Instead we need to be more concerned with the insight they can provide and their ability to describe a variety of experimental results, regardless of their mathematical form. Partial differential equations of interaction and redistribution are ideal for relating pest movement behavior to patterns of damage. These models, together with detailed studies of the mechanistic bases of insect movement, can illuminate much about the heretofore overlooked connection between a pest's trivial movements and crop dynamics.

## ACKNOWLEDGMENTS

This research was supported by a GSRF grant from the University of Washington, National Science Foundation grants DEB 8207117 and BSR 8500343 to P.

Kareiva, and GR/C/63595 to the Centre for Mathematical Biology at Oxford University. I thank J. Bergelson, P. Bierzychudek, N. Cappuccino, E. Evans, D. Pimentel, and D. Strong for comments on the manuscript. Many of the ideas, but none of the mistakes, are due to H. T. Banks, J. Bergelson, M. Cain and G. Odell. The B&B boys of Kincaid Hall provided invaluable energy, entertainment, and harassment during the course of this project.

## REFERENCES

Andow, D. (1983a). The extent of monoculture and its effects on insect pest populations with particular reference to wheat and cotton, *Agric. Ecosyst. Envt.* **9**, 25-35.

Andow, D. (1983b). "Plant diversification and insect populations in agroecosystems," in D. Pimentel, Ed., *Alternative Methods of Pest Control*, Academic, New York, pp. 107-123.

Arcuri, P., and J. D. Murray. (1986). Pattern sensitivity to boundary and initial conditions in reaction-diffusion models, *J. Math. Biol.*, (in press).

Bach, C. E. (1984). Plant spatial pattern and herbivore population dynamics: Factors affecting the abundance of a tropical cucurbit specialist (*Acalymma innubum*), *Ecology*, **65**, 175-190.

Baltensweiler, W., and A. Fischlin. (1979). The role of migration in the population dynamics of the larch bud moth, *Mitt. Schweiz. Entomol. Gesellschaft*, **52**, 259-271.

Banks, H. T., P. M. Kareiva, and P. D. Lamm. (1985). Estimation of temporally and spatially varying coefficients in models for insect dispersal, *J. Math. Biol.*, (in press).

Banks, H. T., and K. Kunisch. (1983). Approximation of feedback controls for parabolic systems, *Proc. IEEE Conf. Decision Control*, **11**, 274-251.

Beddington, J. R., C. A. Free, and J. H. Lawton. (1978). Characteristics of successful natural enemies in models of biological control of insect pests, *Nature*, **273**, 513-519.

Cain, M. L. (1985). Random search by herbivorous insects: A simulation model, *Ecology*, **66**, 876-888.

Cain, M. L., J. Eccleston, and P. M. Kareiva. (1985). The influence of food plant dispersion on caterpillar searching success, *Ecol. Entomol.*, **10**, 1-7.

Cameron, E. A., M. L. McManus, and C. E. Mason. (1979). Dispersal and its impact on the population dynamics of the gypsy moth in the United States, *Mitt. Schweiz. Entomol. Gesellschaft*, **52**, 169-179.

Chesson, P. L., and W. W. Murdoch. (1986). Relationships among host–parasitoid models, *Am. Nat.*, (in press).

Conway, E. D. (1983). Diffusion and the predator–prey interaction: pattern in closed systems, *Tulane University, Comp. & Appl. Math Reports*, **83-2**, 1-47.

Crawley, M. J. (1983). *Herbivory*, University of California Press, Berkeley, California, 437 pp.

Cromartie, W. J. (1981). "Environmental control of insects using crop diversity," in D. Pimentel, Ed., *CRC Handbook of Pest Management*, CRC Press, Boca Raton, Florida, pp. 223-251.

Crook, G. W., P. E. Vezin, and Y. Hardy. (1979). Susceptibility of balsam fir to spruce budworm defoliation as affected by thinning, *Can. J. For. Res.*, **9**, 428-435.

Curtain, R. T. and A. L. Pritchard. (1978). *Infinite Dimensional Linear Systems Theory*, Springer-Verlag, New York, 416 pp.

Dahlsten, D. L., and S. H. Dreistadt. (1984). Forest insect pest management, *Bull. Entomol. Soc. Am.*, **30**, 19-21.

Dempster, J. P. and T. H. Coaker. (1974). "Diversification of crop ecosystems as a means of controlling pests," in D. Price-Jones and M. Solomon, Eds., *Biology in Pest and Disease Control*, Blackwell, London, pp. 106-114.

Dethier, V. G. (1959). Food-plant distribution and density and larval dispersal as factors affecting insect populations, *Can. Entomol.*, **88**, 581–596.

Edelstein, L. (1986). Mathematical theory for plant–herbivore systems, *J. Theor. Biol.*, (in press).

Fletcher, B. S., and E. Kapatos. (1981). Dispersal of the olive fly, *Dacus oleae* during the summer period on Corfu, *Entomol. Exp. Appl.*, **29**, 1–8.

Gentry, C. R., C. E. Yonce, J. L. Blythe, and J. H. Tumlinson. (1979). Lesser peachtree borer: Recovery of marked native males in pheromone baited traps, *Env. Entomol.*, **8**, 218–220.

Gould, F., and R. E. Stinner. (1984). "Insects in heterogeneous environment," in C. Huffaker and R. Rabb, Eds., *Ecological Entomology*, Wiley, New York, pp. 287–301.

Gutierrez, A. P., G. Butler, Y. Wang, and D. Westphal. (1977). The interaction of pink bollworm, cotton, and weather: A detailed model, *Can. Entomol.*, **109**, 1457–1468.

Gutierrez, A. P., Y. Wang, and R. Daxl. (1979). The interaction of coton and boll weevil—a study of co-adaptation, *Can. Entomol.*, **111**, 357–366.

Hallman, G. J., G. L. Teetes, and J. W. Johnson. (1984). Weight compensation of undamaged kernels in response to damage by sorghum midge (Diptera: Cecidomyiidae), *J. Econ. Entomol.*, **77**, 1033–1036.

Hassell, M. P. and R. M. May. (1985). "From individual behavior to population dynamics," in I. M. Sibly and R. H. Smith, Eds., Behavioural Ecology: Ecological Consequences of Adaptive Behaviour. *25th Symp., Br. Ecol. Soc.*, **25**, 3–32.

Hassell, M. P., and T. R. E. Southwood. (1978). Foraging strategies of insects, *Annu. Rev. Ecol. Syst.*, **9**, 75–98.

Hawkes, C. (1972). The estimation of the dispersal rate of the adult cabbage root fly, *J. Appl. Ecol.*, **9**, 617–632.

Heads, P. P., and J. H. Lawton. (1983). Studies on the natural enemy complex of the holly leaf-miner: The effects of scale on the detection of aggregative responses and the implications for biological control, *Oikos*, **40**, 267–276.

Howell, J. F., and A. E. Clift. (1974). The dispersal of sterilized codling moths released in the Wenas Valley, Washington, *Environ. Entomol.*, **3**, 75–81.

Hubbell, S. P. (1979). Tree dispersion, abundance, and diversity in a tropical dry forest, *Science*, **203**, 1299–1309.

Huffaker, C. B. (1958). Experimental studies on predation: dispersion factors and predator–prey oscillations, *Hilgardia*, **27**, 343–383.

Huffaker, C. B., and R. Rabb, Eds. (1984). *Ecological Entomology*, Wiley-Interscience, New York, 800 pp.

Ito, Y. and K. Miyashita. (1961). Studies on the dispersal of leaf and planthoppers, *Jap. J. Ecol.*, **11**, 181–186.

Jones, D. D. (1979). "The budworm site model," in G. Norton and C. S. Holling, Eds., *Pest Management*, Pergamon Press, England, pp. 313–337.

Joyce, R. J. V. (1976). "Insect flight in relation to problems of pest control," in R. C. Rainey, Ed., *Insect Flight*, Blackwell, Oxford, pp. 135–155.

Karban, R., and J. R. Carey. (1984). Induced resistance of cotton seedlings to mites, *Science*, **225**, 53–54.

Kareiva, P. M. (1981). "Non-migratory movement and the distribution of herbivorous insects: Experiments with plant spacing and the application of diffusion models to work-recapture data," Ph.D. dissertation, Cornell University, Ithaca, New York, 292 pp.

Kareiva, P. M. (1982). Experimental and mathematical analysis of herbivore movement; Quantifying the influence of plant spacing and quality on foraging discrimination, *Ecol. Monogr.*, **52**, 261–282.

Kareiva, P. M. (1983). "Influence of vegetation texture on herbivore populations: Resource concentration and herbivore movement," in R. F. Denno and M. S. McClure, Eds., *Variable Plants and Herbivores in Natural and Managed Systems*, Academic, New York, pp. 259–289.

Kareiva, P. M. (1984). Predator-prey dynamics in spatially-structured populations: Manipulating dispersal in a coccinellid-aphid interaction, *Lect. Notes. BioMath*, **54**, 368–389.

Kareiva, P. M. (1985a). Finding and losing host plants by flea beetles: patch size and surrounding habitat, *Ecology*, **66**, 1809–1816.

Kareiva, P. M. (1985b). "Patchiness, dispersal, and species interactions: consequences for communities of herbivorous insects," in J. Diamond and T. Case, Eds., *Community Ecology*, Harper & Row, New York, pp. 192–206.

Kareiva, P. M., and G. M. Odell. (1986). Aggregative behavior of predators and its effect in containing prey outbreaks, *Am. Nat.*, (in press).

Karlin, S., and H. M. Taylor. (1975). *A First Course in Stochastic Processes*, 2nd ed., Academic, New York.

Kemp, W. P., and G. A. Simmons. (1979). Influence of stand factors on survival of early instar spruce budworm, *Environ. Entomol.*, **8**, 993–996.

Kidd, N. A. C. (1982). Predator avoidance as a result of aggregation in the grey pine aphid. *Schizolachnus pineti, J. An. Ecol.*, **51**, 397–412.

Lamb, K. P., E. Hassan, and D. R. Scotter. (1971). Dispersal of scandium-46-labeled *Pantorhytes* weevils in Papual cacao plantations, *Ecology*, **52**, 178–182.

Lampert, E. P., D. C. Cress, and D. L. Haynes. (1984). Temporal and spatial changes in abundance of the asparagus miner, *Ophiomyia simplex, Mich. Environ. Entomol.*, **13**, 733–736.

Levin, S. A. (1981). The role of theoretical ecology in the description and understanding of populations in heterogeneous environments, *Am. Zool.*, **21**, 865–875.

Levin, S. A., and L. A. Segel. (1986). Pattern generation in space and aspect, *SIAM Rev.*, (in press).

Lions, J. L. (1971). *Optimal Control of Systems Governed by Partial Differential Equations*, Springer-Verlag, New York, 593 pp.

Ludwig, D., D. G. Aronson, and H. F. Weinberger. (1979). Spatial patterning of the spruce budworm, *J. Math. Biol.*, **8**, 217–258.

Murdoch, W. W., J. Chesson, and P. L. Chesson. (1985). Biological control in theory and practice, *Am. Nat.*, **125**, 344–366.

Obrycki, J. J., J. R. Nechols, and M. J. Tauber. (1982). Establishment of a European ladybeetle in New York State, *N.Y. Food Life Sci. Bull.*, **94**, 1–3.

Okubo, A. (1980). *Diffusion and ecological problems: Mathematical Models*, Springer-Verlag, New York, 254 pp.

Parker, M. A. (1984). Local food depletion and the foraging behavior of a specialist grasshopper, *Hesperoteffix viridis, Ecology*, **65**, 824–835.

Perrin, R. M. and M. Phillips. (1978). Some effects of mixed cropping on the population dynamics of insect pests, *Entomol. Exp. Appl.*, **24**, 385–393.

Price, P. W. (1976). Colonization of crops by arthropods: Nonequilibrium communities in soybean fields, *Environ. Entomol.*, **5**, 605–611.

Ray, W. H., and D. G. Lainiotis, Eds. (1978). *Distributed Parameter Systems: Identification, Estimation, and Control*, Marcel Dekker, New York, 706 pp.

Rhoades, D. F. (1985). Offensive–defensive interactions between herbivores and plants: Their relevance in herbivore population dynamics and ecological theory, *Am. Nat.*, **225**, 205–238.

Risch, S. J. (1981). Insect herbivore abundance in tropical monocultures and polycultures: An experimental test of two hypotheses, *Ecology*, **62**, 1325–1340.

Risch, S. J. (1985). Multi-lines of corn augment biological control of the European corn borer, *Ecol. Entomol.*, (in press).

Risch, S. J., D. Andow, and M. A. Altieri. (1983). Agroecosystem diversity and pest control: Data, tentative conclusions and new research directions, *Env. Entomol.*, **12**, 625–629.

Root, R. B. and P. M. Kareiva. (1984). The search for resources by cabbage butterflies (*Pieris rapae*): Ecological consequences and adaptive significance of markovian movement in a patchy environment, *Ecology*, **65**, 147–165.

Semtner, P. J. (1984). Effect of early-season infestations of the tobacco flea beetle (Coleoptera: Chrysomelidae) on the growth and yield of flue-cured tobacco, *J. Econ. Entomol.*, **77**, 98–102.

Shoemaker, C. A. (1981). Applications of dynamic programming and other optimization methods in pest management, *IEEE Trans. Aut. Control*, **26**, 1125–1132.

Southwood, T. R. E. (1962). Migration of terrestrial arthropods in relation to habitat, *Biol. Rev.*, **37**, 171–214.

Southwood, T. R. E. (1978). *Ecological Methods*, Chapman & Hall, London, 524 pp.

Stinner, R. E., C. S. Barfield, J. L. Stimac, and L. Dohse. (1983). Dispersal and movement of insect pests, *Ann. Rev. Entomol.*, **28**, 319–335.

Tsubaki, Y., and Y. Shiotsu. (1982). Group feeding as a strategy for exploiting food resources in the burnet moth *Pryeria sinica*, *Oecologia*, **55**, 12–20.

Turchin, P. (1986). Modelling the effect of host patch size on Mexican Bean beetle (*Epilachna varivestis*) emigration, *Ecology*, **67**, 124–132.

Waage, J. K. (1979). Foraging for patchily-distributed hosts by the parasitoid, *Nemeritis canescens*, *J. An. Ecol.*, **48**, 353–372.

Waddill, V., K. Pohronezny, R. McSorley, and H. H. Bryan. (1984). Effect of manual defoliation on pole bean yield, *J. Econ. Entomol.*, **77**, 1019–1023.

Wang, Y., A. Gutierrez, G. Oster, and R. Daxl. (1977). A population model for plant growth and development: Coupling cotton herbivore interactions, *Can. Entomol.*, **109**, 1359–1374.

Way, M. J., and C. J. Banks. (1967). Intra-specific mechanisms in relation to the natural regulation of numbers of *Aphis fabae*, *Ann. Appl. Biol.*, **59**, 189–205.

Way, M. J., and M. Cammell. (1970). "Aggregation behavior in relation to food utilization by aphids," in A. Watson, Ed., *Animal Populations in Relation to Their Food Resources*, Blackwells, Oxford, England, pp. 229–247.

Welter, S. C., M. M. Barnes, I. P. Ting., and J. T. Hayashi. (1984). Impact of various levels of late-season spider mite feeding damage on almond growth and yield, *Environ. Entomol.*, **13**, 52–55.

Wetzler, R. E., and S. J. Risch. (1984). Experimental studies of beetle diffusion in simple and complex crop habitats, *J. Anim. Ecol.*, **53**, 1–19.

# 5

---

# PLANT DEFENSE
# STRATEGIES AND HOST-PLANT
# RESISTANCE

*Marcos Kogan*

## 5.1. INTRODUCTION

The scientific foundations for crop-plant resistance to arthropods were established independently of the development of contemporary concepts on plant ecophysiology, plant–herbivore interactions, and coevolutionary theory. Until recently, host-plant resistance (HPR) has been the endeavor of practically oriented agricultural scientists who felt that, although desirable, knowledge of the mechanisms of resistance was not essential to breeding programs whose ultimate goal was to produce new varieties resistant to insect pests or pathogens. Painter (1951) stated that since "one must frequently deal with a number of causes or mechanisms which result in resistance rather than with a single factor, in attempting to breed resistant varieties a knowledge of these mechanisms may sometimes be of little use." Therefore, researchers in HPR have been mostly concerned with developing techniques to identify rapidly sources of resistance in germ plasm banks and to monitor inheritance of resistance in breeding lines. (Painter, 1951; Chesnokov, 1962; Panda, 1979; Harris, 1980; Maxwell and Jennings, 1980). Intellectual curiosity about biochemical–physiological mechanisms of resistance was usually satisfied by classifications according to the categories of Painter (1951)—antibiosis, antixenosis (nonpreference), or tolerance.

The peculiarities of the relationships of arthropods with their host plants have intrigued humans at least since the silkworm was domesticated in ancient China over 6000 years ago (see Konishi and Ito, 1973). However, scientific inquiry into the mechanisms of host selection goes back perhaps less than 100 years (Errera, 1886), yet only in the last 20 years has there been a real flurry of research in the behavioral, ecological, and physiological bases of insect–plant interactions (IPI).

Just as early HPR researchers showed little enthusiasm for the fundamentals of IPI, researchers in IPI tended to overlook the wealth of detailed information in the HPR literature. To be sure, many IPI researchers have used crop pests and resistant plants as models for their chemical, ecological, and electrophysiological studies. Important agricultural pests such as the tobacco hornworm [*Manduca sexta* (L.)], the Colorado potato beetle [*Leptinotarsa decemlineata* (Say)], the onion fly [*Delia antiqua* (Meigen)], the imported cabbageworm (*Pieris rapae* L.), the black cutworm [*Agrotis ipsilon* (Hufnagel)], the Mexican bean beetle, (*Epilachna varivestis* Mulsant), the European corn borer [*Ostrinia nubilalis* (Hubner)], the corn earworm [*Heliothis zea* (Boddie)], and various species of aphids and tephritid fruit flies have been preferred laboratory animals in some of the most productive IPI research. However, despite a few important exceptions, this research has remained disassociated from the mainstream of HPR.

As a consequence, workers in the two fields use different jargons, and interact infrequently. Most importantly, IPI researchers do not fully benefit from opportunities that HPR offers to test IPI theory. Conversely, HPR researchers do not fully appreciate the implications of the conceptual and theoretical framework generated by IPI investigations in furthering HPR science.

In an attempt to demonstrate the "parallel evolution" of IPI and HPR, I have compiled some landmarks in the historical development of those fields (Table 5.1). Landmark concepts, publications, or events in IPI are listed in the left-hand column, those for HPR in the right-hand column. Events and publications that have contributed equally to the development of both fields are listed in the

**Table 5.1 Landmark Concepts, Publications, or Events in the Development of Insect–Plant Interactions Theory and Host-Plant Resistance Theory and Practice**

| Insect–Plant Interactions (IPI) | Interface | Host-Plant Resistance (HPR) |
|---|---|---|
| | | *E. lanigerum* resistance —apple (1831) |
| Botanical instinct (Fabre, 1880s) | | Hessian fly resistance —wheat (1881/92) |
| Chemical defense (term) (Stahl, 1888) | | Grape Phylloxera |
| Glucosinolates/crucifer/ *Pieris* (Verschaffelt, 1910) | | resistance—American grape roots (1890s) |
| Parallel evolution insects and plants (Brues, 1920) | | |
| Secondary substances (Czapek, 1922–25) | | Biotypes of Hessian fly (Painter, 1930) |

**Table 5.1**— *Continued*

| Insect–Plant Interactions (IPI) | Interface | Host-Plant Resistance (HPR) |
|---|---|---|
| Chemical factors in HPS[a] (Dethier, 1941) | | Snelling (1941) review Bot. Rev. |
| Host-finding—aphids (Kennedy, 1950) | Brues (1946) book | Painter (1951) book |
| Evolution host specificity (Dethier, 1954) | | Isolation of DIMBOA in corn (Beck, 1957) |
| 1st Int. Symp. Ins. and Food Plant (Wageningen, 1957)[b] | First issue Entomol. Exp. Applic. (Holland, 1958) | Chartering of IRRI (1959) |
| "Raison d'etre" Secondary metabolites (Fraenkel, 1959) | Thorsteinson (1960) review Ann. Rev. Entomol. | |
| Coevolution (Ehrlich and Raven, 1964) | Beck (1965) review Ann. Rev. Entomol. | Onset of the Green Revolution (1960s) |
| Chemosensory bases IPI (Schoonhoven, 1968) | Intake and utilization analysis (Waldbauer, 1968) | 1st HYVs of Green Revolution |
| 2nd Int. Symp. Ins. and Food Plant (Wageningen, 1969)[c] | | Genetic biotypes Hessian |
| Allelochemical systematics (Whittaker and Feeny, 1971) | Int. Conf. Ins. & Mite Nutrit. (Rodriguez, 1972) | fly (Hatchett and Gallun, 1970) |
| Sympatric speciation (Bush, 1975) | | Release IR26-BPH[d] resistant rice (1973) |
| 1st Issue J. Chem. Ecol. (1975) | | HPR Newsletter (Gallun, 1975) |
| Apparency and defense (Feeny, 1976; Rhoades and Cates, 1976) | | BPH biotypes (India) 1st ACS Symp. HPR (Hedin (ed.), 1977) Maxwell and Jennings (1979) book |
| 3rd Trophic level effects (Price et al., 1980) | | |
| 70/84 @ 2 books on IPI published per year | 1st Gordon Conf. on IPI (Santa Barbara, 1980) | 1979/1984 3 books and scores of chapters on HPR in IPM books |
| 80/82 @ 2K papers on IPI (Bernays, 1982) | | |

[a]HPS = host plant selection
[b]Grison et al., 1958
[c]de Wilde and Schoonhoven, 1969
[d]BPH = brown planthopper

center, the interface column. The criteria for considering a publication as land-mark use, in part, its impact on the literature, as judged by its heuristic value and frequency of citation. However, inclusion in Table 5.1 also reflects my per-sonal bias. Brief discussion of this table serves as an introductory review of the literature.

### 5.1.1. From Fabre to Fraenkel

The famous French entomologist, Jean Henry Fabre (1890), in his *Souvenirs Entomologiques*, was one of the first to wonder about the botanical instinct of insects. He described, in his colorful language, the frustrations of trying to find an alternative food for the silkworm (*Bombyx mori* L.) when foliage from re-gional mulberry groves had been killed by an early frost. Although he could not overcome the predilection of the silkworm for mulberry leaves, he questioned why. Later generations of sericulturists have attempted to answer that question in what has become one of the best documented relationships of an oligophagous insect with its host plant. The term "chemical defenses" apparently was first employed by Stahl (1888), but the unmistakable association of plants' "second-ary substances" [an expression first used by A. Kossel in 1891 (see Mothes, 1980) and later adopted by Czapek, 1922–25] with feeding and oviposition pref-erences in insects was first documented by Verschaffelt (1910). This work started the line of research on glucosinolate-rich crucifers and pierid butterflies that remains one of the most prolific in the IPI literature (Hovanitz, 1969 and bibli-ography in Miller, 1978–1979; Chew, 1975; Feeny, 1977; van Etten and Tookey, 1979; Rodman and Chew, 1980; among many others).

Brues (1920) ushered in the idea of a linkage between the evolution of green plants and phytophagous insects, which he called "parallel evolution." This concept, redefined as "reciprocal adaptive evolution" in an almost forgotten pa-per by Fraenkel (1958), finally blossomed with the classic work by Ehrlich and Raven (1964) on coevolution. Meanwhile, evidence of the allelochemical role of secondary plant compounds on IPI was rapidly accumulating (Dethier, 1941). The definitive statement on this role was made by Fraenkel (1959). He consid-ered that their performance in defense was the very raison d'etre of secondary compounds in green plants. This extreme view generated considerable contro-versy, particularly among aphidologists, who claimed that primary metabolites, as essential nutrients, were just as important in aphid host selection behavior (Kennedy, 1950, 1953, 1965).

Ecological awareness in the 1960s renewed interest in basic mechanisms of IPI, and generated an impressive succession of reviews, symposia, and articles that attempted to provide a cohesive theoretical interpretation for the proliferat-ing mass of information. National and international meetings promoted the flow of information and stimulated debate. The "School of Wageningen," led by de Wilde and Louis Schoonhoven, pioneered this effort with a series of interna-tional symposia on insects and their food plants, the first of which convened in 1957 (Grison et al., 1958).

Perhaps the most stimulating and general concepts of the period related to the coevolution of insects and plants. Coevolutionary theory and the hypotheses formulated by Feeney (1976), Rhoades and Cates (1976), their extensions discussed by Rhoades (1979), and the notion of optimization of fitness provide the main framework for the rest of this chapter. The influence of secondary plant metabolites on higher trophic levels also attracted the attention of community ecologists interested in predator–prey interactions. An important paper in this area is the review by Price et al. (1980).

The vitality of the field has not abated. Research on the chemical bases of IPI is some of the most active in the burgeoning field of chemical ecology. Between 1970 and 1984, this activity resulted in the publication of over 30 single- and multiauthored books (Grison et al., 1958; Wilde and Schoonhoven, 1969; National Academy of Sciences, 1969; Sondheimer and Simeone, 1970; van Emden, 1971; Rodriguez, 1972; Gilbert and Raven, 1975; Jermy, 1976; Wallace and Mansell, 1976; Hedin, 1977, 1983, 1985; Labeyrie, 1977; Marini-Bettolo, 1977; Shorey and McKelvey, 1977; Chapman and Bernays, 1978; Harborne, 1978, 1982; Edwards and Wratten, 1980; Denno and Dingle, 1981; Nordlund et al., 1981; Thompson, 1982; Visser and Minks, 1982; Ahmad, 1983; Crawley, 1983; Denno and McClure, 1983; Futuyma and Slatkin, 1983; Jolivet, 1983; Whitehead and Bowers, 1983; Bell and Carde, 1984; Price et al., 1984; Strong et al., 1984a,b; Green and Hedin, 1986). Bernays (1982) estimated that about 2000 papers had been published on IPI in the previous 2 years, including many extensive reviews.

### 5.1.2. From Burgundy to Los Baños

While much action and attention seemed to focus on IPI research accomplishments, important events were occurring in HPR. Earliest records of resistance in plants to insect attack were the *Eriosoma lanigerum* (Hausmann) in apple (in England in 1831), and the Hessian-fly [*Mayetiola destructor* (Say)] in wheat (in the United States in 1881–1882). However, the most spectacular early success with HPR in solving a crucial economic problem was the use of resistant American grape rootstocks in grafts with European grape scions to control the outbreak of the grape phylloxera in some of the most famous wine-producing regions of France. This early success was not soon superceded; in fact, the next major event in HPR was a presage of future troubles—the detection of biotypes of Hessian fly capable of colonizing wheat genotypes that up until then had been resistant (Painter, 1930). This event was early evidence of coevolution accelerated through human interference—breeders followed discovery of biotypes with releases of new resistant lines of host plants that were soon overtaken by new biotypes.

The keystone work in this early phase of HPR was *Insect Resistance in Crop Plants* (Painter, 1951, reprinted in 1969). In this book Painter provided the theoretical foundation for HPR work and a superb review of the literature. Symptomatic, however, of the disassociation between HPR and IPI, Verschaffelt's work was not cited in this review, although it was referred by Snelling (1941) in an equally insightful, if not equally famous, review of HPR.

During the 1940s and 1950s, HPR, as well as all other nonchemical methods of insect pest control, took a distant second place to the overwhelming emphasis on control by the novel organo-synthetic insecticides. Failures of insecticidal controls and the ecological imbalances resulting from their overuse (see Metcalf, Chapter 10) rekindled interest in the more "traditional" methods of pest control, including HPR. Combined economical, ecological, and sociological pressures inspired agricultural scientists to formulate the principles of a holistic approach to pest control—integrated pest management (IPM) (Geier and Clark, 1961). Within this scenario, a major event in modern agricultural research took place.

In 1959 the International Rice Research Institute (IRRI) was chartered at Los Baños, Phillipines. It was the first of 15 international centers currently devoted to improving agricultural output in the developing world. Emphasis in these centers is to increase productivity through breeding well-adapted, high-yielding varieties (HYV). The spectacular success achieved with HYVs of wheat and later of rice ushered in the so-called Green Revolution. Concurrent with the breeding programs were efforts to identify sources of resistance to insect pests and pathogens. However, the goals of the Green Revolution often collided with objectives of resistance programs (Wade, 1974a,b; Harlan, 1977). In 1973 IRRI released the very successful IR26, a variety of rice resistant to the brown planthopper [*Nilaparvata lugens* (Stal)]. A few years later, brown planthopper biotypes adapted to the resistant varieties were detected in India and throughout southeastern and eastern Asia (Ikeda and Kaneda, 1982). These biotypes threatened again the stability of rice production in the region; additional cycles of breeding were required to incorporate resistance against the new biotypes. Concern about the rapid upsurge of Hessian fly and brown planthopper biotypes stimulated some of the most elegant studies of the genetics of insect–plant interactions (Hatchett and Gallun, 1970; Gallun and Khush, 1980).

From the early 1960s, HPR researchers grew increasingly aware of the need to investigate the intricacies of resistance mechanisms. The discovery of DIMBOA (Beck, 1957; Beck and Smissman, 1960, 1961) as a chemical resistance factor in young corn plants against the European corn borer was one of the first detailed studies of mechanisms of biochemical resistance, and it remained the sole outstanding work of this nature for nearly 15 years. Some of the successes in this area resulted from increased support by the U.S. Department of Agriculture for research on the physiological bases of plant resistance in laboratories at Brownsville, Texas, and Mississippi State, Mississippi, on cotton; Ankeny, Iowa, on corn; and Albany, California, on the general nature of antibiosis in various crop plants.

As research expanded, new forums for communicating ideas opened. Four symposia sponsored by the American Chemical Society addressed the most current topics on the biochemical basis of HPR (Hedin, 1977, 1983, 1985; Green and Hedin, 1986). The considerable progress achieved in the field of HPR since

Painter's landmark book (1951) was summarized in a volume edited by Maxwell and Jennings (1980) that resulted from a Bellagio conference sponsored by the Rockefeller Foundation. Reflecting the growing scientific and practical interest in mechanisms of resistance, the volume included chapters on the chemical, physical, and sensorial bases of resistance to insects (Beck and Schoonhoven, 1980; Norris and Kogan, 1980); environmental influences on the expression of resistance (Tingey and Singh, 1980); and conceptual considerations in resistance to pathogens (MacKenzie, 1980; Robinson, 1980).

### 5.1.3. The Middleground

Finally, some landmark events and papers had an impact on both IPI and HPR. An example is Brues's (1946) classic book *Insect Dietary*, which is unmatched in its broad coverage of the field. The recent volume by Strong et al. (1984a) exceeds Brues's work in analytical depth but makes no attempt to digest the huge volume of information amassed in the nearly 40 years that intervene.

Another major contribution of the Wageningen school was the publication of *Entomologia Experimentalis et Applicata* (1st issue, 1958). This journal has published research in both HPR and IPI, and as a result has greatly enhanced the flow of information between the two fields.

Thorsteinson's (1960) and Beck's (1965) reviews of host selection in phytophagous insects and host-plant resistance, respectively, have contributed substantially to reconcile concepts and terminology in both areas. One of the most useful papers for both IPI and HPR research is Waldbauer's (1968) review of methods for measuring food intake and utilization. This work provided the foundation for the rapidly expanding field of nutritional ecology which is helping to interpret observed patterns of intraspecific variation of resource exploitation (Scriber and Slansky, 1981; Slansky, 1982; Scriber, 1983). It has also helped elucidate nutritional effects of resistant crop genotypes (Kogan, 1972a). A conference on insect and mite nutrition [Lexington, Kentucky (Rodriguez, 1972)] was an early attempt to convene researchers in all phases of insect alimentary physiology and nutritional ecology.

The dichotomy, however, still seems to persist among "basic" and "applied" researchers of insect–plant relationships. The main thesis of this chapter is that both fields deal with the same fundamental biological phenomenon and, consequently, they share a common scientific foundation. The fundamental difference is that HPR operates within ecological systems in which humans function at both the second and third trophic levels—they are primary consumers and they are a powerful and highly disruptive mortality factor for competing herbivores. In addition, humans perform a fundamental role in the reproductive ecology of the crop plant through normal agricultural practices. As humans acquired this multiple role, they also "coevolved" with crop plants into a complex mutualistic relationship. Under current socioeconomic circumstances neither crop plants, nor, most certainly, humans can survive on this planet without the other.

## 5.2. FUNDAMENTAL CONCEPTS IN INSECT–PLANT INTERACTIONS

Herbivorous insects and green plants have coexisted for at least 250 million years (Strong et al., 1984). The diversity and complexity of insect–plant associations observed today presuppose a long evolutionary process of coadaptation superimposed on processes of adaptation to other ecological factors, both physical and biological. Although it is appealing for entomologists to ascribe to insects a driving role in plant evolution, those other ecological factors must have a preponderant role in most instances.

Gifford et al. (1984) analyzed crop productivity and photosynthate partitioning and concluded that those were rarely limited by single environmental factors. Consequently, it would be unrealistic to rank these factors hierachically so that if the top one were made nonlimiting, productivity would depend on the next one in rank. So the evolution of plant adaptations is the result of the combined selection pressures of multiple environmental factors. If herbivory is not the driving force in the evolution of plants, it certainly contributes a significant share of the total selection pressure.

Plants, in either natural or agricultural communities, are normally visited by many herbivorous insect species. However, only a small fraction of those visitors ever establish enduring trophic associations with the plants. For example, extensive samples of soybean fields in Illinois over a period of 12 years yielded over 400 different species of phytophagous insects (Kogan, 1980, 1981). Yet actual records of colonization of individual soybean fields in the state included only about 40 species (Mayse and Price, 1978). Numbers of transient species would be much greater if more effort were made to collect them. Over 60 species of aphids were trapped in soybean fields in Illinois (Halbert et al., 1981), and yet not a single aphid species is capable of colonizing soybean in either North or South America. Such patterns of faunal adaptation to specific hosts have more than passing interest to IPM, for the major agricultural pests come from this pool of potential colonizers. These diverse patterns of association between insects and plants in time and space are the cardinal phenomena that IPI theory attempts to explain.

Four fundamental concepts and hypotheses in IPI will be discussed, and their significance to IPM in general and to HPR in particular will be evaluated. These concepts and hypotheses are: (1) the chemical defense hypothesis based on the concept of the dual role of secondary plant metabolites; (2) the dual discrimination hypothesis based on concepts of nutritional ecology and optimization of fitness; (3) the risk–cost–value hypothesis based on concepts of optimal defense strategy; and (4) coevolution of insects and plants. A synopsis of these hypotheses and their conceptual sources is presented in Table 5.2.

### 5.2.1. Chemical Defense Hypothesis

Records of the diversity, multiplicity, and unique patterns of occurrence of secondary plant compounds, and the observed patterns of host preferences by phytophagous insects provided the foundation for current IPI and coevolutionary

**Table 5.2 Summary of Fundamental Concepts and Hypotheses in IPI Theory**

| Concept | Hypothesis | Statement or Definition and Corresponding Section in Chapter |
|---|---|---|
| I. Dual role of secondary metabolites | A. Chemical defense | Plants produce many secondary metabolites that act as kairomones to a few and allomones to the majority of the sympatric herbivore fauna (Sect. 5.2.1.) |
| II. Optimization of fitness (herbivore × plant) | B. Dual discrimination | Insects maximize colonization success by responding to both primary and secondary metabolites (Sect. 5.2.2.) |
| III. Optimization of defense (plant × herbivore) | C. Risk–Cost–Value | Plants maximize defense posture in direct proportion to risk of attack and the value of the tissue protected, and in inverse proportion to the cost of defense (Sect. 5.2.3.) |
| | a. Apparency and defense | (Qualitative vs. Quantitative defenses) (Sect. 5.2.3, 5.2.3.1.) |
| | b. Parsimonious allocation of defensive energy | (More protection where the need is greatest) (Sect. 5.2.3.2.) |
| | c. Latent-defense mobilization | (Induced resistance factors) (Sect. 5.2.3.3.) |
| | d. Competitive demands for energy | (Defenses relaxed if there is greater need for energy to overcome life threatening stress) (Sect. 5.2.3.4.) |
| | e. Cost-sharing and spreading of the risk | (Defense guilds) (Sect. 5.2.3.5.) |
| | f. Synergistic effects with other mortality factors | (Interactions with third trophic level) (Sect. 5.2.3.4.) |
| IV. Coadaptation of insects and plants | D. Coevolution or sequential evolution | Plants and herbivores undergo evolutionary changes under mutually exerted selective pressures (Sect. 5.2.4.) |

theories (Dethier 1954, 1970; Fraenkel, 1959, 1969; Thorsteinson, 1960; Ehrlich and Raven, 1964; Kogan, 1977; Berenbaum, 1983a; and volumes cited on page 87).

Although plants may be relatively monotonous in their qualitative profiles of primary metabolite content, they have an incredibly diverse array of secondary compounds (Harborne, 1982). However, the distinction between primary and secondary plant compounds is far from absolute either chemically (Mothes, 1980) or ecologically (Seigler and Price, 1976). Whereas primary metabolites (carbohydrates, amino acids, purine and pyrimidine bases, lipids) are ubiquitous and perform vital roles in all living organisms, the secondary metabolites (alkaloids, terpenoids, flavonoids, tannins, etc.) have rather limited distribution among organisms and frequently have little obvious survival value outside an ecological context. If the physiological role of many secondary plant compounds is obscure, their potential ecological function is currently recognized even by nonecologically oriented organic chemists (see most chapters in Bell and Charlwood, 1980).

Taxonomically related plants often have similar natural product chemistry (Hegnauer, 1962–1969; Swain, 1963; Bell and Charlwood, 1980; Bisby et al., 1980). However, it is not uncommon for unrelated plant species to converge in the nature of their typical secondary metabolites. For instance, glucosinolates occur in practically all species studied in five closely related families (including the Cruciferae, Capparaceae, and Resedaceae) and in several species in six other unrelated families (including Tropeolaceae, Caricaceae, and Euphorbiaceae) (Underhill, 1980), and cucurbitacins have been isolated not only from many species of Cucurbitaceae, but also from a few species of Cruciferae (Curtis and Meade, 1971), Scrophulariaceae (Gmelin, 1967), and Begoniaceae (Doskotch et al., 1969).

Secondary plant compounds perform allelochemical functions (Whittaker and Feeny, 1971) either as allomones or as kairomones. [There are various classifications of the allelochemic effect of secondary plant compounds on insect behavior, development, growth, survival, fecundity, and fertility. The system that I adopted (Table 5.3) combines the classifications proposed by Dethier et al. (1960) and Whittaker and Feeny (1971) with the categories of plant resistance to insects commonly used by workers in HPR (Beck, 1965; Kogan and Ortman, 1978; Horber, 1980)].

On the conceptual basis of coevolution, allelochemicals exist primarily to perform allomonal functions in a defensive mode (Pasteels, 1976). The reciprocal adaptation of specialist herbivores has resulted not only in the evolution of mechanisms that render allomones ineffectual (e.g., detoxification, behavioral avoidance), but also in converting them into kairomones, or cues for host-finding and feeding or oviposition excitation. Consequently, many secondary plant compounds are defensive allomones for the majority of the sympatric herbivorous fauna, and kairomones for a few oligophages associated with the allelochemic-producing plant. Such compounds perform a dual allelochemical role (Norris and Kogan, 1980). For example, the bitter, oxygenated tetracyclic triter-

**Table 5.3  Principal Classes of Chemical Plant Factors (Allelochemics) and the Corresponding Behavioral or Physiological Effects on Insects[a]**

| Allelochemical Factors | Behavioral or Physiological Effects |
| --- | --- |
| **Allomones** | Give adaptive advantage to the producing organism |
|   Antixenotics | Disrupt normal host selection behavior |
|     Repellents | Orient insects away from plant |
|     Locomotory excitants | Start or speed up movement |
|     Suppressants | Inhibit biting or piercing |
|     Deterrents | Prevent maintenance of feeding or oviposition |
|   Antibiotics | Disrupt normal growth and development of larvae; reduce longevity and fecundity of adults |
|     Toxins | Produce chronic or acute intoxication syndromes |
|     Digestibility reducing factors | Interfere with normal processes of food utilization |
| **Kairomones** | Give adaptive advantage to the receiving organism |
|   Attractants | Orient insects toward host plant |
|   Arrestants | Slow down or stop movement |
|   Feeding or oviposition excitants | Elicit biting, piercing, or oviposition; promote continuation of feeding |

[a]Adapted from Dethier et al. (1960); Beck (1965); Whittaker and Feeny (1971) and modified from Kogan (1982).

penoids, cucurbitacins, are effective feeding deterrents or even toxins for most phytophagous insects, but they are powerful feeding excitants for many diabroticine beetles. Nielsen et al. (1977) have isolated cucurbitacins E and I from green parts of the cruciferous plant *Iberis amara* (L.), and have shown that a crucifer specialist, the flea beetle *Phyllotreta nemorum* L., was deterred from feeding on *Iberis* because of its cucurbitacin content. Other species tested performed similarly (Nielsen, 1978) (for exception, see Usher and Feeny, 1983). Two-spotted spider mites (*Tetranychus urticae* Koch) produced fewer nymphs on bitter (high cucurbitacins) cucumbers (*Cucumis sativus* L.) than on a nonbitter genotype (Costa and Jones, 1971). However, these same cucurbitacins are among the most active feeding excitants ever tested with an insect (Chambliss and Jones, 1966; Howe et al., 1976; Metcalf et al., 1982). Metcalf and co-workers (1982) reported aggregations of five species of diabroticites 15 times greater on the high cucurbitacin (1.01 mg/g) *Cucurbita okeechobeensis* Bailey than on several low cucurbitacin species (with <0.02 mg/g total cucurbitacin content). Similar duality has been demonstrated for glucosinolates, which are powerful feeding excitants for crucifer specialists but are feeding deterrents for most other phytophagous insects. *Papilio polyxenes* Fabr. larvae, oligophagous on Umbelliferae; *Spodoptera eridania* (Cramer), a polyphagous species; and *Pieris rapae* L., oligophagous on crucifers, were fed acceptable plants (carrot, lima bean, and collards, respectively) that were cultured in solutions containing various concentrations of allylglucosinolate. Absorption of the compound by the plant cutting was confirmed by gas chromatography. As glucosinolate concentration increased, growth efficiency (regression of relative growth rate on relative consumption rate) sharply declined for *P. polyxenes*, was unchanged for *P. rapae*, and showed an intermediate effect for *S. eridania* (Blau et al., 1978). Thus, glucosinolates have an acute effect on oligophagous insects not associated with crucifers, and a moderate dose-dependent effect on polyphagous species (Lerin, 1980).

As with any generalization in IPI theory, there are many exceptions to this duality rule (Schoonhoven, 1981). Some compounds have considerable allomonal activity but with no obvious kairomonal effect, and some common kairomones have no apparent allomonal roles. For instance, the triterpenoid aldehyde gossypol, found in epidermal glands and in seeds of various malvaceous plants (Lukefahr and Fryxell, 1967), is a feeding deterrent and a toxin for several species of lepidopterous, coleopterous, and hemipterous insects associated with cotton in the United States (Bottger and Patana, 1966), but it is neutral, at best, for the boll weevil (*Anthonomus grandis* Boheman). Although high-gossypol plants seem to be more susceptible to thrips and whiteflies (Niles, 1980), there is no evidence that gossypol is a kairomone for those species.

Many compounds seem to act exclusively as allomones, and insects that feed on plants containing these compounds do so because they can detoxify them or have developed behavioral adaptations that avoid their impact. Examples of the former are insects associated with species of Solanaceae containing tropane or steroidal alkaloids. *Solanum* spp. plants are usually rich in steroidal alkaloids and are hosts for the Colorado potato beetle; *Datura* spp. plants contain tropane alkaloids and are preferred hosts for *Lema trilineata* (Olivier). Fitness of both insect species is greatly reduced when hosts are reversed (Fig. 5.1). However, neither species seems to use the specific alkaloids as kairomones. The dual-role hypothesis does not seem to apply to this system and other plant compounds may serve as mediators of host specificity (Visser, 1983). There are few, if any, absolute allomones whose action cannot be neutralized by some insect (Schoonhoven, 1982). One compound that approaches this category is the triterpenoid azadirachtin, extracted from leaves, fruit, and seed of the neem tree (*Azadirachta indica* A. Juss.) (Reed et al., 1982; Reinhold et al., 1982; Schoonhoven, 1982).

Even more difficult to find are absolute kairomones. It is conceivable that such apparently useless compounds would not be preserved in the genome of green plants, although certain low-molecular-weight alcohols, aldehydes, and monoterpenes seem to form part of a widespread "green leaf odor" in many plants (Visser, 1983). Thus, the estimated 100,000–400,000 kinds of secondary metabolites (Schoonhoven, 1982), qualitatively and quantitatively combined in many complex ways, provide a mind-boggling diversity of secondary plant chemistry, capable of uniquely endowing each plant species with a characteristic

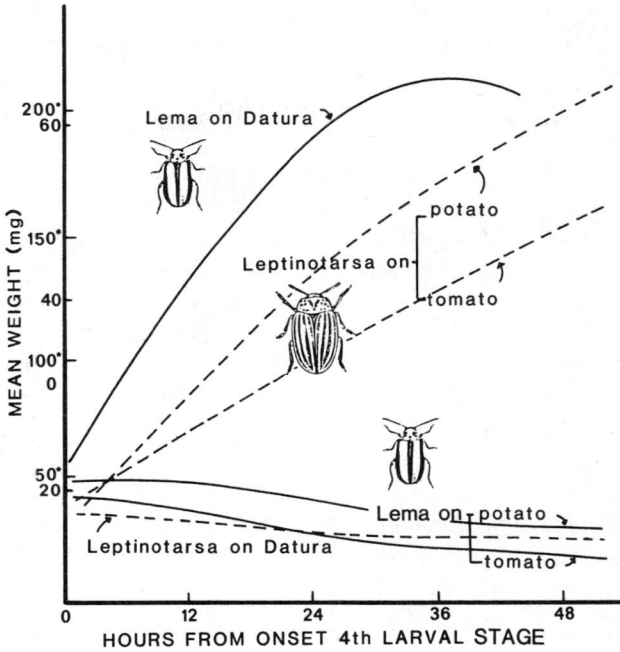

Fig. 5.1 Weight variation of larvae of *Lema trilineata* and *Leptinotarsa decemlineata* fed leaves of *Datura*, potato, or tomato during the fourth instar. The scale numbers on the abscissa followed by an asterisk refer to *Leptinotarsa*. The weight variation curves reflect the relative susceptibility of each species to the prevalent alkaloid complexes in each plant (from Kogan, 1986).

chemical profile. Insects are equally diverse in their ability to decode those profiles and to adjust, behaviorally or physiologically, to specific antixenotic or antibiotic compounds. We can, therefore, understand why so much emphasis has been placed on the key role of secondary metabolites in the mediation of insect–plant interactions [but see Strong and Brodbeck (1986), and Duffey and Bloem (Chapter 6) for a discussion of interactions between primary and secondary metabolites].

In summary, the allelochemical effect of many secondary plant compounds on insects has been confirmed by behavioral, nutritional, and electrophysiological experiments that support field observations and experiments. If antiherbivory is not the sole function of these compounds, it is, nevertheless, an important function acquired in evolutionary time. Most of these compounds are part of the normal chemical makeup of the plants that produce them, whether or not those plants are subjected to herbivory. Plants with such compounds are said to possess constitutive defenses (Levin, 1976). By contrast, some secondary metabolites are synthesized *de novo*, only if the plant is subjected to stress. The best known factors that induce allomone production are plant pathogens and some abiotic environmental factors (temperature, ultraviolet radiation). Mechanical

injury, including that resulting from herbivory, is also capable of inducing production of certain defensive chemicals. Plants displaying these reactions have
induced defenses (Levin, 1976). Thus, allomonal secondary compounds can be
either constitutive or induced. Most kairomonal compounds studied so far are
constitutive.

The allomonal effects on insects range from mild repulsion to reduced fecundity and longevity, or high toxicity and even lethality. The kairomonal effects
range from mild, short-term attraction to powerful, long-term feeding or oviposition excitation. The close host associations of most plant-feeding insects have
been explained by the unique secondary chemical makeup of host plants
(Fraenkel, 1959; see also list of references on page xx). Oligophagous species
tend to respond specifically to certain compounds that elicit feeding or oviposition. These same or other compounds act as repellents, feeding or ovipositional
deterrents, or antibiotics against the vast majority of the rest of the sympatric
phytophagous fauna. Polyphagous insects, on the other hand, require only feeding excitants of ubiquitous occurrence (including nutrients) (but see Ahmad,
1983). Their host range is probably defined by the absence of antibiotic or antixenotic factors with which they cannot cope, rather than the presence of specific
kairomones (Jermy, 1966).

### 5.2.2. Dual Discrimination Hypothesis

Emphasis on the dominant role of secondary metabolites in host plant selection
by phytophagous insects generated considerable reaction by those claiming that
both primary and secondary metabolites played important roles in the host selection process (Kennedy, 1965). Based mainly on evidence from host selection
studies with aphids, Kennedy (1953) postulated the dual discrimination hypothesis (Kennedy, 1953; Thorsteinson, 1960). More recently, studies on the combined effects of nutrients and secondary metabolites led to the emergence of the
field of nutritional ecology (Beck and Reese, 1976; Scriber and Slansky, 1981;
Slansky, 1982; Scriber, 1983). According to the dual discrimination hypothesis,
insects tend to select plants of specific ages or physiological state so as to optimize their nutritional intake. Secondary metabolites act as token stimuli that
signal the adequacy of the plant for feeding or oviposition, and both nutrients
(primary metabolities) and token stimuli (secondary metabolites) coact to regulate the final selection of a suitable host (van Emden, 1978). Perception of a host
plant is thus assumed to be a "Gestalt," rather than a mosaic of discrete stimuli
(Kogan, 1977).

Despite the overall qualitative similarity in primary metabolite content, there
is considerable variation in the nutritional quality of a plant for herbivores
(Denno and McClure, 1983). Therefore, if secondary plant compounds affect
host selection at the initial stages of the process (host finding and recognition),
the actual selection of a feeding site is more likely to be influenced by nutritional
quality (Beck and Reese, 1976; Slansky, 1982). Soil texture, mineral and organic
matter contents, water saturation level, sun exposure, photoperiod, tempera-

ture, and relative humidity are some of the many environmental factors that influence the nutritional quality of the plant for an insect. [See also Caswell and Reed (1976) for a discussion of possible nutritional differences between the $C_3$-$C_4$ plants.]

The relative concentration of nutrients also varies with plant phenology and plant growth patterns. For example, nitrate reductase (NR) is an important and ubiquitous enzyme in plants; it catalyzes a two-electron reduction of nitrate to nitrite. Most NR activity is induced by the presence of $NO_3^-$; the end-product of reduction is ammonia ($NH_3$), which is then used in the biosynthesis of amino acids (Bandurski, 1965). In most species of Fabaceae, NR activity prevails until symbiotic $N_2$ fixation takes over; in soybean each mechanism contributes about 50% of the incorporated nitrogen. The following, taken mostly from Harper et al. (1972), illustrates the intraspecific variability in activity of an enzyme that catalyzes important metabolic processes in living organisms. Furthermore, activity of this enzyme is intimately related to the nitrogen–water index of food quality proposed by Scriber (1984). There is a positive correlation between NR activity in corn (*Zea mays* L.), wheat (*Triticum aestivum* L.), and rye grass (*Lolium perenne* L.) and total reduced plant nitrogen, grain protein production, and dry matter accumulation in varieties with similar nitrogen transport efficiencies grown under adequate environmental conditions. However, NR activity requires high energy input (light) and ceases under low light intensities, resulting in nitrate accumulation in plant tissues. In fully grown corn plants, nitrate accumulates in the lower stalk and lower leaves, which are shaded by the upper canopy; upper leaves have virtually no nitrates and middle stalk and leaves have intermediate concentrations (Aldrich, 1980). Seed protein content among soybean genotypes ranges from 39.7 to 50.6% (Harper et al. 1972), but more important from the perspective of herbivory are the seasonal profiles of NR activity and the distribution of NR within the leaf canopy of soybean (Fig. 5.2 *A*, *B*, adapted from Harper et al., 1972). At this fundamental (primary) level of metabolism, the genetic, seasonal, and intraindividual variation in nitrate concentration and NR activity suggest an extremely variable resource in terms of herbivore efficiency. Reese (1979) and Scriber (1984) provide detailed discussions and examples of the effect of nutrient concentration (particularly N compounds) on insect feeding, and on plant suitability for insect growth and survival. The ability of insects to select suitable food by its nutrient content has been demonstrated with seed feeders (Loschiavo et al., 1969; Waldbauer and Bhattacharya, 1973), and also with a foliage and fruit feeder—the corn earworm in self-selection of optimal artificial diets (Waldbauer et al., 1984). The sensorial capability of phytophagous insects to detect nutrients has been also demonstrated by amputation and electrophysiological methods (Staedler, 1977; Visser, 1983).

In summary, it seems that throughout the catenary process (Thorsteinson, 1960; Kogan, 1977) that leads a dispersing insect population to select a specific host for feeding or oviposition, both primary and secondary metabolites are intricately involved. In the early phases of the process—host finding and recognition—secondary metabolites are likely to prevail with primary metabolites act-

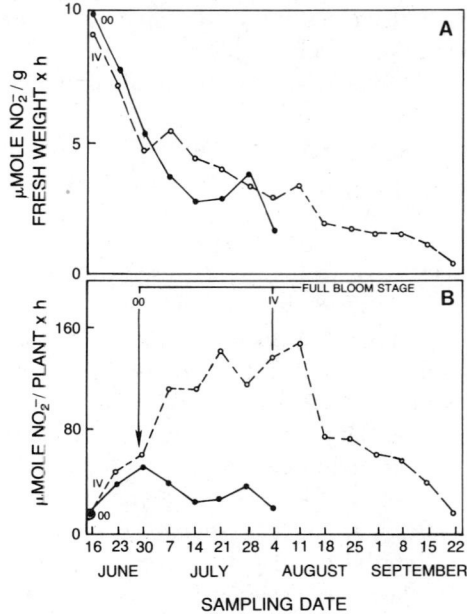

Fig. 5.2 (*A*) Seasonal profiles of mean nitrate reductase activity of entire soybean leaf canopy at respective sampling dates. (*B*) Seasonal profiles of the sum of nitrate reductase activity of all leaf positions within the soybean canopy at respective sampling dates. Solid line curves for one very early maturing variety (Group 00), and dashed line curves represent means for three varieties of intermediate maturity (Group IV). (Redrawn from Harper et al., 1972).

ing as synergists after probing contacts are made. At the ingestive phase, primary metabolites become increasingly important and the synergistic relationship is likely to be reversed, particularly if "token" stimuli are still required by the insect. Finally, at the level of host suitability, the nutritional quality of the host (i.e., its primary metabolite content) takes over, provided there are no antibiotic factors that the insect is unable to neutralize either behaviorally or enzymatically. Again, such a broad generalization must be approached with extreme caution. In oligophagous insects, secondary metabolites are more likely to regulate both pre- and postingestive phases of the process. Polyphagous insects, often endowed with more general detoxification mechanisms (Brattsten, 1979), are capable of handling a broader range of defensive chemicals.

## 5.2.3. Optimal Defense Strategies

Feeny (1976) and Rhoades and Cates (1976) reached similar conclusions concerning patterns of antiherbivore strategies in green plants in papers presented at a meeting of the Phytochemical Society of North America, held at Tampa, Florida (Wallace and Mansell, 1976). Feeny used mainly evidence from his work on oak trees (*Quercus robur* L.) and the winter moth (*Operophtera brumata* L.) (Feeny, 1970) and work on species of Cruciferae and associated and nonassociated arthropods (Feeny et al., 1970); Rhoades and Cates used mostly work on creosote bush (*Larrea* spp.) and associated fauna (Rhoades, 1977). A summary of the basic premises of their hypothesis follows.

**1.** Chemical plant defenses are broadly divided into two categories: (1) toxins or qualitative defenses and (2) digestibility reducing factors or quantitative defenses. Toxins are mainly low-molecular-weight compounds, easily translocated within the plant, active against insects at low concentrations, and generally having a restricted and characteristic distribution among plant taxa (see secondary metabolites above); examples are alkaloids, glucosinolates, cyanogens, and monomeric phenolics. Digestibility reducing factors are usually high-molecular-weight compounds that normally are structurally associated with specific tissues, active at rather high concentrations, and commonly occur among many plants of related and unrelated taxa. Examples are polymeric phenolics such as tannins (Swain, 1979; Harborne, 1980; Zucker, 1983) and proteinase inhibitors (Ryan, 1979).

**2.** Plant resources can be divided into two broad groups: (1) apparent (or predictable), and (2) nonapparent (or unpredictable). The concept of apparency applies equally to whole plants within a community (e.g., early successional versus climax species) or to resources within an individual plant (e.g., young versus old tissue, or top leaves versus bottom leaves).

**3.** Apparent resources, such as trees, are usually long lived, are easily detected by potential insect colonizers, and accumulate quantitative defenses. Nonapparent resources have a greater propensity for escaping in space and time; they resist attack by accumulating qualitative defenses (toxins). The compounds that are typically considered as part of qualitative defenses are likely to be neutralized and used as token stimuli by oligophagous species specialized on the toxin-producing plant.

These general postulates have been the object of considerable testing that yielded both supportive (e.g. Courtney, 1982), as well refutative evidence (e.g. Bernays, 1978). Furthermore, the presence of condensed tannins in an annual primitively nonapparent plant, such as cotton (*Gossypium hirsutum* L.) (Chan et al., 1978), and grasses and wild soybean (see below) or the presence of alkaloids and other typical qualitative defenses in many apparent species of Fagaceae (Hegnauer, 1966), suggest that these hypotheses do not have the generalized scope that, if not intended by the authors, at least was readily assumed by many workers in the field. Though these hypotheses may lack the generality of an ideal comprehensive theory of antiherbivory strategy in plants, they provide the most reasonable body of concepts proposed to date to explain many qualitative IPI patterns. I shall borrow freely from these concepts to examine the implications of IPI under managed agricultural systems. This exercise will use mainly examples from soybean, [*Glycine max* (L.) Merrill], an expanding crop that provides many features amenable to such an analysis and with which I am reasonably familiar. Other crops will be used as needed to complement discussion of the argument. My intent is not a comprehensive review of the subject because whole books have been written and continue to appear at a breath-taking rate (see references on p. 87).

### 5.2.3.1. The Metabolic Cost Hypothesis

#### Statement of the Hypothesis

Biosynthesis of defensive metabolites requires considerable metabolic energy input. As a limited supply of energy must be allocated to all vital functions, better defended plants (those with high investment in defense) have lower fitness than less well defended plants, if attackers are absent and all other environmental conditions are similar.

There is more speculation than data on the metabolic cost of producing secondary metabolites. Chew and Rodman (1979) hypothesized three possible relationships between biosynthetic cost of chemical defense and potential benefits to the plant: (1) a linear relationship, assuming that each incremental allocation to defense corresponds to an increment in antiherbivory efficiency; (2) a threshold relationship, assuming that no antiherbivory effect exists before a critical concentration of the allelochemic is produced and that efficiency increases exponentially above the threshold; and (3) a relationship that assumes efficiency increases toward a maximum as defensive chemicals accumulate, and then declines sharply with further accumulation of chemicals [Chew and Rodman (1979) describe this as a ditonic relationship]. The direct cost of energy allocated for defense, however, is difficult to estimate as few biosynthetic pathways for secondary metabolites have been totally elucidated and the turnover rates of metabolites are not well defined (Swain, 1979). Notwithstanding these reservations, current theory presupposes considerable metabolic costs for defense. Since plant energy resources are finite (Gifford et al., 1984), resources for defense are probably mobilized at the expense of energy otherwise available for reproduction; thus, less well defended organisms would have higher fitness than better defended organisms, when enemies are absent (Rhoades, 1979). This concept, if validated by additional experimental work, is particularly significant for crop plants.

Rice breeders claim to have obtained varieties highly resistant against insect pests and diseases without loss of yield (Khush, 1984). However, soybean genotypes resistant to foliage feeders are agronomically inadequate. These plants are procumbent (commercial varieties should be erect), seeds are small (less than 10 g/100 seeds as opposed to ~18 g/100 seeds in good commercial varieties), pods have a tendency to shatter, and their grain yield is low. One of those resistant genotypes, plant introduction (PI) 229358, has been used in several breeding programs from which about 20 advanced germ plasm lines have been released for further improvement. Despite the considerable effort invested into some of these programs, the best resistant breeding lines still yield less than the best available varieties in the absence of pests (Kogan et al., 1986).

Rice and soybean are very different. While high-yielding rice varieties may have an energy surplus to invest in defense, soybean biosynthesis of defensive chemicals must be closer to an energy threshold (see above and Chew and Rodman, 1979) with less of its energy resources available for defense. One way that soybean seems to compensate for this tight biosynthetic energy budget is through a remarkable tolerance to leaf area reduction. Even commercial cultivars may lose up to 30% of their foliage area without a yield loss, if defoliation occurs prior to pod development. Although cost of defense may not be the single explanation for the yield differential between susceptible and resistant soybean, this is one factor seldom considered by breeders and plant physiologists. But before such consideration can be given, much more must be known about the chemical basis of resistance in plants.

### 5.2.3.2. Parsimonious Allocation of Defense Energy Hypothesis

#### Statement of the Hypothesis

Since there is a metabolic cost for defense, to achieve maximum return on this investment, energy must be parsimoniously allocated for defensive purposes in direct proportion to the risk of attack and the value of target tissues.

Considering the multiple demands on energy resources, it is reasonable to assume that defensive energy should flow most where demand is greatest—plants within the community most vulnerable to attack (more apparent plants), and organs within the plant most needed for survival. Rhoades (1983) expressed this principle in terms of cost, risk, and value. An example is the tendency for apparent plants to accumulate quantitative defenses, for example, tannins in many woody plants (Zucker, 1983). But tannins are not restricted to typically "apparent" plants. A primitive race of cotton contains up to 20% condensed tannins by dry weight in its leaves (Lane and Schuster, 1981) and some $C_3$ grasses have tannin concentrations that reduce acceptibility by the range catepillar (*Hemileuca oliviae* Cocq.) (Capinera et al., 1983). However, Bernays et al. (1983) have shown that certain grasshoppers grow better when tannic acids are added to their diets. The ecological role of tannins, mainly as digestibility reducing factors, has been much debated (Bernays, 1978, 1981; Berenbaum, 1983b). However controversial, there seems to be a tannin role in plant defense in certain systems, and cost of biosynthesis of tannins is likely to be high (Swain, 1983).

Apparency (or predictability) may be reduced by escape in space or time. Soybean as an annual crop escapes outbreaks of various lepidopterous defoliators in

northern United States because migrating colonizers arrive too late in the growing season to build up damaging populations.

In their primitive state, soybean lines (*Glycine soja* Sieb. & Zucc.) are early successional annual vines found in fields, hedgerows, and ditch banks in eastern Asia. The seeds are small and black and the plants are extremely nonapparent, at least to human eyes. Although wild soybeans would qualify as nonapparent (*sensu* Feeny, 1976), still they are found by several soybean specialists, for example, the soybean aphid (*Aphis glycines* Mats.), the soybean pod borer (*Leguminivora glycinivorella* Mats.), and several species of stem, root, and shoot boring flies (Kogan, 1981). However, these herbivores seldom cause extensive injury to the wild hosts, for the soybean seems to possess effective physical and chemical defenses in addition to nonapparency. Resistance to the agromyzid stem fly, *Ophiomyia centrosematis* (de Meij.), has been ascribed, in part, to the presence of several phenolics, including small concentrations of tannins in the stems (Chiang and Norris, 1984).

The risk–cost–value hypothesis postulates also that defenses tend to concentrate in more "valuable" tissues. Although it is difficult to assign "value" to different plant tissues and organs, meristematic tissues, and reproductive organs presumably rank higher than vegetative parts. In crop plants the reproductive potential is measured by the seed yield and its capacity to germinate. In soybean, leaves are vulnerable to herbivory but plants are generally tolerant of defoliation. Plants can tolerate ~30% defoliation without loss in yield, except during the sensitive period of pod and seed development (Turnipseed and Kogan, 1976). However, soybean yields are strongly and negatively affected by insect attack to seeds. Soybean seeds are protected by a hard tegmen and they grow within a thick hairy pod; the seed itself contains high concentrations of proteinase inhibitors (Ryan, 1979), saponins (Applebaum and Birk, 1979), and a recently identified chitinase (Wadsworth and Zikakis, 1984). Although green seeds are vulnerable to some attack by both sucking and chewing insects, dried seeds are highly resistant to most pests of stored grain. In the soybean plant the valuable seeds seem to be better defended than the vegetative organs (Table 5.4).

In summary, although the dichotomy between qualitative and quantitative defenses does not seem to hold within the definition of apparent and nonapparent plants, nonapparent plants seem more apt to escape herbivory both in space and time. Vital reproductive organs seem to be better protected than vegetative organs; compensatory growth and excessive initial production of leaves may allow for a lesser concentration of defensive chemicals in vegetative organs without loss of fitness.

### 5.2.3.3. Mobilization of Defenses Triggered by Attack

#### Statement of the Hypothesis

To reduce investment in defense, allomones are biosynthesized *de novo* as needed upon attack. Initial attack or signals thereof induce the biosynthetic processes.

**Table 5.4 Some Characteristic Secondary Metabolities Found in Various Plant Parts of *Phaseolus* spp. and *Glycine* spp. (based mainly on Harborne et al., 1971).**

| Secondary Plant Compound | *Phaseolus* | *Glycine* |
|---|---|---|
| Flavonoids | | |
|   Flavonols/ones | Robinin | — |
|   | Quercitin-3-glucuronide | — |
|   Isoflavones | Daidzein | Daidzein |
|   | | Genistein |
|   | | Glyceollins |
|   | | Coumestrol |
|   | | Sojagol |
| Alkaloids | Guanidine | — |
|   | Trigonelline | — |
| Carbohydrates | Raffinose | Raffinose |
|   | Stachyose | Stachyose |
|   | Myoinositol | Pinitol[a] |
| Steroids | Stigmasterol | Stigmasterol[b] |
|   | Sitosterol | Sitosterol |
|   | | Campesterol |
|   | Soyasapogenol-C (glycoside) | Soyasapogenols A–E (glycosides) |
|   | Phaseolin | — |
| Lipids | — | Ceramides[c] |
| Phenolic acids | | Gallic[d] |
|   | | Protocatechuic |
|   | | p-Hydroxybenzoic |
|   | | Vanillic |
|   | | Caffeic |
|   | | p-Coumaric |
|   | | Ferulic |
| Miscellaneous | Linamarin | Cerebrosides[c] |
|   | Lotaustralin | — |

[a]Dreyer et al. (1979).
[b]Grunwald and Kogan (1981).
[c]Ohnishi et al. (1982).
[d]Hardin (1979).

The most parsimonious defense strategy would use inducible resistance factors. In the absence of attack, the plant would have invested little energy into the biosynthesis of defensive chemicals.

Although much research on the allelochemical role of secondary metabolites has centered on chemicals that normally accumulate in plants whether attacked by herbivores or not (constitutive defenses), there is mounting evidence that some metabolites are synthesized *de novo* in response to attack (Ryan, 1979, 1983; Kogan and Paxton, 1983; Rhoades, 1985). These are inducible defenses (Levin, 1976). Phytoalexins, a group of inducible compounds with antifungal properties, have been well studied by plant pathologists (Bailey and Mansfield, 1982). Recently, phytoalexins have been shown to be active against herbivores (Haukioja and Niemela, 1977; Russell et al., 1978; Sutherland et al., 1980; McIntyre et al., 1981; Hart et al., 1983).

Resistance in soybean to root and stem rot caused by *Phytophthora megasperma* var. *sojae* results from the accumulation at the point of attack of isoflavonoid phytoalexins, mainly glyceollins. Glyceollins are also a powerful feeding deterrent for the Mexican bean beetle. Soybean cotyledons treated under UV light accumulate large amounts of glyceollin. When offered a choice between glyceollin-rich and normal (untreated) cotyledons, adult Mexican bean beetles accept only the normal tissue, although they make probing markings on the treated cotyledones (Hart et al., 1983). Preliminary evidence indicates that previous herbivory predisposes soybean to increased resistance to subsequent attacks, similar to the effect of previous pathogen infection predisposing plants to resist future infections (Kuć and Caruso, 1977). The involvement of phytoalexins in the herbivory case, however, remains to be proven (M. Kogan, D. Fischer and J. Paxton, University of Illinois, unpublished). Karban and Carey (1984) have shown that cotton seedlings exposed for 5 days to spider mites are more resistant to mite infestations 12 days after the end of the initial exposure, but the mechanisms are unknown.

Not only do plants seem to communicate systemically via chemical signals to mobilize inducible resistance mechanisms, but interplant communication has been claimed as well (Rhoades, 1983, 1985; Baldwin and Schultz, 1983). These claims, however, have been met with considerable skepticism (Fowler and Lawton, 1984). In summary, induction of resistance factors may represent an energy efficient step in the evolution of plant defenses.

### 5.2.3.4. Competitive Demands for Energy

#### Statement of the Hypothesis

During periods of extreme stress, energy normally allocated for defense may be diverted to functions essential for survival; or total energy may be so reduced that not enough is available for defense.

Herbivory is but one of many stresses that affect a plant. Under most normal growing conditions, energy is allocated to various physiological and ecological processes according to a budget, which is genetically regulated and influenced by environmental factors. If the plant is subjected to a temporary imbalance, for example, a period of drought, there will be pressing energy demands to overcome the life-threatening stress. Under such circumstances, defenses are probably relaxed and energy is spent to maintain more vital functions.

It may be difficult to rank such contingencies, although imbalances that result in irreversible injury affecting vital functions presumably rank highest. Few stresses cause such widespread physiological syndromes in living organisms as water deficits. Water stress in the mesophytic soybean causes reduced growth, inhibits photosynthetic rates, increases stomatal closure, reduces transpiration, decreases respiration, reduces nitrogen absorption, reduces overall nitrate reductase activity (Adjei-Twum and Splittstoesser, 1976), and inhibits nodulation (Williams and Mallorca, 1984). In stressed plants, however, nitrate reductase activity per unit dry weight is high. This activity, combined with low rates of photosynthesis and consequent reduced plant size, results in high relative concentration of tissue nitrogen. Of potential importance for herbivory, however, is that water-stressed plants show increased levels of free amino acids, particularly an increase in proline (White, 1976, 1984; Mattson, 1980; Strong and Brodbeck, 1986) and high relative concentrations of soluble carbohydrates (Fukutoku and Yamada, 1982). Whether such profound metabolic changes impose additional demands of energy or whether these changes are actually devices to mobilize energy has not been determined. The overall level of energy in the plant may be reduced by water stress, although this does not preclude higher concentrations of some secondary metabolites (cyanogens, alkaloids) in water-stressed plants (Rhoades, 1979; Mattson, 1980). Since some of these metabolites may be quickly turned over in the primary metabolism, their higher levels in stressed plants do not contradict the hypothesis (see argument in Rhoades, 1979). The fact remains that some water-stressed plants are more susceptible to insect attack (McNeill and Southwood, 1978; White, 1984).

Outbreaks of the two-spotted spider mite (*Tetranychus urticae* Koch) on soybean are highly correlated with periods of dry weather. This relationship may be explained by one or a combination of the following hypotheses: (1) There is a direct positive effect of low relative humidity on spider mite fecundity, fertility, and female longevity (Boudreaux, 1958; Hazan et al., 1973) (Fig. 5.3); (2) there is a differential susceptibility among phytophagous and predaceous mites to low relative humidity, high temperature, or the direct mechanical impact of rain fall (Huffaker et al., 1969); and (3) water-stressed plants provide a more nutritious food for the herbivorous spider mites. The increased quality of the food coupled with greater overall fitness under low RH and high temperature would account for the outbreaks. The nutritional quality hypothesis seems to conform with the suggestion that a higher demand for metabolic energy relaxes plant defenses. For the highly polyphagous two-spotted spider mites, the nutritional status of the host may be the most critical factor in host acceptance and suitability. However, there is no direct information that there are parallel decreases in secondary metabolite composition or concentration. In fact, Gould (1978) provides evidence that apparently conflicts with the above hypothesis. Bitter cucumbers (high cucurbitacins content) subjected to water stress were significantly more resistant to *T. urticae* than nonbitter or bitter but nonstressed plants. Since cucurbitacins were already present in the plants prior to the onset of stress, the water deficit possibly just resulted in a higher concentration of cucurbitacins per unit of sap ingested.

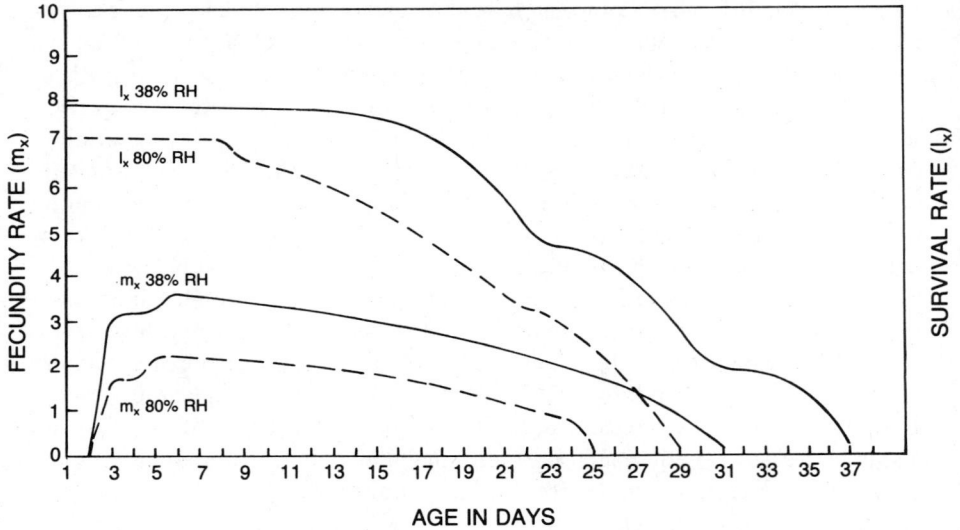

Fig. 5.3  Longevity ($l_x$) and age specific fecundity ($m_x$) of *Tetranychus cinnabarinus* (Boisd.) at 19°C and 38 or 80% RH. Fitness parameters are considerably higher at lower relative humidities. (Redrawn from Hazan et al., 1973.)

### 5.2.3.5. Cost-Sharing and Spreading of the Risk Hypothesis

#### Statement of the Hypothesis

Plants in communities with relatively high species diversity increase their defensive posture at no increase in cost if there is a synergistic effect among allomones and mutual interference among kairomones produced by individual species in the community. Plants within single-species populations reduce cost of defense if high genetic diversity enhances the quantitative and qualitative diversity of allomonal profiles among individuals.

Atsatt and O'Dowd (1976) suggested that plant defense guilds exist within communities, and that their interacting defensive postures were more effective than the defenses of isolated species. The mechanisms of this associational resistance may be: (1) Some plants within the community are reservoirs for natural enemies of the herbivores; (2) diversity interferes with host searching behavior by the herbivore; (3) some "toxic" plants may attract herbivores and reduce their impact on more susceptible hosts (O'Dowd and Williamson, 1979); and (4) certain allelopathic relationships among plants may render unsuitable otherwise adequate hosts. There is evidence to support some of these hypotheses, but more testing is needed.

The question of diversity and stability of agricultural systems has been much debated (Southwood and Way, 1970; Altieri, 1983; see also Kareiva, Chapter 4, and Herzog and Funderburk, Chapter 9). Some studies have shown that increased plant diversity may reduce overall specialist herbivore load (Risch, 1980, 1981; Matteson et al., 1984). Other studies have failed to show differences in herbivore populations in monocultures vis-à-vis polycultures, although small plot sizes may account for negative results in some cases. Under controlled conditions, experimental designs to be manageable may so restrict plot dimensions that the total experiment becomes a polyculture, and effects of monocultures are not detected (see, for example, Nordlund et al., 1984). The choice of plants in the polyculture may also be critical. For instance, Latheef and Irwin (1979) failed to detect differences in eggs and larval populations of several crucifer specialists on patches of cabbage growing alone or with borders of odoriferous plants such as thyme (*Thymus vulgaris* L.) and peppermint (*Mentha piperita* L.). Although differences among treatments were not statistically significant, some associations had more visible injury than the control, which suggests that the companion crop attracted more ovipositing females. Another explanation is that *Pieris rapae* (L.), one of the species included in the study, has a marking pheromone that tends to deter females from ovipositing on leaves already containing another *Pieris* egg (Rothschild and Schoonhoven, 1977; Klijnstra, 1982). The powerful odor of some of these companion crops may have masked the effect of the marking pheromone.

In summary, interactions of multiple plant species may enhance, reduce or have no effect on herbivory. The interactions are always complex and when operating may reduce the risk of herbivory at a lower metabolic cost than that needed to achieve similar levels of defense by plant species growing in pure stands. Thus, plants in the guild share the cost of defense.

As plant communities benefit from diverse defensive postures, so do populations benefit from genetic diversity and individuals from physiological heterogeneity. Allelochemical diversity within populations of a single species would tend to avoid the rapid selection for new biological races of herbivores, in a process that has been compared to the delay in development of insecticide resistance by alternating insecticides with different modes of action (Whitham et al., 1984; see also Gould, 1984).

The enormous genetic potential for intraspecific variability in plants is readily apparent in germ plasm collections of common cultivars (Harlan and Starks, 1980). Variability exists also in time, as plants change with each phenological stage of development. The genetic diversity within plants is the raison d'etre of research in HPR. Screening programs tailored to reveal this diversity identify new sources of resistance (Fig. 5.4). In areas of subsistence farming, local varieties (land races) may have much genetic variability that is usually lost as modern agricultural methods and improved cultivars are introduced. The low-yielding land races frequently produce stable yields (low standard deviations of mean yield over time), under rather variable environmental conditions. These races, often planted as blends, have the necessary genetic makeup to compensate for

Fig. 5.4 Nursery of breeding lines submitted to heavy pressure of *Anticarsia gemmatalis* populations in Southern Brazil. Differential defoliation levels illustrate genetic variability for resistance to larval feeding and probably also oviposition selection by female moths. (Original photo.)

the effects of those environmental fluctuations. Pest impact is dampened by similar mechanisms. Plants within a population may range from highly susceptible to highly resistant to insect pests and disease. In general, there is an inverse relationship between resistance and yield potential among races, suggesting that costly defenses are not equally spread within the population. For the population this means that in the absence of attack, yields will be maximum because the high-yielding, susceptible plants will produce a rich crop. In the presence of attack, the most susceptible plants, and probably the most abundant, will be destroyed, but some will survive to maintain genetic diversity. Most of the resistant will survive and produce the bulk of the seed for the next generation. Within this scenario of fluctuating herbivore impact and relative proportions of susceptibility/resistance within populations, diversity is preserved both in the host and in the herbivore populations, thus maintaining the efficiency of extant defenses and avoiding the rapid evolution of new biological races of herbivores.

Whitham (1983) suggests that the advantages of within-tree heterogeneity result from the following processes: (1) Heterogeneity creates pockets of lower suitability that may result in reduced insect fitness; (2) in pockets of high suitability there may be an increase in competition for the superior resources; and (3) herbivores may concentrate in those pockets of high suitability, making the herbivores more vulnerable to predation. Also by concentrating defenses in pockets, the plant may reduce the overall cost of defense. If attackers are absent, defense energy is wasted only in a segment of the plant; if attack occurs, only the unprotected pockets are seriously affected.

Whitham et al. (1984) see the origins of within-plant heterogeneity in the following ways: (1) Different parts within the same plant are genetically different so the host is a "mosaic of defenses"; (2) the host varies in time and space in inducible defenses produced in response to localized attack; and (3) during the life span of a plant, genes are "turned on and off" differentially in various plant parts. These possible mechanisms are probably superimposed on each other and result in considerable internal diversity of long-lived plants.

In summary, given the high genotypic and phenotypic variability of most plant populations, and the frequent within-individual variability of many plant defenses, the risk of attack is spread among and within individuals. The cost of defense of such a variable resource is substantially lower than that of an homogeneous resource. The outcome is preservation of variability, stability of overall fitness, and relaxation of selection pressure for development of more virulent herbivore biotypes. As more is known about variability of defense in individuals and in populations, it may become apparent that (1) variability of defense is more common than heretofore assumed, and (2) this variability is more important in IPI than heretofore realized.

### 5.2.3.6. Synergism of Plant Defenses with Other Biotic Mortality Factors

#### Statement of the Hypothesis

Nonlethal antibiotic plant allelochemicals act synergistically with entomopathogens, parasitoids, and predators by interfering with mechanisms that provide herbivore immunity to disease or escape from predation. These allelochemicals must be compatible with the biological control agents.

Nonlethal defenses may potentiate the role of predation, parasitism, and pathogen infection in regulating herbivore populations. The interactions between a herbivore's host plants and its predators are fundamental community level processes (Price et al., 1980). Understanding these interactions is essential in developing a comprehensive IPI theory, as well as in evaluating the roles of predators, parasitoids, and diseases in the population dynamics of herbivores (see also Duffey and Bloem, Chapter 6). Studies of interactions of parasitoids (Martin et al., 1981; McCutcheon and Turnipseed, 1981) and a fungal disease (Oliveira, 1981) with noctuid larvae feeding on resistant and susceptible soybean failed to show profound mutual influences but revealed a variability indicative of a subtle nonlethal effect on the herbivore. Similar nonlethal effects are exemplified by the Mexican bean beetle-*Podisus maculiventris* (Say)-soybean PI 80837 interactions described by Bouton et al. (in Price et al., 1980; Thompson, 1982). Optimization of defensive postures by synergism of predators, parasitoids, or diseases with resistance may not occur in systems involving potent toxins that can be sequestered by the herbivores (Duffey, 1980), or may otherwise be incompatible with the natural enemies (Duffey and Bloem, Chapter 6). Other nonselective plant characteristics that may interfere with both herbivore and predator behavior or physiology are glandular trichomes (Price, 1981; Duffey, Chapter 6) and simple trichomes (Treacy et al., 1985).

### 5.2.4. Coevolution, Sequential, or Reciprocal Evolution of Insects and Plants

The evolutionary processes that selected for the biosynthesis of specific defensive natural products in plants, and the counteradaptive reversal of defensive chemicals into attractants and feeding excitants in some phytophagous insects have been defined as coevolution (Ehrlich and Raven, 1964). According to Janzen (1980), in the strict sense, coevolution is the process whereby an evolutionary change in a defensive trait of species A (e.g., a host plant) is followed by a genetic change in species B (e.g., an herbivore) that allows B to overcome the temporarily reduced vulnerability (due to the new trait) to its trophic association with A. The pressure exerted by B on A, in turn, eventually selects for new defensive traits in successive cycles. The close interrelationship between the evolution of herbivores and their host plants had been recognized by Brues (1920), who called it "parallel evolution," and Fraenkel (1958) who called it "reciprocal evolution." Jermy (1984) suggested that "sequential evolution," rather than coevolution, explains extant IPI systems.

The patterns of host specificity observed in contemporary insects certainly result from adaptive evolutionary processes. Also, plants' physical and chemical defenses are the result of evolutionary processes. However, there is a major argument about the coordination of the two processes (Futuyma, 1983; Jermy, 1984; Strong et al., 1984). Some argue that few, if any, insect–plant antagonistic systems observed in nature result from direct coevolutionary interactions. Most contemporary insect–plant associations probably resulted from secondary adaptations of insects to resources with phytochemical characteristics evolved under selection pressures from complexes of antagonists and other environmental forces. As more is known about the similarities of chemical antiherbivory and pathogen resistance mechanisms, it becomes apparent that pathogens may be a preponderant force in plant chemical evolution. For instance, DIMBOA, a classical example of an antibiotic factor in corn against the European corn borer, is an important resistance factor against the fungus *Helminthosporium turcicum* in several inbred lines (Toldine, 1984). Some phytoalexins active against herbivores (Kogan and Paxton, 1983) are, in fact, second lines of defense against plant pathogens (Bailey and Mansfield, 1982). That so many secondary chemicals are broadly biocidal is consistent with the notion that plant chemical evolution is the result of simultaneous selection by many different factors, both biotic and abiotic.

Whether by coevolution, parallel evolution, reciprocal evolution, or sequential evolution, insects and plants have evolved characters that allow insects to exploit most available plant resources and allow plants to defend themselves against the majority of sympatric herbivore fauna. Mutualistic systems of insects and plants are the best examples of coadaptation [e.g., the *Ficus*-Agaonidae system (Janzen, 1979; Wiebes, 1979)].

The patterns of host specificities in antagonistic systems of arthropods and plants do not fit the coevolutionary model, if the association is recent. For example, most of the arthropod fauna of soybean in North, Central, and South America have adapted secondarily to the introduced crop. That coevolution is the model to explain associations established in geological time has been supported by only a few well-tested cases (Benson et al., 1976; Berenbaum, 1983).

## 5.3. INFERENCES FROM IPI THEORY—IMPLICATIONS TO HPR PRACTICE

The body of theory discussed to this point is constantly evolving and hence sweeping generalizations are still risky. As more information accumulates on the ecological and physiological mechanisms operating in the thousands of insect–plant interactive systems, some generalizations may be made. Meanwhile, much can be inferred from IPI theory, even at this early stage of its development. The following is an attempt to extract from this body of theory some elements potentially relevant to practical HPR applications. A synopsis of the HPR inferences and their source IPI hypotheses is presented in Table 5.5.

### 5.3.1. Inferences from the Dual Role of Secondary Metabolites Hypothesis

1. The allomone/kairomone role of many secondary plant metabolites imposes certain limits to HPR options and requires special precautions in the genetic manipulation of allelochemicals.

Glucosinolates are essential kairomones for crucifer-feeding specialists and are effective allomones for most other phytophagous insects. Breeding for reduced or zero glucosinolates, if desirable or possible, would reduce attraction of crucifer-feeding specialists, but would also make plants more vulnerable to potential pests presently deterred by the allomonal glucosinolates. In addition, it would change the flavor of cultivars for human consumption. Breeding for high glucosinolate content, on the other hand, may increase overall resistance, except for a few specialists that may be even more attracted to the plants; it may also change the flavor of foods and pose health hazards for humans.

2. Metabolites with allomone-only effects offer options for HPR against generalist herbivores, but specialists may already have within their populations the genetic makeup needed to detoxify or neutralize the action of the allomone; biotypes are more likely to occur in a situation akin to vertical resistance to plant pathogens (Robinson, 1980).

Alkaloids in the Solanaceae seem to perform only allomonal functions, but specialists are not deterred by specific alkaloids of their preferred host plants, up to a certain concentration; above that concentration developmental rates, pupal weights, and survival may be reduced [see Parr and Thurston (1972) for the effect of nicotine in artificial media for the tobacco hornworm]. In addition, ab-

**Table 5.5 Summary of HPR Applications Inferred from IPI Theory**

| IPI Hypothesis and Corresponding Section in Chapter | Some HPR Inferences |
|---|---|
| A. Chemical defense (Sect. 5.3.1.) | 1. Need to monitor undesirable consequences of disturbances of the allomone/kairomone role of secondary metabolites<br>2. Consideration of high allomone concentrations increasing risk of biotype selection<br>3. Development of multifactor resistance |
| B. Dual Discrimination (Sect. 5.3.2.) | 1. Consideration of cultural practices aimed at optimizing (not maximizing) yields and maintaining high intraspecific genetic diversity<br>2. Selection of breeding lines under pressure of genetically diverse herbivore populations of the target pest or pest complex<br>3. Measurement of persistence of resistance under diverse environments |
| C. Risk–Cost–Value | |
| a. Apparency and defense [Sect. 5.3.3(1), (2)] | 1. "Design" of novel defensive strategies for the neo-apparent crop plants<br>2. Exploitation of multifactor resistance (same as A.3.), including physical, quantitative, qualitative, constitutive and inducible defenses, as well as compensatory growth |
| b. Parsimonious allocation of energy [Sect. 5.3.4(1), (2)] | 1. Consideration of the economic value of attainable levels of resistance, since maximum yield with maximum resistance may be unattainable |
| c. Latent defenses [Sect. 5.3.4(3)] | 1. Exploitation of mechanisms to selectively induce resistance mechanisms |
| d. Competitive demands [Sect. 5.3.4(4)] | 1. Reduction of impact of physical (climatic) and nutritional stresses |
| e. Cost-sharing and spreading of the risk [Sect. 5.3.2(1)–(5)] | 1. Enhancement of crop's genetic and phenotypic diversity |
| f. Synergistic effects [Sect. 5.3.3(3)] | 1. Exploitation of resistance factors compatible and (if possible) synergistic with actions of predators, parasitoids and entomopathogens |
| D. Coevolution [Sect. 5.3.5(1), (2)] | 1. Enhancement of stability of resistance through understanding mechanisms of resistance and their impact on the associated herbivore fauna<br>2. Cautious innovation of defensive strategies (open new cycles of evolution) |

112

normally high concentrations of alkaloids may affect plant fitness (Krischik and Denno, 1983).

> 3. Broad-spectrum resistance may be achieved by manipulating multiple factors that involve constitutive, inducible, and physical defenses in a total ecological context.

Broad-spectrum resistance may be achieved by carefully increasing levels of allomones, reducing levels of specific kairomones (for compounds lacking dual roles), and by enhancing response rates and end-levels of inducible defenses. In addition, maximum benefit should be derived from physical defenses (e.g., trichomes, thickened tissues, waxy secretions) that are nonspecific and unlikely to promote biotype formation. Programs for resistance in rice to a complex of pests has followed this scheme (Khush, 1984) and resistance in soybean to foliage-feeding Lepidoptera, Coleoptera, and Homoptera, and to pod-feeding Hemiptera involves a complex of factors including constitutive, inducible, and physical defenses (pubescence) (Kogan, 1986).

### 5.3.2. Inferences from Postulates of Nutritional Ecology

> 1. Maximizing yields through fertilization and use of HYVs in monocultures of highly uniform genetic background drastically reduces intraspecific host variability and enhances resource utilization by herbivores.

Demand for high yields has led to breeding of cultivars uniform in growth pattern, maturity date, and other desirable agronomic traits. Such improvements often increase genetic homogeneity, which makes plants more vulnerable to insect pests and diseases. Recent catastrophic outbreaks of diseases in corn, oats, and wheat illustrate the risk of narrowing the genetic background of crops (Whitham et al., 1984).

> 2. Genetic diversity and phenotype variability among local herbivore populations may result in differences in response to the same plant genotype growing in different regions.

Insect populations may be morphologically indistinguishable but markedly variable and genetically diverse in major biological traits, including host preferences and fitness on different hosts (Mitter and Futuyma, 1983; Diehl and Bush, 1984). Diehl and Bush (1984) expanded the definition of biotypes and stressed the importance of environmental influences on gene expression. The corn earworm is an example of variability in a major agricultural pest in the United States that affects corn, cotton, soybean, and many horticultural crops. It is the most serious pest of soybean on the Atlantic coast and in several major producing states in the southern United States. In the Midwest, however, it never attacks soybean, even when sweet corn growing adjacent to soybean is heavily infested. Sell et al. (1974, 1975) showed that the esterase-II locus of earworm populations collected on corn in 22 locations, from Brownsville, Texas, to Rochelle, Illinois, was remarkably stable in space and time. One exception was the Brownsville population, which might have been introgressed by migrants developed on cotton. Sell et al. suggested that polymorphism at this locus might be associated with the chemical composition of the host. The inability of the corn earworm to colonize soybean in the northern United States may be linked to enzyme polymorphisms that render the plant unsuitable as host. As a migrant pest, the corn earworm in the northern states may be derived from the corn-bred populations of the South and is thus incompatible with soybean.

   3. Variability due to genotype-by-environment interactions in plants may alter expression of resistance and accentuate effects of herbivore variability.

   Although host plants may be genetically homogeneous, they are greatly influenced by a variable environment. Examples of the effect of environment on plant fitness abound in the agronomic literature. Fig. 5.5 illustrates how the planting

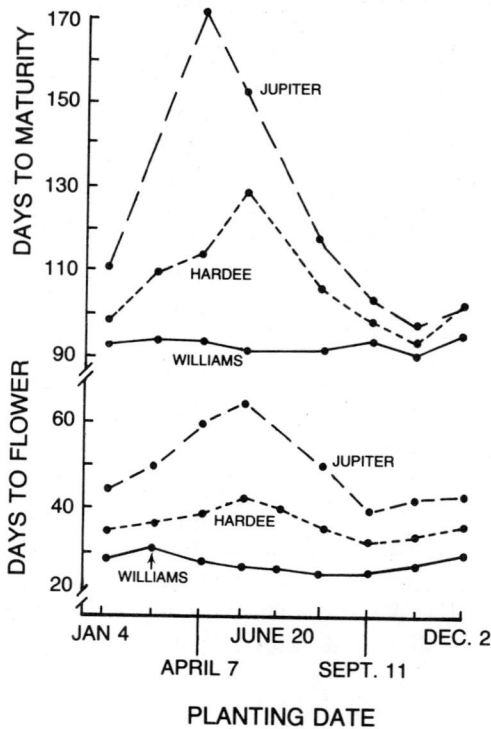

Fig. 5.5 Variation of planting dates of soybean changing environment by genotype interactions expressed in the number of days to flower and days to maturity. Effects are more accentuated in the late maturing soybean variety Jupiter than in the early maturing variety Williams. Hardee is intermediate. Soybeans planted at Isabela, Puerto Rico (18°N). (Redrawn from Minor, 1976.)

date affects days-to-flowering and days-to-maturity in three soybean cultivars growing under subtropical conditions. Varying the planting date subjects plants to environmental changes such as photoperiod, temperature, and precipitation that deeply affect plant physiology. Resistance of cultivars to insect pests and diseases under different environmental conditions may be similarly affected, although it may be difficult to distinguish between geographic variability in virulence of pests or in susceptibility of hosts. One practical consequence of the genotype-by-environment interaction is the limitation of use of breeding nurseries for resistance. Breeders of annual crops commonly use winter nurseries to expedite generation advancement. If selection for resistance is done in the winter nurseries, lines that are resistant in one environment may not be so in another. Resistant soybean lines selected in South Carolina often failed to display similar resistance levels in a winter nursery in southern Brazil (personal observation).

> 4. The use of blends of genotypes and careful rotation of crops or cultivars of the same crop may alleviate the impact of genetic homogeneity in crops.

To compensate for excessive homogeneity in many cultivars, agronomists have tested the use of blends or rotations of varieties. Such a rotation program has been recommended in areas afflicted by the soybean cyst nematode (*Heterodera glycines* Ichinohe). Nematode-resistant, nematode-susceptible, and nonhost crops are rotated over periods of 3–5 years. The use of blends of susceptible and resistant cultivars has also been recommended (Malek and Edwards, 1973). Rotation and blends maintain enough host diversity in the field that selection of biotypes is halted or greatly delayed.

> 5. Polyphagous herbivores probably are less promising targets for successful HPR.

Feeding specialization leads to greater dependence on specific host characteristics. Consequently, genetic manipulations that alter the allomone/kairomome balance of host plants may influence specialists but have little effect on generalists. Generalist herbivores are more likely to require nonspecific feeding stimuli (e.g., sucrose), so they are not affected by reduced kairomone levels. These generalists probably can detoxify a broader spectrum of toxins (Brattsten, 1979; but see Gould, 1984), so they are less likely to be affected by increased allomone levels, particularly toxins that do not interfere with ingestion rates. However, herbivore specialists may tolerate higher concentrations of allomones in their preferred hosts than generalists do (see Blau et al., 1978; Berenbaum, 1985).

### 5.3.3. Inferences from the Concept of Apparency and Defense

> 1. In most annual crop plants grown as monocultures, the apparency/defense equation has been irremeably subverted.

Most annual crops evolved from early successional, typically unapparent plants. Escape in space and time was part of their defense strategies. They probably had qualitative defenses whose metabolic costs were normally high but commensurate with the need to assure the reproductive levels necessary to preserve the gene pool (at least a few viable seeds per plant). Modern agriculture subverted this status to a point of no return (= irremeably). Maximum yields demand a great deal of internal and external energy, some at the expense of energy for defense. Also, plants in monocultures are highly apparent and predictable in time and space (Fig. 5.6A–C). To restore acceptable levels of resistance to modern agricultural crops, HPR researchers will have to compensate for lost segments of the original defense strategy. Understanding what was lost should help design viable alternatives.

> 2. The constitutive or inducible accumulation of quantitative defenses in "neo-apparent" crop plants is limited by time and probably also by high metabolic costs.

Theory postulates the preponderance of quantitative defenses in apparent resources (but see Sect. 5.2.3.2. for severe restrictions to generalizing this concept). However, in neo-apparent, short-lived crop plants, there are serious constraints for the effective use of such defenses. Tannins occur in some resistant cotton varieties (Chan et al., 1978) and have been detected in stems of wild soybean lines resistant to agromyzid stem flies (Chiang and Norris, 1984), but the effectiveness of such defenses has not been determined. Silica, also considered a quantitative defense, is an important resistance factor in many grasses (McNaughton and Tarrants, 1983; Moore, 1984) and it plays a role in the resistance of rice to some of its pests. Proteinase inhibitors (PIs) accumulate in leaves of potato and tomato as a result of mechanical injury; injury releases a PI-inducing factor (PIIF) that is translocated and induces synthesis of PIs in all aerial parts of the plant (Ryan, 1979, 1983; Nelson et al., 1983). An antiherbivory role for PI had been hypothesized for a long time (Green and Ryan, 1972), but only recently has experimental evidence supported this hypothesis. Broadway et al. (1986) freeze-dried tomato leaves with high levels of induced PI and incorporated the leaf powder into an agar base medium. The medium, when fed to beet armyworm [*Spodoptera exigua* (Hubner)] larvae, significantly reduced growth rates.

The effective accumulation of quantitative defenses is limited by the length of the growing season. Short season varieties have little time to accumulate sufficient defenses to deter rapidly developing pests. So neo-apparency and quantitative defenses may not be compatible, but the degree of apparency may be manipulated to achieve phenological asynchronizations between host crops and colonizers (see Herzog and Funderburk, Chapter 9).

> 3. Much HPR research should be directed toward finding effective physical defenses and multiple qualitative chemical defenses compatible with key production traits and with actions of other biological control agents.

Fig. 5.6 (A) Highly unapparent wild soybean growing in the midst of other early successional vegetation. (B) Ditch bank typical of the habitat of wild soybeans in Northern China. (C) Production soybean field in Illinois that makes the resources highly apparent and predictable for arthropod colonizers. [(A) and (B) originals taken at Heilungjong Province, Peoples Republic of China, 1981; (C) courtesy of Agricultural Communications Office, University of Illinois at Urbana-Champaign.]

The impact of qualitative defenses may be potentiated if those are compatible with the actions of parasitoids, predators, and insect pathogens. One of the inadequacies of modern breeding programs in HPR is that rarely, if ever, are actions of natural enemies evaluated in conjunction with screening procedures. Experiments designed to evince the potential synergistic effects of resistance and natural enemies should become an integral part of the variety development routine. Attention should also be given to exploring effective physical defenses that are proven compatible with natural enemy actions (see also Duffey and Bloem, Chapter 6).

### 5.3.4. Inferences from the Risk–Cost–Value Hypothesis

   1. Maximum yield and maximum resistance must be reconciled based on risk and value criteria.

   Fig. 5.7 attempts to show the relationship between level of defense and yield. As a crop approaches its maximum yield potential, its level of defense is greatly reduced. Conversely, as concentration of allelochemicals increases, the metabolic cost may be such that the yield potential cannot be realized. There must be a ceiling for concentration of allelochemicals above which plant fitness itself is affected. Consequently, to achieve the goal of maximum yield with optimal defense, it is necessary to evaluate risks of economic injury and value of tissues to be protected. Risk varies among regions and among years within a region. Most HPR researchers have an intuitive perception of risk, as assessed by the probability of injury from a given pest in a certain region. Priorities are set by selecting dominant species in the community, or that species most likely to do injury. It is, however, more appropriate to attempt an analysis of pest impact (Kogan, 1980)

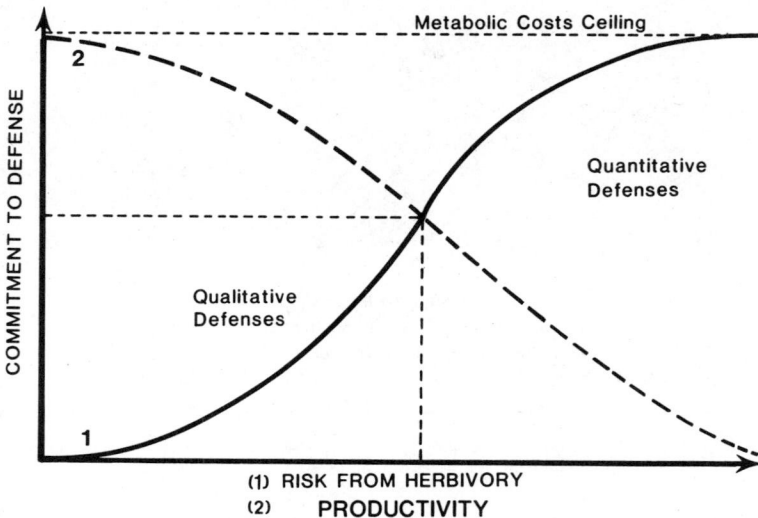

Fig. 5.7 Theoretical relationship between a plant's energetic investment in the biosynthesis of defense phytochemicals and productivity (= yield). If commitment to defense is low, productivity reaches a maximum in the absence of pests. However, as plants increase their investment in defense in proportion to risk from attack, they may reach a metabolic cost ceiling above which there is no reproduction and the plants, with the genes for such extreme defenses, are eliminated.

and to prioritize the breeding effort on the basis of more measurable criteria. Modeling may also help determine the directions for HPR programs by predicting the potential impact of certain defense strategies on the pests and on the crop (Zavaleta and Kogan, 1984).

Interseasonal variation in risk is much more difficult to incorporate in HPR planning. However, if IPM programs provide early forecasts of pest outbreaks, farmers may select a less productive but more resistant variety if the probability of outbreak is high, and vice-versa if the probability is low. This option, becoming available to farmers through reliable forecasting, may allow breeders to release varieties with lower yield potential but higher resistance for use in those years of greater pest risk.

> 2. Compensatory growth (= tolerance) probably is a cost-effective defense alternative for lower value tissues.

Many annual plants have a remarkable capacity to compensate for losses due to pest injury. Soybean may lose up to 30% of its foliage prior to the fruiting stages without a significant yield decrease. Stands thinned by 50% may produce as much as fields with the full original stand. This is due to the compensatory growth of the remaining plants (Kogan and Turnipseed, 1980). As plants are bred to be more efficient at converting energy into grain yield, some of this compensatory capacity may be lost. HPR researchers will be well advised to run cost–benefit analyses before they increase plant yields to a point that every green leaf counts.

> 3. Inducible defenses may offer an alternative solution to the "yield-defense" dichotomy.

The inverse correlation between yield and defense assumes that constitutive defenses accumulate whether the plant is attacked or not. Since inducible defenses are triggered by the attacker, there is no energy expenditure if an attack does not occur. The physiological bases of inducible defenses are still obscure [but see Ryan (1979) and Nelson et al. (1983) for discussions of the nature of the proteinase inhibitor inducing factor], but the few authenticated cases suggest that there are several possible avenues for use of such defenses in IPM programs (Kogan and Paxton, 1983; Rhoades, 1985).

> 4. Cropping systems that reduce the impact of physical and nutritional stresses on plants should help avoid the relaxation of plant commitment to defense.

An old tenet of agriculturists is that a vigorous plant (= stress free) is less susceptible to pests. Good crop management practices usually result in greater plant vigor. There is a vast literature on the effect of fertilization, particularly with sources of nitrogen, on susceptibility to herbivores. Scriber (1984) reviewed this literature and observed that high nitrogen in a host plant often resulted in greater fitness of arthropods feeding on that plant. These cases contradict the assumption that vigor is always associated with resistance. Crop management, however, may profoundly affect pest populations through changes in the habitat or through maintenance of vigor that increases the plant's ability to resist or to compensate for losses due to herbivory (see Herzog and Funderburk, Chapter 9).

### 5.3.5. Inferences from Coevolutionary Theory

1. Stability of resistance depends on the physiological bases of resistance factors and the genetic plasticity of associated herbivores.

Breeding for resistance against herbivores starts with an attempt to uncover genotypic variability for resistance within plant populations. Preferred sources of resistance are cospecific with the recurrent cultivar, but congeneric and even heterogeneric sources have also been used. Genetic incompatibilities often limit successful crossings with noncospecific lines. As these sources of resistance are uncovered and incorporated into crop cultivars, selection pressure switches to the associated herbivore populations. Low-frequency genotypes may be selected and produce new host races (biotypes) at rates proportional to the frequency of these genotypes in native populations. Whether release of resistant cultivars accelerates cycles of coevolution is a matter of definition. No mutations may be involved either within the plant or within the herbivore populations. However, it is important to understand the homeostatic processes that maintain stability within natural plant–herbivore systems. The nature of the resistance mechanism (antibiosis or antixenosis), the complexity of resistance factors and their genetic control, and the plasticity of the herbivores may greatly influence stability (Gould, 1983).

Plant defense strategies are seldom based on a single highly effective mechanism, either physical (e.g., hairiness), biochemical (e.g., presence of allomones), phenological (e.g., asynchrony of life cycles), or physiological (e.g., compensatory growth). Greater stability of resistance results from multilayered defense strategies. Multilayered defenses have been identified in crop plants, but resistance programs often focus on only a few obvious traits from one or another layer.

A first layer of antiherbivory defense in soybean is a dense vesture of simple trichomes that protects against leafhoppers (Lee, 1983). A second layer is the suspected release of low-molecular-weight volatiles acting as powerful feeding deterrents for the Mexican bean beetle (Fisher and Kogan, 1986; M. Kogan et al., in preparation). A third layer is as yet undefined constitutive antibiotic factors that greatly reduce fitness of herbivores (Kogan, 1972a; Dreyer et al., 1979; Smith and Fischer, 1983; Binder and Waiss, 1984). Finally, a fourth layer is the inducible glyceollins that accumulate in tissues and operate as powerful feeding inhibitors for the Mexican bean beetle (Hart et al., 1983). The details of the antiherbivory system of soybean are still obscure, but the emerging picture shows complex interactive factors that involve compounds of primary and secondary metabolism, both constitutive and inducible defenses, and physical barriers. Only by mobilizing all these factors can current levels of resistance to multiple pests be increased over the geographic range of the crop.

2. Future cycles of directed coevolution of crop plants and herbivores may attempt to incorporate novel secondary chemistry into the crop genotype through genetic engineering.

The extant reservoirs of genes for resistance may not be sufficient to produce economically viable varieties resistant to multiple pests under a wide range of ecological conditions. To create biochemical novelty within crop plants may require only a single-step change in a biosynthetic pathway, but even more complex alterations may become possible through techniques of gene transfer. For instance, since nitrogenous compounds are efficiently accumulated in leguminous plants, it might be possible to divert the pathway of ornithine to produce tropane alkaloids. A leguminous plant containing such alkaloids (if they could be kept out of seeds of grain legumes) would have an enormous advantage over its specialist herbivores since they probably would be unable to detoxify such compounds. How far this scenario unfolds depends on how rapidly breeders exhaust the available genetically compatible sources of resistance. It will depend also on the progress and support of research in genetic engineering focused on these resistance processes.

## 5.4. CONCLUSIONS

Current IPI theory predicts the antagonism between HPR goals of maximizing resistance while maintaining maximum yields. The metabolic threshold for this antagonism is likely to differ among crops, but present high-yielding varieties of wheat and rice are quickly approaching the threshold, if they have not yet surpassed it. As this threshold is surpassed, added costs for protection (not only production inputs, but also the social and environmental costs) must be computed in a realistic economic analyses of the benefits of those high-yielding varieties.

Applications of IPI information in innovative pest control approaches, other than HPR, have been rather limited. Staedler (1983) discussed possible practical uses of plant attractants and feeding and oviposition excitants. Methods for mass trapping have been attempted with some measure of success, mainly when plant attractants are combined with aggregation pheromones (Wood and Bedard, 1977; Payne, 1981) or with visual cues (Prokopy and Owens, 1983). Staedler (1983) suggested also the use of specific attractants to mask nonhosts, thus enticing insects to feed and oviposit on toxic plants. Encouraging results have been achieved with cucurbitacins, a potent feeding excitant for diabroticites. In Brazil the subterranean fruit of the cucurbit *Caiaponia tayuya* is cut in half and laced with insecticides to control populations of *Diabrotica speciosa* (personal communication from the National Rice and Bean Center, Goiania, Brazil). Special plastic traps loaded with pure cucurbitacins are being used in Illinois to monitor populations of *Diabrotica* spp. Most of these uses, however, imply some degree of human manipulation and interference.

It is more appealing to exploit leads offered by IPI theory that minimize manipulation beyond the stages of development of the techniques. Essentially, the approach deduced from current IPI theory is the development of a multifaceted strategy involving physical and biochemical defenses operating in an ecological context that favors the plant by allowing it to escape in space and time. Concurrently, cultural practices that minimize nonpest-related stresses are adopted.

The apparent ability of plants to mobilize defenses as needed offers intriguing possibilities. One immediate consequence of the discovery of defenses induced by previous attack is the need to reassess economic injury levels of early season pests. If not catastrophic, early injury may predispose plants to higher levels of resistance later in the season when herbivore populations are usually higher and plants are past the stage of compensatory growth. Operational "guild defenses" may also be devised even without greatly infringing established routines of large-scale, mechanized agriculture. The concept of crop mosaics, using blends of susceptible and resistant cultivars, may increase the chances for economic use of resistant cultivars, even if they are outyielded by more susceptible cultivars in the absence of herbivory. Finally, partial resistance may be potentiated by enhancing the actions of other biological control agents.

Although the treatment of IPI theory in this chapter is incomplete, I hope that it is sufficient to illustrate that the extant body of information on IPI offers a reasonable conceptual framework for strengthening HPR research and IPM practice. Much of this theory has omitted human interference because it was conceived to explain natural systems. However, humans and crop plants exist now in a tight, mutualistic relationship, and it is impossible to make a realistic analysis of herbivore–crop interactions without including human interference. The breakdown of civilized values because of hunger is evident in recent history (e.g., the prolonged droughts in the Sahel countries of Africa) and in the imagination of writers of science fiction (e.g., John Christopher's 1957 novel *No Blade of Grass*). The dependence of annual crops on humans for protection, preservation of seed, and multiplication is demonstrated by the many species known today only from domesticated cultivars (soybean and corn, to mention only two of the best known).

It may be impossible to maintain high production of food and fiber by relying solely on plant natural defenses. Human interference in crop protection has become indispensible. However, the intensity and nature of this interference can be economically and ecologically optimized if we learn how to exploit the potential of natural plant defenses as they operate alone or in concert with other natural control agents.

## ACKNOWLEDGMENTS

This work is a contribution of the Section of Economic Entomology, Illinois Natural History Survey and College of Agriculture, University of Illinois at Urbana-Champaign. It was supported in part by Hatch funds through project ILLU-12-0324, Regional funds through Project S-157, and from U.S. Department of Agriculture grant CR806277-03 through the Consortium for Integrated Pest Management, Texas A&M University. The opinions expressed herewith are those of the author and not of the funding institutions.

Valuable comments on early versions of this chapter were made by May Berenbaum, Sean Duffey, Don Peters, and Don Strong. Charles Helm and Dan Fischer were my first forum for discussion of some of the ideas, and Jenny Kogan helped me with the references. Jo Ann Auble and Joan Traub handled the word processing of numerous versions and Eva Steger edited the final version of this chapter. To all, my deepest appreciation.

## REFERENCES

Adjei-Twum, D. C., and W. E. Splittstoesser. (1976). The effect of soil water regimes on leaf water potential, growth and development of soybeans, *Physiol. Plant*, **38**, 131–137.

Ahmad, S. (Ed.). (1983). *Herbivorous Insects—Host-Seeking Behavior and Mechanisms*, Academic, New York, 257 pp.

Aldrich, S. R. (1980). *Nitrogen in Relation to Food. Environment and Energy*, Univ. of Illinois. Ag. Exp. Sta., Urbana, Illinois, Sp. Pub. 61, 452 pp.

Altieri, M. A. (1983). *Agroecology—The Scientific Basis of Alternative Agriculture*, private publication, Berkeley, California, 173 pp.

Applebaum, S. W., and Y. Birk. (1979). "Saponins," in G. A. Rosenthal and D. H. Janzen, Eds., *Herbivores: Their Interaction with Secondary Plant Metabolites*, Academic, New York, pp. 539–566.

Atsatt, P. R., and D. J. O'Dowd. (1976). Plant defense guilds, *Science*, **193**, 24–29.

Bailey, J. A., and J. W. Mansfield, Eds. (1982). *Phytoalexins*, Wiley, New York, 334 pp.

Baldwin, I. T., and J. C. Schultz. (1983). Rapid changes in tree leaf chemistry induced by damage: Evidence for communication between plants, *Science*, **221**, 277–279.

Bandurski, R. S. (1965). "Biological reduction of sulfate and nitrate," in J. Bonner and J. E. Varner, Eds., *Plant Biochemistry*, Academic, New York, pp. 467–490.

Beck, S. D. (1957). The European corn borer, *Pyrausta nubilalis* (Hbn.), and its principal host plant. VI. Host plant resistance to larval establishment, *J. Insect Physiol.*, **1**, 158–177.

Beck, S. D. (1965). Resistance of plants to insects, *Annu. Rev. Entomol.*, **10**, 207–232.

Beck, S. D., and J. C. Reese. (1976). "Insect–plant interactions: nutrition and metabolism," in J. W. Wallace and R. L. Mansell, Eds., *Biochemical Interaction between Plants and Insects*, *Recent Adv. Phytochem. 10*, Plenum, New York, pp. 41–92.

Beck, S. D., and L. M. Schoonhoven. (1980). "Insect behavior and plant resistance," in F. G. Maxwell and P. R. Jennings, Eds., *Breeding Plants Resistant to Insects*, Wiley, New York, pp. 115–135.

Beck, S. D., and E. E. Smissman. (1960). The European corn borer, *Pyrausta nubilalis*, and its principal host plant. VIII. Laboratory evaluation of host resistance to larval growth and survival, *Ann. Entomol. Soc. Amer.*, **53**, 755–762.

Beck, S. D., and E. E. Smissman. (1961). The European corn borer, *Pyrausta nubilalis*, and its principal host plant. IX. Biological activity of chemical analogs of corn resistance factor A (6-methoxybenzoxazol inone). *Ann. Entomol. Soc. Amer.*, **54**, 53–61.

Bell, E. A., and B. V. Charlwood. (1980). *Secondary Plant Products*, Springer-Verlag, Berlin, 674 pp.

Bell, W. J., and R. T. Carde, Eds. (1984). *Chemical Ecology of Insects*, Sinauer Assoc., Sunderland, Massachusetts, 524 pp.

Benson, W. W., K. S. Brown, Jr., and L. E. Gilbert. (1976). Coevolution of plants and herbivores: Passion flower butterflies, *Evolution*, **29**, 659–680.

Berenbaum, M. (1983a). Coumarins and caterpillars: A case for coevolution, *Evolution*, **37**, 163–179.

Berenbaum, M. R. (1983b). Effects of tannins on growth and digestion in two species of papilionids, *Entomol. Exp. Appl.*, **34**, 245–250.

Berenbaum, M. (1985). Brementown revisited: Interactions among allelochemicals in plants, *Rec. Adv. Phytochem.*, **19**, 139–169.

Bernays, E. A. (1978). Tannins: An alternative viewpoint, *Entomol. Exp. Appl.*, **24**, 44–53.

Bernays, E. A. (1981). Plant tannins and insect herbivores: An appraisal, *Ecol. Entomol.*, **6**, 353–360.

Bernays, E. A. (1982). "The insect on the plant—a closer look," in J. H. Visser and A. K. Minks, Eds., *Proceedings 5th International Symposium on Insect–Plant Relationship*, Centre for Agricultural Publishing and Documentation, Wageningen, Holland, pp. 3–17.

Bernays, E. A., D. J. Chamberlain, and S. Woodhead. (1983). Phenols as nutrients for a phytophagous insect *Anacridium melanarhodon*, *J. Insect Physiol.*, **29**, 535–539.

Binder, R. G., and A. C. Waiss, Jr. (1984). Effects of soybean leaf extracts on growth and mortality of bollworm (Lepidoptera: Noctuidae) larvae, *J. Econ. Entomol.*, **77**, 1585–1588.

Bisby, F. A., J. G. Vaughan, and C. A. Wright, Eds. (1980). *Chemosystematics: Principles and Practice*, Academic, London, 449 pp.

Blau, P. A., P. Feeny, L. Contardo, and D. S. Robson. (1978). Allylglucosinolate and herbivorous caterpillars: A contrast in toxicity and tolerance, *Science*, **200**, 1296–1298.

Bottger, G. T., and R. Patana. (1966). Growth, development and survival of certain Lepidoptera fed gossypol in the diet, *J. Econ. Entomol.*, **59**, 1166–1168.

Boudreaux, H. B. (1958). The effect of relative humidity on egg laying, hatching and survival in various spider mites, *J. Insect Physiol.*, **2**, 65–72.

Brattsten, L. B. (1979). "Biochemical defense mechanisms in herbivores against plant allelochemicals," in G. A. Rosenthal and D. H. Janzen, Eds., *Herbivores—Their Interactions with Secondary Plant Metabolites*, Academic, New York, pp. 200–270.

Broadway, R. M., S. S. Duffey, G. Pearce, and C. A. Ryan. (1986). Plant proteinase inhibitors: A defense against herbivorous insects? *Entomol. Exp. Appl.* **41**, 33–38.

Brues, C. T. (1920). The selection of food-plants by insects, with special reference to lepidopterous larvae, *Am. Nat.*, **54**, 313–332.

Brues, C. T. (1946). *Insect Dietary—An Account of the Food Habits of Insects*, Harvard University Press, 466 pp.

Capinera, J. L., A. R. Renaud, and N. E. Roehrig. (1983). Chemical basis for host selection by *Hemileuca oliviae*: Role of tannins in preference of $C_4$ grasses, *J. Chem. Ecol.*, **9**, 1425–1437.

Caswell, H., and F. C. Reed. (1976). Plant–herbivore interactions. The indigestibility of $C_4$ bundle sheath cells by grasshoppers, *Oecologia (Berlin)*, **26**, 151–156.

Chambliss, O. L., and C. M. Jones. (1966). *Cucurbitacins*: Specific insect attractants in Cucurbitaceae, *Science*, **158**, 1392–1393.

Chan, B. G., A. C. Weiss, Jr., and M. Lukefahr. (1978). Condensed tannin, an antibiotic chemical from *Gossypium hirsutum*, *J. Insect Physiol.*, **29**, 113–118.

Chapman, R. F., and E. A. Bernays, Eds. (1978). *Insect and Food Plant*, Proc. 4th Int. Symp., Slough, England, Nederlandse Entomologische Vereniging, Amsterdam, 566 pp.

Chesnokov, P. G. (1962). *Methods of Investigating Plant Resistance to Pests*, Sh. Shoshan and S. Nemchonok, translators, published for NSF and USDA by the Israel Prog. for Scientific Translations (Jerusalem), Off. Tech. Serv., U.S. Dept. of Commerce, Washington, D.C., 107 pp.

Chew, F. S. (1975). Coevolution of pierid butterflies and their cruciferous food plants. I. The relative quality of available resources, *Oekologia (Berlin)*, **20**, 117–127.

Chew, F. S., and J. E. Rodman. (1979). "Plant resources for chemical defenses," in G. A. Rosenthal and D. H. Janzen, Eds. *Herbivores: Their Interaction with Secondary Metabolites*, Academic, New York, 718 pp.

Chiang, H. S., and D. M. Norris. (1984). "Purple Stem," a new indicator of soybean stem resistance to bean flies (Diptera: Agromyzidae), *J. Econ. Entomol.*, **77**, 121–125.

Costa, C. P. da, and C. M. Jones. (1971). Cucumber beetle resistance and mite susceptibility controlled by the bitter gene in *Cucumis sativus* L., *Science*, **172**, 1145–1146.

Courtney, S. P. (1982). Coevolution of pierid butterflies and their cruciferous foodplants. IV. Crucifer apparency and *Anthocharis cardamines* (L.) oviposition, *Oekologia (Berlin)*, **52**, 258–265.

Crawley, M. J. (1983). *Herbivory: The Dynamics of Animal Plant Interactions*, University of California Press, Berkeley, California, 437 pp.

Curtis, P. J., and P. M. Meade. (1971). Cucurbitacins from the Cruciferae, *Phytochemistry*, **10**, 3081–3082.

Czapek, F. (1921). *Biochimie der Pflanzen*, 2 Aufl, 3 Baud., G. Fischer, Zena.

Denno, R. F., and H. Dingle, Eds. (1981). *Insect Life History Patterns: Habitat and Geographic Variation*, Springer-Verlag, Berlin, 225 pp.

Denno, R. F., and M. S. McClure, Eds. (1983). *Variable Plants and Herbivores in Natural and Managed Systems*, Academic, New York, 717 pp.

Dethier, V. G. (1941). Chemical factors determining the choice of food plants by *Papilio* larvae, *Am. Nat.*, **75**, 61–73.

Dethier, V. G. (1954). Evolution of feeding preferences in phytophagous insects, *Evolution*, **8**, 33–54.

Dethier, V. G. (1970). "Chemical interactions between plants and insects," in E. Sondheimer and J. B. Simeone, Eds., *Chemical Ecology*, Academic, New York, pp. 83–102.

Dethier, V. G., L. Barton Brown, and C. N. Smith. (1960). The designation of chemicals in terms of the response they elicit from insects, *J. Econ. Entomol.*, **53**, 134–136.

Diehl, S. R., and G. L. Bush. (1984). An evolutionary and applied perspective of insect biotypes, *Ann. Rev. Entomol.*, **29**, 471–504.

Doskotch, R. W., N. Y. Malik, and K. C. Beal. (1969). Cucurbitacin B, the cytotoxic principle of *Begonia tuberhybrida* var. *alba*, *Lloydia*, **32**, 115–122.

Dreyer, D. L., R. G. Binder, B. G. Chan, A. C. Waiss, Jr., E. E. Hartwig, and G. L. Beland. (1979). Pinitol, a larval growth inhibitor for *Heliothis zea* in soybean, *Experientia*, **35**, 1182–1183.

Duffey, S. S. (1980). Sequestration of plant natural products by insects, *Ann. Rev. Entomol.*, **25**, 447–477.

Edwards, P. J., and S. D. Wratten. (1980). *Ecology of Insect–Plant Interactions*, Studies in Biology 121, The Institute of Biology, Edward Arnold, London, 60 pp.

Ehrlich, P. R., and P. H. Raven. (1964). Butterflies and plants: A study in coevolution, *Evolution*, **18**, 586–608.

Errera, L. (1886). *Rev. Sci. Soc. Bot. Real. Belgique*, **5**, cited in Hellen C. de S. Abbott. (1887). Comparative chemistry of higher and lower plants, *Am. Nat.*, **21**, 800–810.

Fabre, J. H. (1890). *Souvenirs Entomologiques*, C. Delagrave, Paris, 2nd ed., 3rd vol., 433 pp.

Feeny, P. (1970). Seasonal changes in oak leaf tannins and nutrients as a cause of spring feeding by winter moth caterpillars, *Ecology*, **51**, 565-581.

Feeny, P. (1976). "Plant apparency and chemical defense," in J. W. Wallace and R. L. Mansell, Eds., *Biochemical Interactions between Plants and Insects*, *Rec. Adv. Phytochem.*, Plenum, New York, pp. 1-40.

Feeny, P. (1977). Defensive ecology of the cruciferae, *Ann. Missouri Bot. Gard.*, **64**, 221-234.

Feeny, P., K. L. Paauwe, and N. J. Demong. (1970). Flea beetles and mustard oils: Host plant specificity of *Phyllotreta cruciferae* and *P. stimolata* adults (Coleoptera: Chrysomelidae), *Ann. Entomol. Soc. Amer.*, **63**, 832-841.

Fischer, D., and M. Kogan. (1986). Chemoreceptors of adult Mexican bean beetles: Structure and role in food preference, *Entomol. Exp. Appl.*, **40**, 3-12.

Fowler, S. V., and J. H. Lawton. (1984). Trees don't talk: Do they ever murmur? *Antenna*, **8**, 69-71.

Fraenkel, G. (1958). The basis of food selection in insects which feed on leaves, Invited paper, 8th Ann. Meetings, Entomol. Soc. Japan, Hokkaido Univ., Sapporo, 5 pp.

Fraenkel, G. (1959). The "raison d'etre" of secondary plant substances, *Science*, **129**, 1966-1970.

Fraenkel, G. (1969). Evaluation of our thoughts on secondary plant substances, *Entomol. Exp. Appl.*, **12**, 473-486.

Fukutoku, Y., and Y. Yamada. (1982). Accumulation of carbohydrates and proline in water-stressed soybean (*Glycine max* L.), *Soil Sci. Plant Nutr.*, **28**, 147-151.

Futuyma, D. J. (1983). "Evolutionary interactions among herbivorous insects and plants," in D. J. Futuyma and M. Slatkin, Eds., *Coevolution*, Sinauer, Sunderland, MA, pp. 207-231.

Futuyma, D. J., and S. C. Petersen. (1985). Genetic variation in the use of resources by insects, *Ann. Rev. Entomol.*, **30**, 217-238.

Futuyma, D. J., and M. Slatkin, Eds. (1983). *Coevolution*, Sinauer, Sunderland, MA, 555 pp.

Gallun, R. L., and G. S. Khush. (1980). "Genetic factors affecting expression and stability of resistance," in F. G. Maxwell and P. R. Jennings, Eds., *Breeding Plants Resistant to Insects*, Wiley, New York, pp. 63-85.

Geier, P. W., and L. R. Clark. (1961). An ecological approach to pest control, *8th Proc. Tech. Meet. Int. Union Conserv. Nature Nat. Res.*, 1960, Warsaw, pp. 10-18.

Gifford, R. M., J. H. Thorne, W. D. Hitz, and R. T. Giaquinta. (1984). Crop productivity and photoassimilate partitioning, *Science*, **225**, 801-808.

Gilbert, L. E., and P. H. Raven, Eds. (1975). *Coevolution of Animals and Plants*, University of Texas Press, Austin, Texas, 246 pp.

Gmelin, R. (1967). Analysis of active compounds in *Gratiola officinalis*, *Arch. Pharm.*, **300**, 234-240.

Gould, F. (1978). Resistance in cucumber varieties to *Tetranychus urticae*: Genetic and environmental determinants, *J. Econ. Entomol.*, **71**, 680-683.

Gould, F. (1983). "Genetics of plant-herbivore systems: Interactions between basic and applied studies," in R. F. Denno and M. S. McClure, Eds., *Variable Plants and Herbivores in Natural and Managed Systems*, Academic, New York, pp. 599-653.

Gould, F. (1984). Mixed function oxidases and herbivore polyphagy: The devil's advocate position, *Ecol. Entomol.*, **9**, 29-34.

Green, M. B., and P. A. Hedin, Eds. (1986). *Natural Resistance of Plants to Pests: Roles of Allelochemicals*, ACS Symposium Ser. 296, American Chemical Society, Washington, D.C., 244 pp.

Grison, P., D. J. Kuenen, P. A. van der Laan, K. Mellanby, H. J. Muller, and J. de Wilde, Eds. (1958). (Issue dedicated to 1st Symposium "Insect and Foodplant") Wageningen, *Entomol. Exp. Appl.*, **1**, 1-72.

Grunwald, C., and M. Kogan. (1981). Sterols of soybeans differing in insect resistance and maturity group, *Phytochemistry*, **20**, 765-768.

Halbert, S. E., M. E. Irwin, and R. M. Goodman. (1981). Alate aphid (Homoptera: Aphididae) species and their relative importance as field vectors of soybean mosaic virus, *Ann. Appl. Biol.*, **97**, 1–9.

Harborne, J. B., Ed. (1978). *Biochemical Aspects of Plant and Animal Coevolution*, Academic, New York, 435 pp.

Harborne, J. B. (1980). "Plant phenolics," in E. A. Bell and B. V. Charlwood, Eds., *Secondary Plant Products*, Springer-Verlag, Berlin, pp. 329–402.

Harborne, J. B. (1982). *Introduction to Ecological Biochemistry*, Academic, London, 278 pp.

Harborne, J. B., D. Boulter, and B. L. Turner, eds. (1971). *Chemotaxonomy of the Leguminosae*, Academic, London, 612 pp.

Hardin, J. M. T. (1979). "Phenolic acids of soybeans resistant and non-resistant to leaf feeding larvae," M.Sc. Thesis, University of Arkansas, 50 pp.

Harlan, J. R. (1977). "How green can a revolution be," in D. S. Seigler, Ed., *Crop Resources*, Academic, New York, pp. 105–110.

Harlan, J. R., and K. J. Starks. (1980). "Germplasm resources and needs," in F. G. Maxwell and P. R. Jennings, Eds., *Breeding Plants Resistant to Insects*, Wiley, New York, pp. 253–273.

Harper, J. E., J. C. Nicholas, and R. H. Hageman. (1972). Seasonal and canopy variation in nitrate reductase activity of soybean (*Glycine max* L. Merr.) varieties, *Crop Sci.*, **12**, 382–386.

Harris, M. K., Ed. (1980). *Biology and Breeding for Resistance to Arthropods and Pathogens in Agricultural Plants*, Texas Ag. Exp. Sta. MP-1451, Texas A&M University, College Station, Texas, 605 pp.

Hart, S. V., M. Kogan, and J. D. Paxton. (1983). Effect of soybean phytoalexins on the herbivorous insects Mexican bean beetle and soybean looper, *J. Chem. Ecol.*, **9**, 657–672.

Hatchett, J. R., and R. L. Gallun. (1970). Genetics of the ability of the Hessian fly, *Mayetiola destructor*, to survive on wheats having different genes for resistance, *Ann. Entomol. Soc. Amer.*, **63**, 1400–1407.

Haukioja, E., and P. Niemela. (1977). Retarded growth of a geometrid larva after mechanical damage to leaves of its host tree, *Ann. Zool. Fenn.*, **14**, 48–52.

Hazan, A., V. Gershon, and A. S. Tahori. (1973). Life history and life tables of the carmine spider mite, *Acarologia*, **3**, 414–440.

Hedin, P. A., Ed. (1977). *Host Plant Resistance to Pests*, ACS Symposium Ser. 62, American Chemical Society, Washington, D.C., 286 pp.

Hedin, P. A., Ed. (1983). *Plant Resistance to Insects*, ACS Symposium Ser. 208, American Chemical Society, Washington, D.C., 375 pp.

Hedin, P. A., Ed. (1985). *Bioregulators for Pest Control*, ACS Symposium Ser. 276, American Chemical Society, Washington, D.C., 540 pp.

Hegnauer, R., Ed. (1962–1969). *Chemotaxonomie der Pflanzen*, 5 vols., Birkhaauser Verlag, Basel. Vol. 1, 517 pp.; Vol. 2, 540 pp.; Vol. 3, 743 pp.; Vol. 4, 551 pp.; Vol. 5, 506 pp.

Herzog, D. C., and J. W. Todd. (1980). "Sampling velvetbean caterpillar on soybean," in M. Kogan and D. C. Herzog, Eds., *Sampling Methods in Soybean Entomology*, Springer-Verlag, New York, pp. 107–140.

Horber, E. (1980). "Types and classification of resistance," in F. G. Maxwell and P. R. Jennings, Eds., *Breeding Plants Resistant to Insects*, Wiley, New York. pp. 15–21.

Hovanitz, W. (1969). Inherited and/or conditioned changes in host-plant preference in *Pieris*, *Entomol. Exp. Appl.*, **12**, 729–735.

Howe, W. L., J. R. Sanborn, and A. M. Rhodes. (1976). Western corn rootworm adult and spotted cucumber beetle associations with *Cucurbita* and cucurbitacins, *Environ. Entomol.*, **5**, 1043–1048.

Huffaker, C. B., M. van de Vrie, and J. A. McMurtry. (1969). The ecology of tetranychid mites and their natural control, *Ann. Rev. Entomol.*, **14**, 125–174.

Ikeda, R., and C. Kaneda. (1982). Genetic studies on brown planthopper resistance of rice in Japan, *JARG*, **16**, 1–5.

Janzen, D. A. (1979). How to be a fig, *Ann. Rev. Ecol. Syst.*, **10**, 13–15.

Janzen, D. A. (1980). When is it coevolution, *Evolution*, **84**, 611–612.

Jermy, T. (1966). Feeding inhibitors and food preferences in chewing phytophagous insects, *Entomol. Exp. Appl.*, **9**, 1–12.

Jermy, T. (1976). *The Host-Plant in Relation to Insect Behavior and Reproduction*, Plenum, New York, 322 pp.

Jermy, T. (1984). Evolution of insect/host plant relationships, *Am. Nat.*, **124**, 609–630.

Jolivet, P. (1983). *Insectes et Plantes. Evolution Parallèle et Adaptations*, Num. Spec. Bull. Soc. Lin. Lyon, 148 pp.

Karban, R., and J. R. Carey. (1984). Induced resistance of cotton seedlings to mites, *Science*, **225**, 53–54.

Kennedy, J. S. (1950). Host-finding and host alternation in aphids, *8th Int. Congr. Entomol.* (Stockholm), 423–426.

Kennedy, J. S. (1953). Host plant selection in Aphididae, *Trans. 9th Int. Congr. Entomol.* (Amsterdam), **2**, 106–113.

Kennedy, J. S. (1965). Mechanisms of host plant selection, *Ann. Appl. Biol.*, **56**, 317–322.

Khush, G. S. (1984). Breeding of rice resistant to insects, *Protec. Ecol.*, **7**, 147–165.

Klijnstra, J. W. (1982). "Perception of the oviposition deterrent pheromone in *Pieris brassicae*," in J. H. Visser and A. K. Minks, Eds., *Proc. 5th Int. Symp. Insect-Plant Relationships*, Centre for Agricultural Publishing and Documentation, Wageningen, Holland, pp. 145–152.

Kogan, M. (1972a). Feeding and nutrition of insects associated with soybeans. 2. Soybean resistance and host preferences of the Mexican bean beetle, *Epilachna varivestis*, *Ann. Entomol. Soc. Amer.*, **65**, 675–683.

Kogan, M. (1972b). "Intake and utilization of natural diets by the Mexican bean beetle, *Epilachna varivestis*—A multivariate analysis," in J. G. Rodriguez, Ed., *Insect and Mite Nutrition: Significance and Implication in Ecology and Pest Management*, North-Holland, Amsterdam, pp. 107–126.

Kogan, M. (1977). The role of chemical factors in insect/plant relationships, *Proc. 15th Int. Congr. Entomol.* (Washington, D.C.), 211–227.

Kogan, M. (1980). "Insect problems of soybean in the United States," in F. T. Corbin, Ed., *World Soybean Res. Conf. II Proceedings, Raleigh, North Carolina*, Westview Press, Boulder, CO, pp. 303–325.

Kogan, M. (1981). Dynamics of insect adaptations to soybean: Impact of integrated pest management, *Environ. Entomol.*, **10**, 263–271.

Kogan, M. (1982). "Plant resistance in pest management," in R. L. Metcalf and W. H. Luckmann, Eds., *Introduction to Insect Pest Management*, Wiley, New York, pp. 93–134.

Kogan, M. (1986). Natural chemicals in plant resistance to insects, *Iowa St. J. Res.*, **60**, 501–527.

Kogan, M., and E. F. Ortman. (1978). Antixenosis—A new term proposed to define Painter's "nonpreference" modality of resistance, *Bull. Entomol. Soc. Amer.*, **24**, 175–176.

Kogan, M., and J. Paxton. (1983). "Natural inducers of plant resistance to insects," in P. Hedin, Ed., *Plant Resistance to Insects*. ACS Symposium Ser. 228, American Chemical Society, Washington, D.C., pp. 152–171.

Kogan, M., and S. G. Turnipseed. (1980). "Soybean growth and assessment of damage by arthropods," in M. Kogan and D. C. Herzog, Eds., *Sampling Methods in Soybean-Entomology*, Springer Verlag, New York, pp. 3–29.

Kogan, M., W. V. Campbell, E. E. Hartwig, and M. J. Sullivan. (in press). "Resistance in soybean to insect pests," in L. D. Newsom, Ed., *New Technology in Pest Control for Soybean.* Wiley, New York.

Konishi, M., and Y. Ito. (1973). "Early entomology in East Asia," in R. F. Smith, T. E. Mittler, and C. N. Smith, Eds., *History of Entomology*, Annual Reviews, Palo Alto, CA, pp. 1-20.

Krischik, V. A., and R. F. Denno. (1983). "Individual, population, and geographic patterns in plant defense," in R. F. Denno and M. S. McClure, Eds., *Variable Plants and Herbivores in Natural and Managed Systems*, Academic, New York, pp. 463-512.

Kuć, Z., and F. L. Caruso. (1977). "Activated coordinated chemical defense against disease in plants," in P. Hedin, Ed., *Host Plant Resistance to Pests*. ACS Symposium Ser. 62, American Chemical Society, Washington, D.C., pp. 78-89.

Labeyrie, V., Ed. (1977). *Comportement des Insectes et Millieu Trophique*, Ed. Centre Nat. Recherche Scientifique, Paris, France, 493 pp.

Lane, H. C., and M. F. Schuster. (1981). Condensed tannins of cotton leaves, *Phytochemistry*, **20**, 425-427.

Latheef, M. A., and R. D. Irwin. (1979). The effect of companionate planting on lepidopteran pests of cabbage, *Can. Entomol.*, **111**, 863-864.

Lee, Y. I. (1983). "The potato leafhopper, *Empoasca fabae*, soybean pubescence and hopperburn resistance," Ph.D. dissertation, University of Illinois at Urbana-Champaign, 119 pp.

Lerin, J. (1980). Influence des substances allelochimiques des cruciferes sur les insectes, *Acta Oecol.*, **1**, 215-235.

Levin, D. A. (1976). The chemical defenses of plants to pathogens and herbivores, *Ann. Rev. Ecol. Syst.*, **7**, 121-159.

Loschiavo, S. R., A. J. McGinnis, and D. R. Metcalfe. (1969). Nutritive value of barley varieties assessed with the confused flour beetle, *Nature*, **224**, 288.

Lukefahr, M. J., and P. A. Fryxell. (1967). Content of gossypol in plants belonging to genera related to cotton, *Econ. Bot.*, **21**, 128-131.

MacKenzie, D. R. (1980). "The problem of variable pests," in F. G. Maxwell and P. R. Jennings, Eds., *Breeding Plants Resistant to Insects*, Wiley, New York, pp. 183-213.

Malek, R. B., and D. I. Edwards. (1973). Battling the soybean cyst nematode. *Ill. Res.*, (winter): 12-13.

Marini-Bettolo, G. B., Ed. (1977). *Natural Products and the Protection of Plants*, *Pontif. Acad. Sci. Scripta Varia 41*, Elsevier, Amsterdam, 864 pp.

Martin, P. B., P. D. Lingren, G. L. Greene, and E. E. Grissell. (1981). The parasitoid complex of three noctuids [Lep.] in northern Florida cropping system: Seasonal occurrence, parasitization, alternate hosts, and influence of host-habitat, *Entomophaga*, **26**, 401-419.

Matteson, P. C., M. A. Altieri, and W. C. Gagne. (1984). Modification of small farmer practices for better pest management, *Ann. Rev. Entomol.*, **29**, 383-402.

Mattson, Jr., W. J. (1980). Herbivory in relation to plant nitrogen content, *Ann. Rev. Ecol. Syst.*, **11**, 119-161.

Maxwell, F. G., and P. R. Jennings, Eds. (1980). *Breeding Plants Resistant to Insects*, Wiley, New York, 683 pp.

Mayse, M. A., and P. W. Price. (1978). Seasonal development of soybean arthropod communities in East Central Illinois, *Agro-Ecosystems*, **4**, 387-405.

McCutcheon, G. S., and S. G. Turnipseed. (1981). Parasites of lepidopterous larvae in insect resistant and susceptible soybeans in South Carolina, *Environ. Entomol.*, **10**, 69-74.

McIntyre, J. L., J. A. Kodds, and J. D. Haare. (1981). Effects of localized infections of *Nicotiana tabacum* to tobacco mosaic virus on systemic resistance against diverse pathogens and an insect, *Phytopathology*, **71**, 297-301.

McNeill, S., and T. R. E. Southwood. (1978). "The role of nitrogen in the development of insect/plant relationships," in J. B. Harborne, Ed., *Biochemical Aspects of Plant and Animal Coevolution*, Academic, London, pp. 77-98.

McNaughton, S. J., and J. L. Tarrants. (1983). Grass leaf silification: Natural selection for an inducible defense against herbivores, *Proc. Natl. Acad. Sci. USA.*, **80**, 790–791.

Metcalf, R. L., A. M. Rhodes, R. A. Metcalf, J. Ferguson, E. R. Metcalf, and P. Y. Lee. (1982).
Cucurbitacin content and diabroticite (Coleoptera: Chrysomelidae) feeding upon *Cucurbita*
sp., *Environ. Entomol.*, **11**, 931–937.

Miller, S. E. (1978-1979). Publications of William Hovanitz, *J. Res. Lep.*, **17** (suppl.), 66–76.

Minor, H. C. (1976). "Planting date and plant spacing in soybean production," in R. M. Goodman,
Ed., *Expanding Use of Soybeans*, University of Illinois, Urbana-Champaign, INTSOY ser. 10,
pp. 56–62.

Mitter, C., and D. J. Futuyma. (1983). "An evolutionary-genetic view of host-plant utilization by
insects," in R. F. Denno and M. S. McClure, Eds., *Variable Plants and Herbivores in Natural
and Managed Systems*, Academic, New York, pp. 427–459.

Moore, D. (1984). The role of silica in protecting Italian rye grass (*Lolium multiflorium*) from attack
by dipterous stem-boring larvae (*Oscinella frit* and other related species), *Ann. Appl. Biol.*,
**104**, 161–166.

Mothes, K. (1980). "Historical Introduction," in E. A. Bell and B. V. Charlwood, Eds., *Secondary
Plant Products. Encyclopedia of Plant Physiology*, Vol. 8, Springer-Verlag, Berlin, pp. 1–10.

National Academy of Sciences. (1969). *Insect-Plant Interactions. Report of a Work Conference*,
National Academy of Sciences, Washington, D.C., 93 pp.

Nelson, C. E., M. Walker-Simmons, D. Makus, G. Zuroske, J. Graham, and C. A. Ryan. (1983).
"Regulation of synthesis and accumulation of proteinase inhibitors in leaves of wounded tomato
plants," in P. A. Hedin, Ed., *Plant Resistance to Insects*, Symposium Ser. 208, American
Chemical Society, Washington, D.C., pp. 103–122.

Nielsen, J. K. (1978). Host plant discrimination within Cruciferae: Feeding responses of four leaf
beetles (Coleoptera: Chrysomelidae) to glucosinolates, cucurbitacins and cardenolides, *Entomol. Exp. Appl.*, **24**, 41–54.

Nielsen, J. K., L. M. Larsen, and H. Sorensen. (1977). Cucurbitacin E and I in *Ibens amara*: Feeding inhibitors for *Phyllotreta nemorum*, *Phytochemistry*, **16**, 1519–1522.

Niles, G. A. (1980). "Breeding cotton for resistance to insect pests," in F. G. Maxwell and P. R.
Jennings, Eds., *Breeding Plants Resistant to Insects*, Wiley, New York, pp. 337–369.

Nordlund, D. A., R. B. Chalfant, and W. J. Lewis. (1984). Arthropod populations, yield and damage in monocultures and polycultures of corn, beans and tomatoes, *Agric. Ecosyst. Environ.*,
**11**, 253–367.

Nordlund, D. A., R. L. Jones, and W. J. Lewis, Eds. (1981). *Semiochemicals: Their Role in Pest
Control*, Wiley, New York, 306 pp.

Norris, D. M., and M. Kogan. (1980). "Biochemical and morphological bases of resistance." in F.
G. Maxwell and P. R. Jennings, Eds., *Breeding Plants Resistant to Insects*, Wiley, New York,
pp. 23–61.

O'Dowd, D. J., and G. B. Williamson. (1979). Stability conditions in plant defense guilds, *Am.
Nat.*, **114**, 379–383.

Ohnishi, M., S. Ito, and Y. Fujino. (1982). Chemical characterization of ceramide and cerebroside
in soybean leaves, *Agric. Biol. Chem.*, 46, 2855–2856.

Oliveira, E. B. de. (1981). "Effect of resistant and susceptible soybean genotypes of different phenological stages on development, leaf consumption, and oviposition of *Anticarsia gemmatalis*
Hubner," M.Sc. thesis, University of Florida, Gainesville, 163 pp.

Painter, R. H. (1930). The biological strains of Hessian fly, *J. Econ. Entomol.*, **23**, 322–329.

Painter, R. H. (1951). *Insect Resistance in Crop Plants*, Macmillian, New York, 520 pp.

Panda, N. (1979). *Principles of Host-Plant Resistance to Insect Pests*, Allenheld, Osmun & Co. and
Universe Books, New York, 386 pp.

Parr, J. C., and R. Thurston. (1972). Toxicity of nicotine in synthetic diets to the larvae of the tobacco hornworm, *Ann. Entomol. Soc. Amer.*, **65**, 1185-1188.

Pasteels, J. M. (1976). Evolutionary aspects in chemical ecology and chemical communication, *Proc. 15th Int. Congr. Entomol.* (Washington, D.C.), 281-293.

Payne, T. L. (1981). "Disruption of southern pine beetle infestations with attractants and inhibitors," in E. R. Mitchell, Ed., *Management of Insect Pests with Semiochemicals: Concepts and Practices*, Plenum, New York, pp. 565-586.

Price, P. W. (1981). "Semiochemicals in evolutionary time," in D. A. Nordlund, R. L. Jones, and W. J. Lewis, Eds., *Semiochemicals: Their Role in Pest Control*, Wiley, New York, pp. 256-279.

Price, P. W., C. E. Bouton, P. Gross, B. A. McPheron, Jr., J. N. Thompson, and A. E. Weis. (1980). Interactions among three trophic levels: Influence of plants on interactions between insect herbivores and natural enemies, *Ann. Rev. Ecol. Syst.*, **11**, 41-65.

Price, P. W., C. N. Slobodchikoff, and W. S. Gaud, Eds. (1984). *A New Ecology. Novel Approaches to Interactive Systems*, Wiley, New York, 515 pp.

Prokopy, R. J., and E. D. Owens. (1983). Visual detection of plants by herbivorous insects, *Ann. Rev. Entomol.*, **28**, 337-364.

Reed, D. K., J. D. Worthen, Jr., E. C. Vebel, and G. L. Reed. (1982). Effects of two triterpenoids from neem on feeding by cucumber beetles (Coleoptera: Chrysomelidae), *J. Econ. Entomol.*, **75**, 1109-1112.

Reese, J. (1979). "Interactions of allelochamicals with nutrients in herbivore food," in G. A. Rosenthal and D. H. Janzen, Eds., *Herbivores—Their Interaction with Secondary Plant Metabolites*, Academic, New York, pp. 309-330.

Rembold, H., G. K. Sharma, Ch. Czoppelt, and H. Schmutterer. (1982). Azidarachtin: A potent growth regulator of plant origin, *Zeit. Ang. Entomol.*, **93**, 12-17.

Rhoades, D. F. (1977). "The antiherbivore chemistry of *Larrea*," in T. J. Mabry, J. H. Hunziker, and D. R. Difeo, Jr., Eds., *Creosote Bush*, Dowden, Hutchinson & Ross, Stroudsburg, Pennsylvania, pp. 135-175.

Rhoades, D. F. (1979). "Evolution of plant chemical defense against herbivores," in G. A. Rosenthal and D. H. Janzen, Eds., *Herbivores—Their Interactions with Secondary Plant Metabolites*, Academic, New York, pp. 3-52.

Rhoades, D. F. (1983). "Responses of alder and willow to attack by tent caterpillars and webworms: Evidence for pheromonal sensitivity of willows," in P. Hedin, Ed., *Plant Resistance to Insects*, ACS Symposium Series 208, American Chemical Society, Washington, D.C., pp. 55-68.

Rhoades, D. F. (1985). Offensive-defensive interactions between herbivores and plants: Their relevance in herbivore population dynamics and ecological theory, *Am. Nat.*, **125**, 205-238.

Rhoades, D. F., and R. G. Cates. (1976). "Toward a general theory of plant antiherbivory chemistry," in J. W. Wallace and R. L. Mansell, Eds., *Biochemical Interactions Between Plants and Insects. Rec. Adv. Phytochem. 10*, Plenum, New York, pp. 168-213.

Risch, S. J. (1980). The population dynamics of several herbivorous beetles in a tropical agroecosystem: The effect of intercropping corn, beans and squash in Costa Rica, *J. Appl. Ecol.*, **17**, 593-612.

Risch, S. J. (1981). Insect herbivore abundance in tropical monocultures and polycultures: An experimental test of two hypotheses, *Ecology*, **62**, 1325-1340.

Robinson, R. A. (1980). "The pathosystem concept," in F. G. Maxwell and P. R. Jennings, Eds., *Breeding Plants Resistant to Insects*, Wiley, New York, pp. 157-181.

Rodman, J. E., and F. S. Chew. (1980). Phytochemical correlates of herbivory in a community of native and naturalized Cruciferae, *Biochem. System. Ecol.*, **8**, 43-50.

Rodriguez, J. G., Ed. (1972). *Insect and Mite Nutrition—Significance and Implications in Ecology and Pest Management*, North Holland, Amsterdam, 702 pp.

Rosenthal, G. A., and P. H. Janzen, Eds. (1979). *Herbivores—Their Interactions with Secondary Plant Metabolites*, Academic, New York, 718 pp.

Rothschild, M., and L. M. Schoonhoven. (1977). Assessment of egg load by *Pieris brassicae* (Lepidoptera: Pieridae), *Nature*, **266**, 352–355.

Russell, G. B., O. R. W. Sutherland, R. F. N. Hutchins, and P. E. Christmas. (1978). Vestitol: A phytoalexin with insect feeding-deterrent activity, *J. Chem. Ecol.*, **4**, 572–579.

Ryan, C. A. (1979). "Proteinase inhibitors," in G. A. Rosenthal and D. H. Janzen, Eds., *Herbivores: Their Interaction with Secondary Plant Metabolites*, Academic, New York, pp. 599–618.

Ryan, C. A. (1983). "Insect-induced chemical signals regulating natural plant protection responses," in R. F. Denno and M. S. McClure, Eds., *Variable Plants and Herbivores in Natural and Managed Systems*, Academic, New York, pp. 43–60.

Schoonhoven, L. M. (1981). "Chemical mediators between plants and phytophagous insects," in D. A. Nordlund et al., Eds., *Semiochemicals: Their Role in Pest Control*, Wiley, New York, pp. 31–50.

Schoonhoven, L. M. (1982). Biological aspects of antifeedants, *Entomol. Exp. Appl.*, **31**, 57–69.

Scriber, J. M. (1983). "Evolution of feeding specialization, physiological efficiency, and host races in selected Papilionidae and Saturniidae," in R. F. Denno and M. S. McClure, Eds., *Variable Plants and Herbivores in Natural and Managed Systems*, Academic, New York, pp. 373–412.

Scriber, J. M. (1984). "Host-plant suitability," in W. J. Bell and R. T. Carde, Eds., *Chemical Ecology of Insects*, Sinauer, Sunderland, MA, pp. 159–202.

Scriber, J. M., and F. Slansky, Jr. (1981). The nutritional ecology of immature insects, *Annu. Rev. Entomol.*, **26**, 183–211.

Seigler, D., and P. W. Price. (1976). Secondary compounds in plants: Primary functions, *Am. Nat.*, **110**, 101–105.

Sell, D. K., G. S. Whitt, R. L. Metcalf, and L. K. Lee. (1974). Enzyme polymorphism in the corn earworm, *Heliothis zea* (Lepidoptera: Noctuidae): Hemolymph esterase polymorphism, *Can. Entomol.*, **106**, 701–709.

Sell, D. K., G. S. Whitt, and W. H. Luckmann. (1975). Esterase polymorphism in the corn earworm, *Heliothis zea* (Boddie): A survey of temporal and spatial allelic variation in natural populations, *Biochem. Genet.*, **13**, 885–898.

Shorey, H. H., and J. J. McKelvey, Jr., Eds. (1977). *Chemical Control of Insect Behavior: Theory and Application*, Wiley, New York, 414 pp.

Slansky, Jr., F. (1982). Insect nutrition: An adaptationist's perspective, *Florida Entomol.*, **65**, 45–71.

Smith, C. M., and N. H. Fischer. (1983). Chemical factors of an insect resistant soybean genotype affecting growth and survival of the soybean looper, *Entomol. Exp. Appl.*, **33**, 343–345.

Snelling, R. O. (1941). Resistance of plants to insect attack, *Bot. Rev.*, **7**, 543–586.

Sondheimer, E., and J. A. B. Simeone, Eds. (1970). *Chemical Ecology*, Academic, New York, 336 pp.

Southwood, T. R. E., and M. J. Way. (1970). "Ecological background to pest management," in R. L. Rabb and F. E. Guthrie, Eds., *Concepts of Pest Management*, North Carolina State University, Raleigh, NC, pp. 6–29.

Staedler, E. (1977). Sensory aspects of insect plant interactions, *Proc. XV Intern. Cong. Entomol.* (Washington, D.C.) 228–248.

Staedler, E. (1983). "Attractants, arrestants, feeding and oviposition stimulants in insect plant relationships: Application for pest control," in D. L. Whitehead and W. S. Bowers, Eds., *Natural Products for Innovative Pest Management*, Pergamon, Oxford, pp. 243–258.

Stahl, E. (1888). Pflanzen und Schnecken. Biologische Studie ueber die Schutzmittel der Pflanzen gegen Schneckenfrass. Jenaische Zeits, *Naturwis*, **22**, 555–684.

Strong, D. and B. Brodbeck. (in press). "Amino acid nutrition of herbivorous insects and stress to host plants," in P. Barbosa and J. Schultz, Eds., *Insect Outbreaks*, Academic, New York.

Strong, D. R., J. H. Lawton, and R. Southwood. (1984a). *Insects on Plants: Community Patterns and Mechanisms*, University Press, Cambridge, MA, 313 pp.

Strong, Jr., D. R., D. Simberloff, L. G. Abele, and A. B. Thistle, Eds. (1984b). *Ecological Communities. Conceptual Issues and the Evidence*, Princeton University Press, Princeton, NJ, 614 pp.

Sutherland, O. R. W., G. B. Russell, D. R. Biggs, and G. A. Lane. (1980). Insect feeding deterrent activity of phytoalexin isoflavonoids, *Biochem. Syst. Ecol.*, **8**, 73–75.

Swain, T., Ed. (1963). *Chemical Plant Taxonomy*, Academic, London, 543 pp.

Swain, T. (1979). "Tannins and lignins," in G. A. Rosenthal and D. H. Janzen, Eds., *Herbivores: Their Interaction with Secondary Plant Metabolites*, Academic, New York, pp. 657–700.

Thompson, J. N. (1982). *Interaction and Coevolution*, Wiley, New York, 179 pp.

Thorsteinson, A. J. (1960). Host selection in phytophagous insects, *Annu. Rev. Ent.*, **5**, 193–218.

Tingey, W. M., and S. R. Singh. (1980). "Environmental factors influencing the magnitude and expression of resistance," in F. G. Maxwell and P. R. Jennings, Eds., *Breeding Plants Resistant to Insects*, Wiley, New York, pp. 87–113.

Toldine, T. E. (1984). Relationship between DIMBOA content and *Helminthosporium turcicum* resistance in maize. (in Hungarian with English abstract), *Novenytermeles*, **33**, 213–218.

Treacy, M. F., G. R. Zummo, and J. H. Benedict. (1985). Interactions of host-plant resistance in cotton with predators and parasites, *Agric. Ecosyst. Environ.*, **13**, 151–157.

Turnipseed, S. G., and M. Kogan. (1976). Soybean entomology, *Annu. Rev. Entomol.*, **21**, 25–60.

Underhill, E. W. (1980). "Glucosinolates," in E. A. Bell and B. V. Charlwood, Eds., *Secondary Plant Products*, Springer-Verlag, Berlin, pp. 493–511.

Usher, B. F. and P. Feeny. (1983). Atypical secondary compounds in the family Cruciferae: tests for toxicity to *Pieris rapae*, an adopted crucifer-feeding insect, *Entomol. Exp. Appl.*, **34**, 257–265.

Van Duyn, J. W., S. G. Turnipseed, and J. D. Maxwell. (1971). Resistance in soybeans to the Mexican bean beetle. I. Sources of resistance, *Crop Sci.*, **11**, 572–573.

van Emden, H. F., Ed. (1971). *Insect/Plant Relationships*, Symp. Roy. Entomol. Soc. London, 6, Blackwell, Oxford, 215 pp.

van Emden, H. F. (1978). "Insects and secondary plant substances—An alternative viewpoint with special reference to aphids," in J. B. Harborne, Ed., *Biochemical Aspects of Plant and Animal Coevolution*, Academic, London, pp. 310–323.

van Etten, C. H., and H. L. Tookey. (1979). "Chemistry and biological effects of glucosinolates," in G. A. Rosenthal and D. H. Janzen, Eds., *Herbivores: Their Interaction with Secondary Plant Metabolites*, Academic, New York, pp. 471–500.

Verschaffelt, E. (1910). The cause determining the selection of food in some herbivorous insects, *Proc. Royal Acad. Amsterdam*, **13**, 536–542.

Visser, J. H. (1983). "Differential sensory perceptions of plant compounds by insects," in P. Hedin, Ed., *Plant Resistance to Insects*, ACS Symposium Ser. 208, American Chemical Society, Washington, D.C., pp. 215–230.

Visser, J. H., and A. K. Minks, Eds. (1982). *Insect–Plant Relationships*, Proc. 5th Int. Symposium, Wageningen, Holland. Centre for Agric. Pub. and Documentation, Wageningen, 464 pp.

Wade, N. (1974a). Green revolution (I): A just technology often unjust in use, *Science*, **186**, 1093–1096.

Wade, N. (1974b). Green revolution (II): Problems of adapting a western technology, *Science* **186**, 1186–1192.

Wadsworth, S. A., and J. P. Zikakis. (1984). Chitinase from soybean seeds: Purifications and some properties of the enzyme system. *J. Agric. Food Chem.*, **32**, 1284–1288.

Waldbauer, G. P. (1968). The consumption and utilization of food by insects, *Adv. Insect Physiol.*, **5**, 229–288.

Waldbauer, G. P., and A. K. Bhattacharya. (1973). Self-selection of an optimum diet from a mixture of wheat fractions by the larvae of *Tribolium confusum*, *J. Insect Physiol.*, **19**, 407–416.

Waldbauer, G. P., R. W. Cohen, and S. Friedman. (1984). Self-selection of an optimal nutrient mix from defined diets by larvae of the corn earworm, *Heliothis zea* (Boddie), *Physiol. Zool.*, **57**, 590–597.

Wallace, J. W., and R. L. Mansell, Eds. (1976). *Biochemical Interactions between Plants and Insects*, Plenum, New York, 425 pp.

White, T. C. R. (1984). The abundance of invertebrate herbivores in relation to the availability of nitrogen in stressed food plants, *Oecologia*, **63**, 90–105.

White, T. C. R. (1976). Weather, food and plagues of locusts, *Oecologia*, **22**, 119–134.

White, D. L., and W. S. Bowers, Eds. (1983). *Natural Products for Innovative Pest Management*, Pergamon, Oxford.

Whitham, T. G. (1983). "Individual trees as heterogeneous environments: Adaptation to herbivory or epigenetic noise," in R. F. Denno and M. S. McClure, Eds., *Variable Plants and Herbivores in Natural and Managed Systems*, Academic, New York, pp. 9–27.

Whitham, T. G., A. G. Williams, and A. M. Robinson. (1984). "The variation principle: Individual plants as temporal and spatial mosaics of resistance to rapidly evolving pests," in P. W. Price, C. N. Slobodchikoff, and W. S. Gaud, Eds., *A New Ecology: Novel Approaches to Interactive Systems*, Wiley, New York, pp. 15–51.

Whittaker, R. H., and P. P. Feeny. (1971). Allelochemicals: Chemical interactions between species, *Science*, **171**, 757–770.

Wiebes, J. T. (1979). Co-evolution of figs and their insect pollinators, *Ann. Rev. Ecol. Syst.*, **10**, 1–12.

Wilde, J. de, and L. M. Schoonhoven, Eds. (1969). Insect and host plant, Proc. 2nd Int. Symp. Wageningen, Holland, *Entomol. Exp. Appl.*, **12**, 471–810.

Williams, P. M., and M. S. de Mallorca. (1984). Effect of osmotically induced leaf moisture stress on nodulation and nitrogenase activity of *Glycine max*, *Plant and Soil*, **80**, 267–283.

Wood, D. L., and W. D. Bedard. (1977). The role of pheromones in the population dynamics of the western pine beetle, *Proc. 15th Int. Cong. Entomol.*, Washington, D.C., pp. 643–682.

Zavaleta, L., and M. Kogan. (1984). Economic benefits of breeding host plant resistance: A simulation approach, *No. Cent. J. Ag. Econ.*, **6**, 28–35.

Zucker, W. V. (1983). Tannins: Does structure determine function? An ecological perspective, *Am. Nat.*, **121**, 335–365.

# 6

# PLANT DEFENSE-HERBIVORE-PARASITE INTERACTIONS AND BIOLOGICAL CONTROL

*Sean S. Duffey and Ken A. Bloem*

*Great fleas have little fleas upon*
*their backs to bite 'em,*
*And little fleas have lesser fleas,*
*and so ad infinitum.*
*And the great fleas themselves, in turn*
*have greater fleas to go on;*
*While these in turn have greater still,*
*and greater still, and so on.*
AUGUSTUS DE MORGAN

## 6.1. INTRODUCTION

Prior to the advent of ecology, the extensive feed-forward/feed-backward nature of food chains was nonscientifically realized by the use of metaphors such as the above quotation, or by more fanciful tales such as "The House That Jack Built." By the brunt of modern scientific methods our understanding of the multi-trophic nature of food chains or webs has become more sophisticated than merely an empirical description of flea upon flea. In an entomological vein, the study of insect parasitoids has contributed considerably to our modern views on this topic (Price, 1975a,b). However, despite a moderately large and sophisti-cated body of theoretical information on host–parasitoid relationships (Price et al., 1980; Waage and Hassel, 1982; Hassel and Waage, 1984), the more practi-cal aspects of how to apply this theory accurately so that the "lesser flea" can be used to control the "greater flea" still remains only partially defined and even controversial (Way, 1973; Ehler, 1976; Ehler and Andres, 1983). This volume of symposial papers deals specifically with an attempt to better bridge the gap between ecological theory and practice in integrated pest management (IPM). Our contribution deals specifically with a mechanistic and somewhat theoretical approach to understanding how host-plant resistance to insects can be linked compatibly with the use of biological agents in IPM programs. The major aim of this chapter is to demonstrate that a fundamental knowledge of the mechanisms of host-plant resistance to insects can aid in formulating valid theory as well as in facilitating practice.

The literature cited in this chapter does not represent an exhaustive review. The authors have attempted to cite, in majority, a moderate number of primary research articles that will give the new student a grasp of the wealth of literature available on any topic, as well as provide the opportunity to penetrate further into specific topics by referral to references cited in these primary references. The authors have also tried to cite, where relevant or available, modern textbooks and/or review articles that summarize in a broad and integrative context research and theories on specific topics.

### 6.1.1. Definition of the Problem

It is usually assumed that host-plant resistance (HPR) is "generally" or "highly" compatible with IPM and biological control (BC) strategies (Kogan, 1975; Bergman and Tingey, 1979; Adkisson and Dyck, 1980). Although the majority of cases investigated point to such compatibility, this assumption has its limitations (Bergman and Tingey, 1979; Hare, 1983). The limitations are primarily two: (1) that compatibility has usually been assessed on a short-term basis so that the long-range effects upon BC are unknown, and (2) that the generalizability of the concept of compatibility of resistance is both vague and speculative unless one specifies the nature of the resistance. Since detailed information on the nature or mechanism(s) of most crops' resistance to a given insect is usually either limited or lacking, it is not always possible to depict meaningfully what comprises compatibility. The need for a knowledge of cause-and-effect relationships concerning the compatibility of HPR with BC is a central issue in this paper.

For the sake of simplicity, our discussion of the interaction of HPR and BC agents will be developed with reference to two concepts—*compatibility* and *implementability*. By the use of these two general concepts, we will be able to focus more directly at describing the nature of HPR and its impact upon the use of BC agents in IPM programs. This directedness arises from the necessity to define these concepts in more detail than is generally given for a definition of resistance (see Robinson, 1976, 1980; Ortman and Peters, 1980). By this it is meant that general definitions of HPR, for sake of both clarity and brevity, do not specify *cause, effect*, or *target*.

The term "resistance" is a statement about biological activity with the same degree of predictiveness as the term "pesticide." The ability to utilize pesticides predictively, as with resistance, requires information on at least chemical identity, dose, target, and mode and time of application. Hence, the ability to utilize HPR with a consideration for compatibility with BC agents requires minimally a definition of (1) the causal factor(s) of a plant's resistance, (2) the target organism(s), and (3) a consideration of plant- and insect-related factors that determine the degree and mode of resistance against the target(s) and (4) the BC agents and their mode and degree of control. Without these minimal specifications, the implementability of a compatible HPR program may be difficult. The concept of implementability also entails the availability of other bodies of knowledge about the nature of resistance that are not mutually exclusive from those above: (1) the ability to manipulate genetically or culturally the degree and/or quality of the causal factors in the plant, (2) the degree to which such manipulation affects the plant's interaction with other organisms and the environment, and (3) the degree to and manner in which these factors affect the BC agents.

This chapter is not intended as a comprehensive review of the interaction of HPR with biological control agents. Instead, it will deal in depth with the theory and practice of the simultaneous use of chemically based antibiotic HPR and parasitoids against specific lepidopterous pests.

### 6.1.2. An Overview of Tritrophic Interactions

Our understanding of the role of plant natural products as a partial regulator of plant–insect interactions has greatly expanded in the last decade (see Hedin et al., 1974; Gilbert and Raven, 1975; Friend and Threlfall, 1976; Levin, 1976; Nickol, 1979; Rosenthal and Janzen, 1979; Maxwell and Jennings, 1980; Nordlund et al., 1981; Harborne, 1982; Ahmad, 1983; Denno and McClure, 1983; Futuyma and Slatkin, 1983; Hedin, 1983; Whitehead and Bowers, 1983; Bell and Cardé, 1984; see also Kogan, Chapter 5). Because of the relatively tractable nature of plant–insect systems for experimental analysis, these systems have been particularly attractive to ecologists studying the manner in which biotic factors act as determinants in regulating animal abundance. A particularly prominent avenue of investigation has been on the role of plant quality (e.g., architecture, defensive characteristics, nutritive qualities) as a factor in determining the population dynamics of herbivorous insects and their coevolution with plants (Levin, 1971; Atsatt and O'Dowd, 1976; White, 1976, 1978, 1984; Haukioja, 1980; Fox, 1981; Mooney and Gulmon, 1982; Slansky, 1982; Denno and McClure, 1983; Kareiva, 1983; Price, 1983; Rhoades, 1983). Despite this body of information, attempts to derive unifying theories about the role of plant natural products as defensive agents against herbivorous insects have been overly speculative (Feeny, 1975, 1976; Fox and Maculey, 1977; Janzen, 1979; Bernays, 1981; Rhoades, 1983; Zucker, 1983; Broadway et al., 1986).

Currently, the study of plant chemical defences against insects seems to be divided into two arbitrary and unnecessarily antipodal disciplinary approaches—that with a more holistic ecological–evolutionary emphasis (Levin, 1971; McKey, 1974; Feeny, 1975, 1976; Atsatt and O'Dowd, 1976; Cates and Rhoades, 1977; Janzen, 1979; Rhoades, 1979, 1983; Cates, 1980, 1981; Matt-

son, 1980; Fox, 1981; Slansky, 1982; Schultz, 1983a,b; White, 1984), and that
with a more reductionistic behavioral-physiological-biochemical emphasis
(Feeny and Bostock, 1968; Green and Ryan, 1972; Dahlman and Rosenthal,
1975; Rosenthal et al., 1976, 1977, 1978; Dahlman, 1977; Fox and Macauley,
1977; Berenbaum, 1978a,b; Blau et al., 1978; Tingey and Gibson, 1978; Dryer
et al., 1979; Jones and Firn, 1979a,b; Reese, 1979, 1981; Bernays and Chamberlain, 1980; Berenbaum and Feeny, 1981; Dryer and Jones, 1981; Duffey and
Isman, 1981; Waiss et al., 1981; Dimock et al., 1982; Isman and Duffey, 1983;
Rosenthal, 1982; Dimock and Kennedy, 1983; Kennedy and Dimock, 1983;
Zucker, 1983; Gregory et al., 1984; Tingey et al., 1984; Broadway et al., 1986).

Although both these approaches have contributed greatly to our theoretical
and factual knowledge of the role of plant natural products as modifiers of
plant-insect interactions, the integration of both sets of knowledge into mutually consistent theories or generalizations has been difficult because of disparate
methodologies. The more holistic approach tends to be reliant on inference from
correlation between chemical content of a given plants tissue(s) and some measurable effect upon test organism(s) in field conditions; whereas, the more reductionistic approach tends to be reliant upon extrapolation of the effect of isolated pure chemicals upon test insects in laboratory conditions. Both approaches
share several common weaknesses: (1) neither usually adequately test for cause-
and-effect relationships between chemical and effect, and (2) both attempt to
develop theory based on an essentially two-level interacting trophic system.

It is this last criticism that is our main concern here. Price et al. (1980) and
Whitham (1983) clearly point out that an assessment of plant and insect fitness
and the development of realistic theory on plant defensive strategies against insects will only be possible if, at a minimum, the third trophic level is considered—the interaction between the "plants, herbivores, and natural enemies of
herbivores." The salience of this viewpoint becomes apparent when one wishes
to use a plant chemical as a basis for resistance against a given herbivorous insect, and yet at the same time one does not wish to adversely affect BC agents
such as parasitic wasps. The knowledge to do this predictively is lacking mainly
because a dearth of information in both the holistic and reductionistic fields
prevents the development of implementable strategies.

Our contribution to the development of a more realistic theory is to gather
fundamental physiological-toxicological information about the manner in
which antibiotic plant natural products, with the potential to serve as bases of
HPR, express biological activity not only against the pest insect, but also against
parasitoids. From an ecologist's viewpoint, this question might be phrased in
terms of how plant quality, particularly chemical quality, affects the fitness of
host-insects and their parasitoids.

### 6.1.2.1. Evidence of Effect of Plant Quality on Third Trophic Level Organisms

What evidence is there that plant quality affects the third trophic level in a fashion of consequence to HPR and BC, or to the population dynamics of herbivo-

rous insects and their parasitoids? The notion of plant quality embraces many attributes of a plant from the microscopic to the macroscopic, for example, its community, its phenological characteristics, its physiological state, and its physical and chemical properties. At the macroscopic level, the structure of plant communities, plant phenology, and the gross architectural properties of individual plants are known to have an influence upon the host-seeking ability of parasitoids (Arthur, 1962; Eikenbary and Fox, 1968; Streams et al., 1968; Cheng, 1970; Vinson, 1976, 1984; Harrington and Barbosa, 1978; Waage, 1979; Price et al., 1980; Jones, 1981; Weseloh, 1981; Kareiva, 1983; Price, 1983; Prokopy, 1983; Hassel and Waage, 1984). At a slightly finer level, physical attributes such as the color and/or shape of leaves, the presence of trichomes, and the existence of refugia (e.g., galls, hollow stems) have a major effect upon the efficacy with which parasitoids can locate and utilize hosts (Vinson, 1976; Bergman and Tingey, 1979; Price et al., 1980; Vinson and Iwantsch, 1980; Jones, 1981; Mueller, 1983; Obrycki et al., 1983; Price, 1983; Prokopy, 1983).

Often operating in conjunction with the above factors, plant volatiles, host genotype, and/or host-insect-produced kairomones also strongly influence the efficacy of parasitoids (Greany et al., 1977; Lewis et al., 1977; Prokopy and Webster, 1978; McCutcheon and Turnipseed, 1981; Nordlund and Sauls, 1981; Nordlund et al., 1981; Pair et al., 1982; Roth et al., 1982; Loke et al., 1983; Obrycki et al., 1983; Vinson, 1984). Additional influences upon efficacy arise from the ability of the parasitoid to discriminate between or respond physiologically/behaviorally to diseased, parasitized, and/or otherwise plant-stressed host insects (Holmes and Boethel, 1972; Hagstrum and Smittle, 1977; Prokopy and Webster, 1978; Levin et al., 1981; Weseloh, 1981; Fritz, 1982; Versoi and Yendol, 1982; Beckage and Riddiford, 1983). At the microscopic level of plant chemicals ingested by the host insect, one finds the least information on modulation of efficacy. Based on extrapolations from analyses of ditrophic systems, the intake of organic nitrogen (= nutrients) and/or plant toxins are thought to be a prominent factor determining a parasitoid's fitness (Slansky and Feeny, 1977; McNeill and Southwood, 1978; Scriber, 1978, 1979a,b; Slansky, 1978, 1982; White, 1978, 1976, 1984; Reese, 1979; Rosenthal and Janzen, 1979; Mattson, 1980; Price et al., 1980; Schultz and Baldwin, 1982; Deno and McClure, 1983; McClure, 1983; Price, 1983; Rhoades, 1983; Schultz, 1983b; Whitham, 1983).

The above information sets the stage for understanding the diverse ways in which plant natural products influence parasitoids, but does not directly address the question as to whether or not HPR is compatible with the use of BC agents.

### 6.1.2.2. Compatibility of HPR with BC
Most of the information available demonstrates that generally there is compatibility (Kogan, 1975; Bergman and Tingey, 1979; Adkisson and Dyck, 1980; Ortman and Peters, 1980). At the theoretical level, the two strategies are considered to be compatible because they can simultaneously introduce unrelated mortality factors in a density-dependent or -independent manner upon the pest population (van Emden and Wearing, 1965; Bergman and Tingey, 1979; Adkisson and

Dyck, 1980; MacKenzie, 1980). Presumably, this control results because the resistant variety facilitates searching behavior of the enemy, reduces the vigor of the host to avoid parasitization, delays development of the host so that the pest and enemy populations are temporally synchronized, and/or among other things modifies the behavior of the host so that it is more easily parasitized (Bergman and Tingey, 1979; Jenkins, 1980; Price et al., 1980). In practice, this theory has its variants. Consider a few examples.

The morphological (e.g., frego bract, glandless, and nectariless) and chemical bases (e.g., tannins, gossypol, and flavonoids) of resistance of cotton to a variety of insect pests (e.g., bollworms, weevils, plant bugs, and spider mites) have been extensively studied (Scott and O'Keeffe, 1976; Benedict et al., 1977; Maxwell, 1977; Niles, 1980; Waiss et al., 1981; Adkisson et al., 1982; Reese et al., 1982; Hedin, 1983). In particular, the use of nectariless cotton lines has been useful in controlling lygus bugs, leafhoppers, and the pink bollworm (Scott and O'Keeffe, 1976; Maxwell, 1977; Niles, 1980). Most of this research has centered on assessing the compatibility of the simultaneous use of multiple resistant traits against a complex of pests; yet, little attention has been given to assessing the degree of compatibility of these traits with BC agents. However, it has been demonstrated that some nectariless lines of cotton were compatible with the use of the parasitoid *Campoletis sonorensis* (Cameron) against the tobacco budworm *Heliothis virescens* (F.) (Lingren and Lukefahr, 1977). Certain crops may be more amenable to the efficacy of parasitoids, for a study by Mueller (1983) showed that larvae of *Heliothis* spp. were parasitized more when reared on cotton than on bean or tomato plants. Certain lines of resistant cotton with the frego condition caused the young larvae of bollworms to spend a longer time on the terminal tissues of the plant than on susceptible plants. The extra degree of exposure doubly facilitated the action of predators against these larval pests (see Panda, 1979, Chapter 9).

The resistance of small grains to a variety of pests, particularly to biotypes of the greenbug *Schizaphis graminum* (Rondani), has also been studied in some detail (Pathak and Saxena, 1980; Teetes, 1980). A good example of the complementary nature of HPR and BC was demonstrated by the ability of the parasitic wasp *Lysiphlebus testaceipes* (Cress.) to reduce population levels of the greenbug more effectively on resistant varieties of sorghum and oats than on less-resistant varieties (Todd et al., 1971; Starks et al., 1972; Teetes et al., 1974; Schuster and Starks, 1975; Burton and Starks, 1977; Starks and Burton, 1977). It has also been found that part of the greater degree of parasitism by wasps and flies on the resistant varieties of plants resulted from an increased degree of movement of the aphids, which made them several times more susceptible to parasitization.

The use of antibiotic resistance can be particularly useful in enhancing the efficacy of parasitoids. This is most clearly demonstrated on the alfalfa plant where the control of the first- and second-instar larvae of the alfalfa weevil *Hypera postica* (Gyl.) by the ichneumonid *Bathyplectes curculionis* (Thoms.) is enhanced by antibiotic plant varieties (Panda, 1979, Chapter 9). The basis of the enhanced control resides in the fact that in normal years on susceptible plants the weevil population of first- and second-instar larvae peaks about 2 weeks before the maximal population level of parasitoid, whereas, on antibiotic plants the rate of growth and development of the beetle larvae is sufficiently slowed so that the two populations of insects peak simultaneously. The asynchrony of populations that is observed on susceptible alfalfa leads to a relatively low level of biological control (5%) compared with that achievable through the use of antibiotic plants (70%). In contrast, Pimentel and Wheeler (1973) showed that the control of populations of the aphid *Acyrthosiphon pisum* (Harris) by resistant lines of alfalfa, although compatible, did not enhance the degree of parasitism, nor was there any means found to predict how the resistance would affect the acton of other crop-associated predators and herbivores in controlling populations of the aphid.

A number of other studies on different pests on a variety of crops mostly indicate that the resistant varieties have either neutral or enhancing effects upon the degree of parasitism of aphids and lepidopteran larvae by parasitic wasps (Wyatt, 1970; Curry and Pimentel, 1971; McCutcheon and Turnipseed, 1981).

Plant trichomes are finding a more prominent role as bases of resistance against a number of small insects and mites (Levin, 1973; Norris and Kogan, 1980; Stipanovic, 1983). In the potato plant species and hybrids, multicellular trichomes have been shown to confer resistance against the green peach aphid *Myzus persicae* (Sulzer), the potato aphid *Macrosiphum euphorbiae* (Thomas), the potato leafhopper *Empoasca fabae* (Harris), and the potato flea beetle *Epitrix cucumeris* (Harris) (Gibson, 1978; Tingey and Sinden, 1982; Tingey et al., 1982; Gregory et al., 1984; Tingey et al., 1984). More recent work by Obrycki et al. (1983) has demonstrated that the combination of the sticky potato trichomes in hybrids of *Solanum tuberosum* × *S. berthaultii* and the action of aphid predators and parasitoids (coccinelids, chrysopids, and aphidids) was more effective in reducing aphid populations than either enemies or trichomes alone. However, in other crops the simultaneous use of trichomes with parasitoids has not proved to be a compatible combination. In tobacco, the secretory hairs on foliage were found to interfere significantly with the ability of *Telenomus sphingis* (Ashmead) and *Trichogramma minutum* Riley to parasitize the eggs of the tobacco hornworm, *Manduca sexta* (L.) (Rabb and Bradley, 1968) because the sticky trichomal exudate restricted the movement of the parasitoids. Similar effects of tobacco trichomes on other parasitoid–host interactions have been reported (Milliron, 1940; Elsey and Chaplin, 1978), and in cucumber the efficacy of searching by the parasitoid *Encarsia formosa* Gaham for its host, the white fly *Trialeurodes vaporarium* (Westwood), is greatly reduced by the presence of stiff foliar hairs that slow the parasitoid's walking, and also force the parasitoid to expend more time grooming if the hairs are coated with sticky aphid honey dew (see Price et al., 1980; van Lenteren et al., 1980).

One popular strategy for HPR against insects is to utilize antibiotic or toxic plant natural products (Rosenthal and Janzen, 1979; Maxwell and Jennings, 1980; Denno and McClure, 1983; Hendin, 1983). A large body of literature exists describing the effects of such toxins upon herbivorous insects in both natural and agricultural situations (see above references), yet surprisingly very little is known about the impact of these "toxins" on the third trophic level (Price et al., 1980; Price, 1983). At the more basic level, ecologists are concerned with developing theories about plant chemical quality as it relates to the fitness and population dynamics of herbivores and their enemies. Theories, unfortunately still based on meager evidence, implicate not only allelochemicals (= "toxins"), but also soluble organic nitrogen (= "nutrient") as prime factors dictating the success and abundance of herbivorous insects and their enemies (van Emden, 1969; van Emden and Bashford, 1969; McNeill and Southwood, 1978; Scriber, 1978, 1979a,b; Bergman and Tingey, 1979; Mattson, 1980; Price et al., 1980; Faeth et al., 1981; Prestidge, 1982; Slansky, 1982; Stiling et al., 1982; Al-Zubaidi et al., 1983; McClure, 1983; Price, 1983; Rhoades, 1983; Schultz, 1983a,b; Hilliard and Keeley, 1984; Brewer et al., 1985).

### 6.1.3. Effects of Host's Diet upon Third-Trophic-Level Organisms

Such theorizing is not unreasonable in view of the fact that experimental laboratory manipulation of the quality of the host's diet can have a profound effect upon the growth, development, and survival of the parasitoid via modification of the quality of the host (Shapiro, 1956; Smith, 1957; Cheng, 1970; Zohdy, 1976). The specific manner(s) in which the host's diet modifies its quality for the parasitoid is unknown at a biochemical–physiological level, although it has been clearly shown at a gross physiological level that the type of food, in addition to the presence of a parasitoid, can modify (increase or decrease) the ability of the host to digest and assimilate food (Dahlman, 1977; Brewer and King, 1981; Greenblatt and Barbosa, 1981; Huebner and Chiang, 1982). There is reciprocity, for the parasitoid has the ability to cause changes in the physiology of its host as evidenced by alterations in blood proteins, hormones, blood carbohydrates, and enzyme activities (Vinson, 1975; Dahlman and Vinson, 1976, 1980; Smilowitz and Smith, 1977; Beckage and Riddiford, 1982a,b, 1983; Baronio and Sehnal, 1980; Dahlman and Green, 1981).

Currently, our knowledge of how the nutritive quality of food affects the fitness of a parasitoid does not permit us to construct valid theories about how these complicated mutual interactions arise. Yet, studies on the role of host nutrition indicate that the addition or subtraction of specific chemicals from artificial diets provides a powerful, but highly reductionistic, tool to probe how "food quality" affects the tritrophic interrelationship. Bouletreau and David (1981) found that the ability of several species of parasitic wasps of *Drosophila melanogaster* (Meig.) to survive in their host was dependent on the ability of the wasps to tolerate high concentrations of ethanol, which the ethanol-tolerant fly sequestered from its rotting food. A different set of experiments demonstrated that intake of fatty acids by the host could also lead to overt toxicity to the parasitoid. If the parasitic wasp *Exeristes roborator* (F.), was reared on chemically defined diets that contained free fatty acids (mainly linoleic, linolenic, oleic), most of the larvae died in the first instar; however, if the diets contained no free fatty acids but only free amino acids, the larvae developed normally (Thompson, 1977). This suggests that the inadvertent sequestration of these free fatty acids by the host and/or the ability of the host to regulate free fatty acids in the hemolymph to be a major consequence to the survival of the host. House and Barlow (1961) demonstrated that a high amino acid/dextrose ratio in the food of the fly *Agria affinis* (Fall.) was detrimental to the survival of the parasitoid *Aphaereta pallipes* (Say) when compared with diets having a lower amino acid/dextrose ratio. Nutritionally insufficient diets for the host may not necessarily be detrimental for the parasitoid, but instead can enhance survivorship. Along this line, El-Shazly (1972a,b) noted that when the aphid *Neomyzus circumflexus* (Buck.) was reared on suboptimal diets, it was less able to resist parasitization because of a reduced ability to encapsulate the parasitoid.

### 6.1.4. Effects of Plant Natural Products upon Third-Trophic-Level Organisms

The above kinds of observations point to proximate mechanisms as to how plant natural products modulate tritrophic interactions. These observations also confirm the feasibility of attempting to understand in more detail the interrelationship between the uptake of essential nutrients and putative allelochemicals by the host, the effect of this uptake on the quality of the host, and the nature of the subsequent direct and/or indirect effects of these sequestrates upon the fitness of the parasitoid. A knowledge of the effects of natural products upon the fitness of parasitoids is certainly essential if one is to compatibly implement HPR programs against insect pests on the basis of the use of plant antibiotics.

Most of the examples above deal with interactions that are remote from the agricultural situation. Indeed, very few studies have been done that permit one to make predictions about the compatibility of antibiotic plant natural products that are serving as bases of HPR. One of the first clear statements about the potential of such products to affect entomophagous parasitoids adversely centered on the suspicion that nicotine ingested by the hornworms *Manduca sexta* and *M. quinquemaculata* (Howarth) from the foliage of tobacco reduced the percentage of parasitization by the endoparasitic wasp *Apanteles* spp. because of a direct toxification of the parasitoid (Morgan, 1910). More critical studies of this phenomenon were carried out by Gilmore (1938a,b) from which it was found that if larvae of *M. sexta* were reared upon foliage of varieties of tobacco containing nicotine as well as upon weedy species of solanaceous plants lacking in nicotine, the degree of parasitization by *A. congregatus* was much lower in the insects reared upon nicotine-bearing plants than upon the nicotine-lacking plants. These results suggested that nicotine may be the causal agent in reducing the degree of successful parasitism. Many years later, Thurston and Fox (1972) demonstrated that when nicotine was incorporated into an artificial diet at "high" and "low" levels and fed to tobacco hornworm larvae, the survivorship of the parasitoid *A. congregatus* was lower in larvae reared on the "high" nicotine diet.

More recently Barbosa et al. (1982) have reconfirmed the above results, and have also done more detailed experiments that demonstrate that the parasitoid possesses the ability not only to metabolize nicotine acquired from the host's haemolymph to the less toxic cotinine, but also to compartmentalize the nicotine (supposedly a detoxicative process) into its meconium and the casing of the cocoon (Barbosa and Saunders, 1984). Field experiments with specified high- and low-nicotine-containing lines of tobacco showed that the fitness of the above parasitoid was reduced more when its host fed upon the high-nicotine-containing line (Barbosa et al., 1982). Thus, the use of nicotine as a chemical basis of resistance of tobacco against insect attack may prove to compromise adversely the success of BC agents such as *A. congregatus*. Further work needs to be done to assess this potential incompatibility on an agricultural scale.

Another demonstration of a potential incompatibility of a plant natural product with the activity of an endoparasitic wasp concerns the ability of the glycoalkaloid tomatine to poison the larval ichneumonid *Hyposoter exiguae* (Vier.). When tomatine is ingested by larvae of either *Heliothis zea* or *Spodoptera exigua* (Hubner) (Campbell and Duffey, 1979b, 1981; Duffey et al., 1985), the glycoalkaloid, depending upon dose, can significantly affect the development of the parasitoid, and even cause severe developmental abnormalities and/or death. Although such a chemical is considered a candidate for a basis of resistance against noctuid larval pests of the tomato plant (Isman and Duffey, 1982a; Juvik and Stevens, 1982a,b), its ultimate utility in HPR needs to be seriously considered for a number of reasons that go beyond merely its potential toxicity to the above parasitoid.

Although the above two examples clearly demonstrate that antibiotic HPR based on such natural products need not lead to benign effects upon nontarget organisms, these studies are far from providing sufficient information to begin to understand how to use these specific chemicals compatibly, let alone provide a broad enough platform of knowledge from which one can perhaps build valid generalizations about the compatible use of antibiotic natural products as a bases of HPR in any crop plant. Such generalizations in the end may or may not be derivable; nevertheless, at the present more detailed information, at both the basic and applied levels, is required about the chemical bases of resistance of specific crops so that valid concepts and facts can be used predictively to assess the compatibility and implementability of natural products in IPM tactics, even if they only apply to a particular insect-crop system. Our knowledge of how to use natural products in IPM systems is limited (Nordlund et al., 1981; Ahmad, 1983; Bell and Cardé, 1984; Hedin, 1983; Whitehead and Bowers, 1983).

## 6.2. THE SYSTEM

Our attempt to gather such information is based on a basic research approach to understanding if selected antibiotics from the tomato plant *Lycopersicon esculentum* L. can be used to control several pests, the tomato fruitworm *Heliothis zea* and the beet armyworm *Spodoptera exigua*, and yet be compatible with the simultaneous activity of the endoparasitic ichneumonid wasp *Hyposoter exiguae*.

### 6.2.1. General Details about the System

Currently, the control of these insect pests is predominated by the use of synthetic insecticides (Lange and Kishiyama, 1978; Wyman et al., 1979; Lange and Bronson, 1981; Oatman et al., 1983). A limited number of attempts have been made to incorporate biological control agents such as the egg parasitoid *Trichogramma pretiosum* Riley or the bacterial agent *Bacillus thuringiensis* (Burleigh and Farmer, 1978; Lange and Bronson, 1981; Oatman et al., 1983) into IPM programs. The use of HPR in the control of these and other pests of tomatoes is also negligible (Kennedy, 1978; Lange and Bronson, 1981; Kennedy et al., 1983); although, more recently, advances in our knowledge of potential chemical bases of resistance of tomatoes (phenolics, glycoalkaloids, and 2-tridecanone) against noctuid larvae and a variety of other pests has increased greatly (Sinden et al., 1978; Isman and Duffey, 1982a,b; Juvik and Stevens, 1982a,b; Kennedy and Dimock, 1983). What we cannot currently do is utilize these chemical bases of resistance in a predictive and compatibly integrative manner in IPM programs. In part, this lack of knowledge exists because not enough information is known about how and what plant–insect-related factors affect the action of these chemicals against these pests. By the use of our tomato–noctuid–ichneumonid interaction, we will attempt to develop some guidelines to aid in the future deployment of these chemicals as bases of resistance against noctuid larvae.

Some general information about this interactive system is relevant prior to a detailed discussion of how natural products affect the interaction. The females of *H. zea* and *S. exigua* lay their eggs on the young foliage of tomato plants near inflorescences (Nilakhe and Chalfant, 1981; Pencoe and Lynch, 1982; Snodderly and Lambdin, 1982; Burkett et al., 1983). The neonates initiate feeding on the foliage, and then, as a result of a number of undetermined factors, migrate as older larvae to the fruit where toughness and antixenotic factors (Cosenza and Green, 1979; Juvik and Stevens, 1982a,b) regulate entrance into the fruit. The entrance of a larva into the fruit and the insect's subsequent development in the fruit renders the fruit commercially useless. Therefore, factors that restrict or prevent the larval migration or entrance into the fruit can reduce the degree of damage and hence significantly contribute to resistance. One means of increasing resistance of tomato plants to these noctuid larvae is to employ foliar properties or chemicals that interfere with the searching/ovipositional behavior of the female moths and/or prevent normal growth and development of the larva, thereby reducing the rate and incidence of fruit damage. Since we are confining our argument to the use of antibiotic chemicals, let us consider what advantages such chemicals might have as a tactic in IPM programs.

### 6.2.2. Rationale for Antibiotically Based HPR

One objective is to avoid using plant natural products as alternative lethal agents to synthetic insecticides against insect pests. The concept is to use natural products as nonlethal adjuncts to current IPM practices, thereby reducing the chance of resistance against the action of these chemicals developing in the insect population. An effective and somewhat theoretical means of accomplishing this is to rely on plant-produced antibiotics that exert their effect(s) upon the insect in a nonlethal fashion via significantly reducing the rate of larval growth ($\sim 50\%$). Such an inhibitory action has the potential not only to reduce the vigor of the insect so that it is more susceptible to secondarily lethal biotic (e.g., bacteria, viruses, fungi, and parasitoids) and abiotic factors (inclement climate), but also to increase the probability of death from these (a)biotic agents because of the longer periods of time the insect spends in each larval stage. Hence, with the appropriate degree of growth-reducing antibiosis, coupled with the more effective action of the secondarily acting (a)biotic agents, the number of larval insects that are able to develop and migrate to fruit should be reduced.

This approach is also theoretically sound with respect to the population dynamics of the insects, because one of the most effective means of controlling the growth and proliferation of an insect population is to reduce the growth rate, as opposed to strictly increasing mortality or decreasing birth rate (Dahms, 1969; MacKenzie, 1980; Gould, 1983, 1984; Schultz, 1983a). Another theoretical advantage of such an approach is that the chance of resistance developing in the insect population against the antibiotic factor(s) is greatly reduced because the selection pressure on the insect population is primarily arising from the supposed simultaneous action of the above (a)biotic factors and not directly from the nonlethal antibiotic. The use of antibiosis in this fashion is consistent with increasing the durability or permanence of resistance (Buddenhagen and de Ponti, 1983; Gould, 1983, 1984). However, this line of reasoning is dependent upon the assumption that the joint action of antibiosis and BC agents do not synergize the evolution of resistance. It is possible for the opposite to happen; Gould (1986a,b) presents arguments via genetic modules that the evolution of resistance in the insect population can be enhanced under the influence of multiple selective factors. This acceleration of development of resistance is most likely to occur, for example, where antibiotic factors and parasitoids exert strong selective forces upon one or several major insect genes. Such acceleration is less likely to occur when these factors act upon multiple unrelated gene systems and their products. Obviously, a much more detailed body of knowledge is needed about the genetic responses of insects to multiple selective forces such as those used in IPM programs (i.e., insecticides, pathogens, plant chemicals) before the durability and compatibility of resistance can be accurately assessed.

## 6.3. THE TRITROPHIC INTERACTION

### 6.3.1. Potential Chemical Bases of HPR

On the basis of the above type of argument, the use of such antibiosis for the integration of HPR into IPM programs seems like an advantageous form of pest control, particularly since there is a tendency for HPR to be considered generally compatible with the action of BC agents and other modes of pest control (Kogan, 1975; Bergman and Tingey, 1979; Adkisson and Dyck, 1980). The question is then, are there chemicals in tomato plants that have the ability to serve as antibiotic bases of resistance against our candidate noctuid larvae, as well as to be compatible with the action of BC agents?

The tomato plant contains a number of natural products that have the ability to serve as growth reducing-antibiotics against both the tomato fruitworm (*Heliothis zea*) and the beet armyworm (*Spodoptera exigua*). These chemicals (Fig. 6.1) include several catecholic phenolics (rutin and chlorogenic acid), the glycoalkaloid tomatine, and proteinase inhibitors, all of which to varying degrees of experimental analysis are thought to have the potential to serve as bases of resistance (Green and Ryan, 1972; Campbell and Duffey, 1979b, 1981; Elliger et al., 1980, 1981; Duffey and Isman, 1981; Waiss et al., 1981; Isman and Duffey, 1982a,b; Juvik and Stevens, 1982a,b; Shukle and Murdock, 1983; Broadway et al., 1986).

To understand how these chemicals can be used as bases of resistance against noctuid larvae and yet be compatible with the simultaneous use of BC agents, it is first necessary to calibrate the response of the unparasitized insect to these agents. For simplicity, we will restrict ourselves to a discussion of the biological effects of phenolics and tomatine. Accordingly, it has been observed, via dose-response analyses using artificial diet, that the levels of catecholic phenolics (primarily rutin and chlorogenic acid) and tomatine required to inhibit the growth

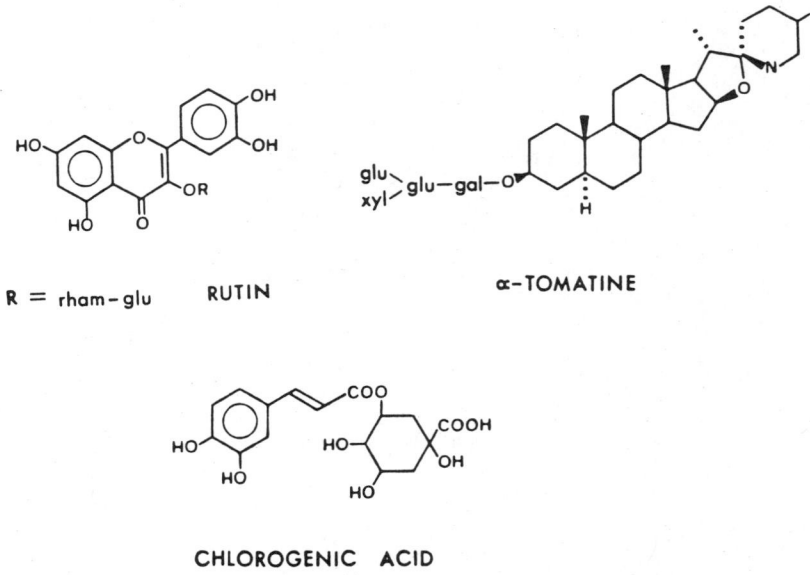

R = rham–glu     RUTIN                    α–TOMATINE

CHLOROGENIC ACID

Fig. 6.1 Chemical structures of some antibiotics in the foliage of the tomato plant *Lycopersicon esculentum*: gal = galactose; glu = glucose; rham = rhamnose; xyl = xylose.

of either larval noctuid species (Fig. 6.2) by 50% is approximately equivalent to the average levels found in tomato foliage near inflorescences. Tomatine is more toxic to both species than rutin or chlorogenic acid, and the larvae of *S. exigua* are several times more sensitive to these chemicals than are those of *H. zea*. There are significant differences in the average foliar content of tomatine and rutin between varieties and/or species of tomato plants to support the contention that a breeding program could select for plants with elevated levels of one or both of these types of chemicals to enhance the degree of antibiosis (Isman and Duffey, 1982b; Juvik and Stevens, 1982a,b; Juvik et al., 1982; Duffey, 1985). On the basis of this limited amount of toxicological data, is it reasonable to suppose initially that one could simply proceed to breed for resistance based on such high chemical-content lines of tomato plants that would on the average, say, effect 75% inhibition of larval growth (an $ED_{75}$)?

It is theoretically possible that the use of an $ED_{75}$ level of tomatine or rutin in an HPR program against *H. zea* would place too great a direct chemically based

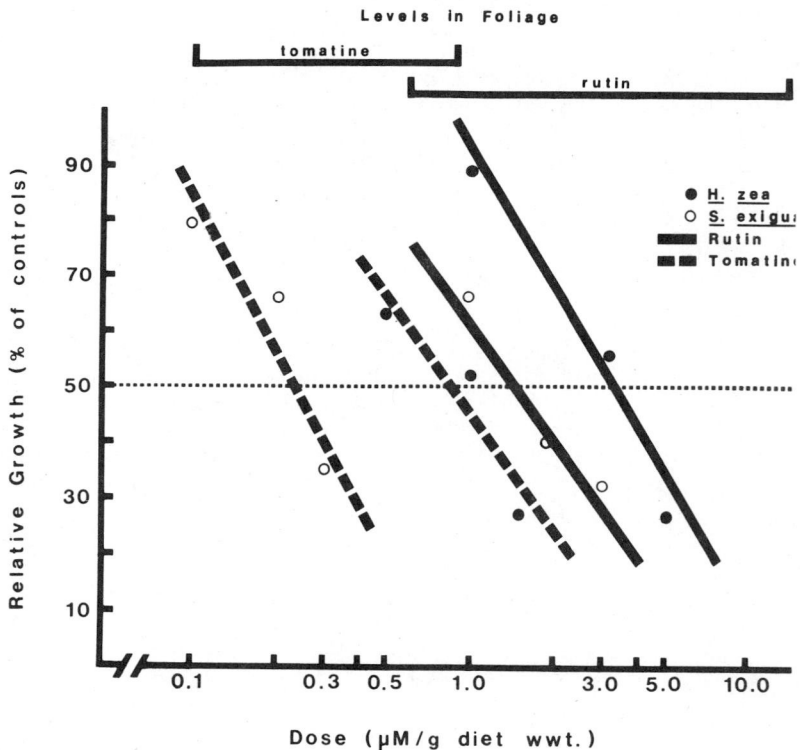

Fig. 6.2 Dose–response relationships showing the relative abilities of rutin and tomatine in artificial diet to reduce the growth of larval *Heliothis zea* and *Spodoptera exigua*. Response to chemicals determined in artificial diet containing 2.4% casein. Relative growth is determined by comparing growth to controls after 10 days of feeding from the neonate stage.

selective force upon the larvae of both *H. zea* and *S. exigua*, particularly the more sensitive larval *S. exigua*, potentially resulting in the development of resistance in the populations to these antibiotics (see Fig. 6.2). The basis of selection for resistance could be upon genes that regulate metabolism, growth, development, and/or behavior in fashions that would confer differential survival to specific genotypes. The potential for the development of such resistance is unknown, since we do not currently understand the genetic basis of the response of these insect species to these chemicals, let alone understand what selective pressure these chemicals exert in field conditions upon the short- and long-term responses of the population. Hence, an important decision in entering a breeding program for elevating chemical content would be to establish what levels of what chemicals can be used most effectively and permanently to suppress populations of key pests.

However, what is not apparent is whether such levels of either chemical would, in fact, be operative in the plant against these pests in the fashion or to the degree predicted from laboratory toxicological studies. It is most often assumed that the action of such antibiotics is specificable by knowing their levels in the appropriate portions of the plant without regard to other plant chemicals that the insect coingests. In the tomato system, we have reason to expect that the operativity will not be predictable because other plant chemicals have the potential to modify the toxicity of rutin and tomatine. In other words, when these larvae eat tomato foliage they do not merely ingest a simple mixture of rutin and tomatine. The insect is also ingesting a variety of other plant natural products such as proteins, phytosterols, cellulose, amino acids, and sugars. The levels of several of these chemicals are shown in Fig. 6.3. Our discussion will be confined to the effects arising from the coingestion of proteins and phytosterols.

### 6.3.2. Interaction of Protein with Phenolics

In light of ecological theory on the role of nitrogen as a major factor in regulating the fitness of an individual and the dynamics of a population of insects and their parasitoids (Mattson, 1980; Faeth et al., 1981; Price et al., 1980; Prestidge, 1982; Stiling et al., 1982; Al-Zubaidi et al., 1983; Price, 1983; Whitham, 1983; Strong et al., 1984; White, 1984; Al-Zubaidi and Capinera, 1984; Hillard and Keeley, 1984; Brewer et al., 1985), it was necessary to establish first how different qualities and quantities of protein affected the growth of unparasitized insects, prior to assessing the impact of these chemicals upon the parasitoid *Hyposoter exiguae*. Likewise, it seemed essential to determine if there was any interaction between phenolics and dietary proteins in affecting the growth of unparasitized hosts, considering the information (1) on the ability of various phenolics to bind (non)covalently to protein (van Sumere et al., 1975; Davies et al., 1978; Singleton, 1981; Pierpoint, 1982, 1983; Barbeau and Kinsella, 1983; Igarashi et al., 1983; Telek and Graham, 1983; Matheis and Whitaker, 1984), and (2) the information demonstrating that phenolics, particularly tannins, have the potential to reduce the digestibility and/or utilization of dietary protein for a

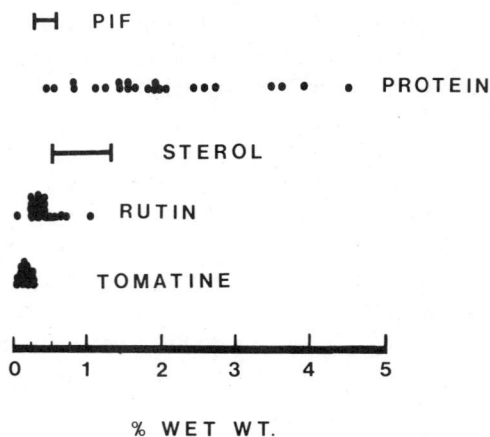

Fig. 6.3 Levels of various natural products within the leaflets of an average plant of a commercial variety of the tomato plant *Lycopersicon esculentum*. PIF = proteinase inhibitors I and II; protein = equals bulk foliar protein; sterol = total foliar sterol content after saponification; rutin = total catecholic phenolic content; tomatine = total tomatine content. Values were derived by estimating chemical content of individual leaflets from single plants of Ace 55 or Castlemart varieties. Bars represent unreplicated determinations. Points represent sample distribution from replicated determinations.

variety of animals including insects (Feeny, 1968, 1975; Horigome and Kandatsu, 1968; Fox and Macauley, 1977; Chan et al., 1978; McKey et al., 1978; Bernays, 1981; Fox, 1981; Hurrell et al., 1982; Martin and Martin, 1982; Reese et al., 1982; Ishii et al., 1983; Rhoades, 1983; Telek and Graham, 1983; Zucker, 1983).

To approximate how plant protein might affect the growth of *H. zea* and *S. exigua*, we assessed the ability of these two larval species to grow on purified soy protein or casein (the protein source in many insect artificial diets) as the major dietary proteins. In the foliage of a given tomato plant during the flowering and setting of fruit, protein levels range between approximately 0.6 and 8.0% wet weight; the nutritive properties of the two model proteins were assessed within this range (Fig. 6.4) by using six concentrations in artificial diet (0.0, 0.2, 0.6, 2.4, 4.8, and 8.0%; at 0.0% added protein the diet contains ~0.7% soluble protein). If these two proteins were equally nutritious to a given species at all paired test concentrations, one would expect, by plotting the relative growth on one protein against the relative growth on the other, a straight line relationship (with a slope = 1; the straight line in Fig. 6.4). However, it is found that the two proteins are not equivalently nutritious over the range of concentrations tested. Maximal growth for both species of larvae occurs between about 2.4 and 4.8% of either protein, although casein is more nutritious than soy protein to both species at these two concentrations (displaced to left of line showing equivalence; Fig. 6.4). At concentrations below 2.4%, growth of both larval species is signifi-

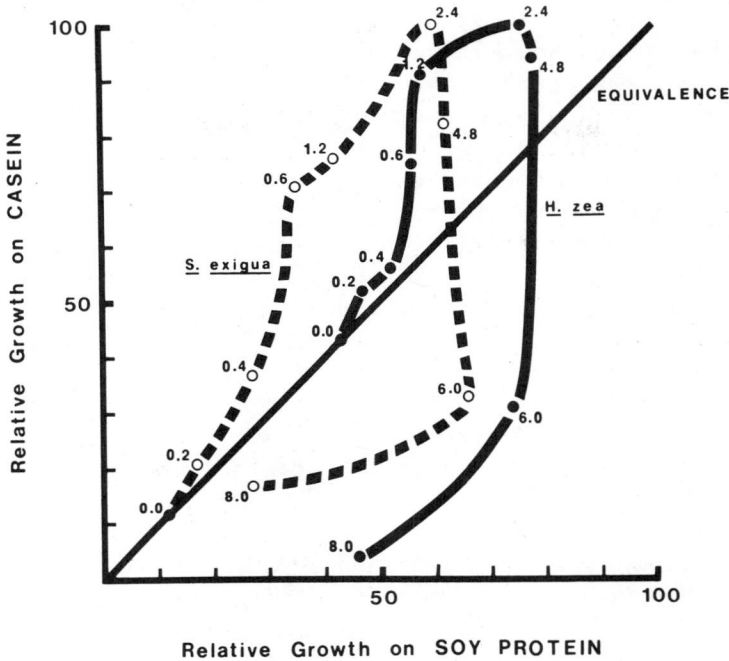

Fig. 6.4 The relative ability of different regimes of dietary protein to support the growth of the two noctuid larvae, *Heliothis zea* and *Spodoptera exigua*. The percentage of protein in diet is on a wet weight basis. Larval growth on the various protein regimes is relative to the growth, after 10 days from the neonate stage, on a diet with 2.4% wet weight of casein (the amount in standard synthetic diet and that supports maximal growth of the two larval species). On standard diet, larvae of *H. zea* and *S. exigua* reach average weights of ~305 and 175 mg, respectively. At about 0.2% and 0.5% casein the growth of both larval species is reduced by 50%. The solid lines indicate theoretical nutritional equivalence of both proteins to either larval species. Displacement of points to the left of the line of equivalence indicates casein is more nutritious; correspondingly, to the right soy protein is more nutritious.

cantly reduced, although casein still remains the most nutritious protein of the two. At concentrations above 2.4%, casein becomes relatively "toxic" to both larval species, as evidenced by the severe inhibition of growth at 8.0% (points displaced to right of line of equivalence; Fig. 6.4); soy protein is much less "toxic" at high levels (8.0%) because it supports a relatively greater level of growth.

The differential ability of these two proteins to support larval growth is not surprising, since it had been previously established by Brewer and King (1979) that different proteins were not equally nutritious to larval *H. zea*, and the growth of *S. exigua* on foliage is strongly influenced by foliar nitrogen (Al-Zubaidi et al., 1983, 1984). However, what these results do point to is a need to understand in more detail how plant protein or total nitrogen affects such larvae. Are the plant proteins eaten by polyphagous insects such as *H. zea* and *S.*

*exigua* generally similar to either of our model proteins in their ability to support larval growth? Even though casein is a milk protein, perhaps it can serve as a valid model protein. Certainly, many studies on the effects of natural plant products upon herbivorous insects have relied upon the use of this protein. Currently, we are investigating the ability of total foliar tomato protein as well as other plant proteins (zein and glutein) to support the growth of these two larval species. As one might expect, these proteins are not equivalently nutritious (casein > soy protein > tomato protein > glutein > zein). Yet, what is more interesting is that the toxicity of catecholic phenolics, as well as proteinase inhibitors, varies with the quality and quantity of each of these proteins (S. S. Duffey et al., unpublished data). Moreover, unless the optimal level of protein is present, the insect's growth can be significantly restricted by either a deficit or surfeit of a given protein. It follows then that the level and type of protein in the plant may be a critical factor determining the toxicity of many phytochemicals, and as a consequence jointly have a profound effect upon the fitness of an insect.

In a theorteical vein, our studies provide evidence that it may be insufficient to talk in general terms (based on total $N$ analysis) about high or low plant protein levels favoring the fitness of insects, unless quality, quantity, and interactivity with other natural products are more carefully specified. In a more practical sense, it may be feasible to consider breeding for either high- or low-foliar-protein-content lines of plants to facilitate antibiosis against these two species of larval noctuids.

Since our original interest was in using natural products such as phenolics and tomatine as basis of resistance, it would be useful to understand how the antibiotic action of these chemicals is affected by different protein regimes. Our results clearly demonstrate that there is a strong interaction between the dietary protein and the toxicity ($=$ growth-reducing properties) of the phenolic rutin to both larval species (Fig. 6.5). Consider *H. zea* first, where it can be noted (Fig. 6.5) that at 2.4% dietary casein the $ED_{50}$ value of rutin is 3.5 $\mu$mol/g of diet (on the $X$ axis), and that growth is optimal (100% on the $Y$ axis) if rutin is not present in the diet. When the level of dietary casein is raised to 4.8% (still a level of protein that provides optimal growth), rutin becomes more toxic because the $ED_{50}$ decreases to 2.0. In contrast, at low-level dietary protein (0.6%), the toxicity of rutin is decreased to an $ED_{50}$ value of 4.0. A similar effect is also observed in *S. exigua*, which is innately more sensitive to rutin, when the levels of casein are altered. Thus, for both larval species, the toxicity of rutin can vary up to fivefold, depending upon the level of protein in the diet. If one extrapolates these results to the real world, then it may only be possible to specify the toxicity of phenolics such as rutin to these larvae when one has specified the level of protein in the foliage that is ingested.

Protein quality is also an important factor that dictates the toxicity of rutin to these noctuid larvae. Again, consider *H. zea* first. When larval *H. zea* are reared on the less nutritious soy protein, the maximal level of growth obtained is 60% of that obtained on casein (Fig. 6.5 and $Y$ axis in Fig. 6.4). At 4.8% soy protein the $ED_{50}$ value of rutin is 3.6, approximately 1.5 times more than that observed on

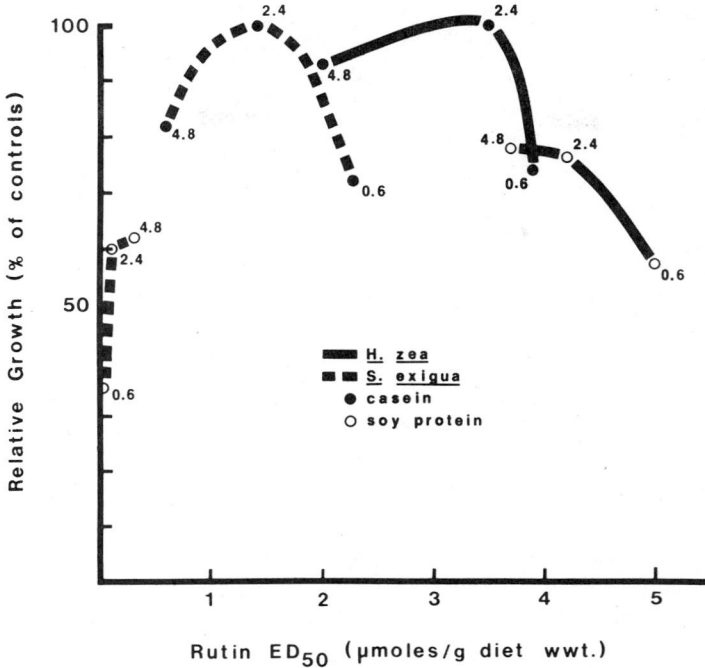

Fig. 6.5 The effect of different regimes of dietary protein on the $ED_{50}$ value of rutin to larval *Heliothis zea* and *Spodoptera exigua*. The *Y* axis indicates the relative growth of the two larval species at the various protein regimes, with respect to 2.4% casein, without the addition of rutin (see Fig. 6.4). The *X* axis indicates the $ED_{50}$ value of rutin when larvae are grown on a given protein regime.

4.8% casein or equivalent to that obtained on 2.4% casein. Thus, rutin is less toxic to larval *H. zea* when soy protein, rather than casein, is the major source of protein. The modulating effect of soy protein on the toxicity of rutin to *S. exigua* is more complicated because *S. exigua* grows only poorly on soy protein (Fig. 6.4). At 0.6% soy protein larval growth is already reduced by greater than 50% so the $ED_{50}$ of rutin is "0." Only at levels above 2.4%, where growth is more than 50% of the controls, is it possible to obtain $ED_{50}$ values for rutin. Thus, the apparent decrease in rutin toxicity associated with increasing soy protein levels (Fig. 6.5) arises from increased growth at increased levels of protein, rather than from an interaction between the protein and rutin. Additionally, studies not described here demonstrate that the toxicity of chlorogenic acid to both species of larvae is similarly affected by the level and type of dietary protein. The toxicity of tomatine is influenced to a much lesser degree by dietary protein.

If one extrapolates these results to the real world, then it may not be possible to specify the antibiotic potential of phenolics like rutin unless one has specified not only the absolute level of plant protein, but also the quality of protein and the herbivore ingesting the mixture. High protein levels in plants have been theorized to benefit herbivorous insects (Mattson, 1980; Price et al., 1980; Price,

1983; Strong et al., 1984; White, 1984). This may be the case if such high levels of protein do not accentuate and/or perhaps alleviate the toxicity of certain plant allelochemicals. However, on the basis of our noctuid–tomato plant interaction, it can be theorized that high plant protein could have a defensive value if it were toxic (as casein) and/or were able to accentuate *in planta* the toxicity of putative defensive agents such as rutin. Our results suggest that the quantity and quality of protein can play a pivotal role in dictating the toxicity of certain plant defenses to herbivorous insects. If such "nutrient–toxin" interactions occur as a general phenomenon in plant–insect interactions, the development of universal theories about the defensive value of plant natural products against insects may be premature until a better biochemical–physiological–ecological understanding of these interactions is gained.

From a more applied viewpoint, the use of catecholic phenolics as bases of resistance of tomato plants against *H. zea* and *S. exigua* may involve more than just breeding for higher levels of phenolics. It may be possible to breed for accentuated resistance by breeding for both high phenolic and foliar and/or fruit protein. However, to use phenolics in a predictive fashion against such pests more information is needed about the mechanism of accentuation of toxicity.

### 6.3.3. Interaction of Tomatine with Phytosterols

Likewise, the independent action of tomatine *in planta* needs to be seriously questioned. Tomatine is a more potent ($\sim 4\times$) growth reducer of larval *H. zea* and *S. exigua* than is rutin in an artificial diet (Elliger et al., 1981; Isman and Duffey, 1982b; Juvik and Stevens, 1982a; Fig. 6.2 and Table 6.1). Interestingly, Juvik and Stevens (1982b) found a strong correlation between total foliar and fruit tomatine content and resistance to larval *S. exigua* but not to *H. zea*. Larval *S. exigua* are much more sensitive to poisoning by tomatine than is *H. zea* (Fig. 6.2). Why then is there no correlation for *S. exigua*? One can derive a partial explanation for this in a roundabout way, if one considers the ability of tomatine to interact with phytosterols.

Tomatine, as well as a number of other plant saponins, are able to form insoluble and/or biologically inactive complexes with certain phytosterols, for example, cholesterol, stigmasterol, and sitosterol (Roddick, 1974, 1979, 1980; Birk and Peri, 1980; Mahadevan, 1982). In fact, cholesterol is often added to soymeal to alleviate the toxicity of soy saponins to livestock (Birk and Peri, 1980), and it has been demonstrated that cholesterol added to an artificial diet will also alleviate the toxic effects of soybean and alfalfa saponins upon the growth and development of seed eating beetles (Gestetner et al., 1970; Shany et al., 1970a,b; Birk and Peri, 1980). Campbell and Duffey (1979b, 1981) were able to demonstrate that equimolar levels of cholesterol, cholesterol oleate, or a sterol fraction isolated from tomato foliage were each able to nullify completely the toxicity of tomatine to larval *H. zea* fed on artificial diet. Moreover, it was found that the foliage of tomato plants contained two to seven times more phytosterols than tomatine. Additional studies have shown that sterols, such as cholesterol and

**Table 6.1 The Ability of Certain Sterols to Alleviate the Toxicity of an $ED_{50}$ Dose of Tomatine in Laval *Heliothis zea* and *Spodoptera exigua***

| Larval species | $ED_{50}$ Level of Tomatine ($\mu$mol/g diet)[a] | Percent Alleviation of Toxicity at Equimolar Level of Sterol[b] | | | | | |
|---|---|---|---|---|---|---|---|
| | | CHOL | CHOLOL | SIT | STIG | TMD | LAN |
| *Heliothis zea* | 0.9 | 100 | 100 | 88 | 84 | 85 | 0 |
| *Spodoptera exigua* | 0.2 | 100 | ? | 96 | 62 | ? | ? |

[a] $\mu$mol/g based on wet weight of diet with 2.4% casein.

[b] CHOL = Cholesterol; CHOLOL = cholesterol oleate; SIT = sitosterol; STIG = stigmasterol; TMD = tomatidine; LAN = lanosterol.

sitosterol, are also able to alleviate the toxicity of tomatine to larval *S. exigua* and *H. zea*. Yet, not all phytosterols are equivalent in their ability to alleviate tomatine-induced toxicity. For example, cholesterol is more alleviatory than stigmasterol or ergosterol in *H. zea*, but stigmasterol is less effective in *S. exigua* than in *H. zea* (K. C. Kelley, K. A. Bloem, and S. S. Duffey, unpublished data; Table 6.1).

Thus, the potential is there, when an insect such as *H. zea* or *S. exigua* is feeding upon tomato foliage, to have the antibiotic effects resulting from the ingestion of tomatine completely quenched because of the coningestion of greater levels of specific phytosterols. This alleviatory phenomenon can be invoked to partially explain Juvik and Stevens' (1982a) results. There may be no correlation between tomatine content and resistance against *S. exigua* because the activity of tomatine was completely quenched by the alleviatory action of phytosterols; the correlation for *H. zea* may perhaps be based on the presence of some other character or phenomenon associated with the presence of tomatine but not the direct result of tomatine-induced toxicity.

Although some phytosterols may function as alleviatory agents against saponins such as tomatine, it is worthwhile to consider that specific types of phytosterols (e.g., cholesterol, sitosterol) are essential nutrients for these insects, the sterols can also be toxic in inappropriately high doses (e.g., cholesterol; Ritter and Nes, 1981a,b). Hence, as pointed out by Kogan (Chapter 5), there is no distinct boundary between a kairomone and an allomone, or a nutrient and a toxin.

An important consideration about the use of natural products in HPR programs arises from the above information on tomatine–sterol interactions. If one were to know merely that tomatine was an effective reducer of larval growth based on studies with artificial diet, one might recommend that a breeding program be begun to produce tomato plants with high levels of tomatine. Theoretically, this approach may be doomed to failure, because one has not accounted for the presence of alleviatory phytosterols. The plants with high levels of tomatine might, therefore, be found to be not resistant to these noctuids. Hence, a breeding program designed to utilize tomatine to control larval *H. zea* and *S. exigua* would have to consider two groups of chemicals rather than one, in which one might attempt to breed for high tomatine content and correspondingly low phytosterol content. Because of our lack of knowledge about the genetic basis of control of phytosterol biosynthesis and the physiological role of phytosterols in the plant, this may be easier said than done. Nevertheless, to complete the scenario, we find that our high-tomatine–low-phytosterol plants are resistant to the insect pests, but more susceptible to attack by fungal pathogens such as *Phytophthora infestans* and *Pythium* species. Why is this the case?

Work on the antibiotic action of tomatine against certain pathogenic fungi has demonstrated that high levels of phytosterols are essential for promoting the entrance of tomatine into fungal hyphae (Mahadevan, 1982; Callow, 1983; Defago and Kern, 1983; Defago et al., 1983; Nes et al., 1983). This apparent paradox demonstrates the value of basic research in facilitating more applied research and also indicates that our ability to utilize tomatine predictively as a basis of resistance against both insects and pathogens will be, to an important degree, dependent upon our biochemical and ecological understanding of the biological consequences of phytosterol–tomatine interactions.

### 6.3.4. Compatibility of Tomato Plant Natural Products with Parasitoids

Our major goal is to understand how HPR, particularly based on the use of anti-biotic natural products, can be used compatibly with a biological control agent such as parasitoids. Since we have calibrated the response of the two noctuid hosts to our candidate antibiotics, we can now make a first-order assessment of whether these antibiotics might be compatible with the use of the endoparasitic ichneumonid wasp *Hyposoter exiguae* to control larval *H. zea* and *S. exigua*.

#### 6.3.4.1. General Aspects of Action of Parasitoid on Host
The endoparasitic wasp *H. exiguae* is most effective at ovipositing in the early instars of both above species of noctuids (Puttler, 1961; Jowyk and Smilowitz, 1978; Campbell and Duffey, 1979a; Browning and Oatman, 1981). The para-sitoid deposits only one egg in the hemolymph of its host. Once the host is parasi-tized (1st or 2nd instar), there is dramatic reduction in the growth rate of the host compared with unparasitized larvae because of the parasitoid's feeding upon the host's tissue. In the first three instars, the larval parasitoid feeds upon the host's hemolymph, but during the remaining instars primarily feeds upon the fat body. This internal feeding so severely restricts the growth of the host that an unparasitized larva will grow to approximately 500 mg and 250 mg prior to pupation (*H. zea* and *S. exigua*, respectively); yet, when parasitized in the sec-ond instar, neither host will reach more than 25 mg. Finally, just prior to the parasitoid's pupation, when it has grown to a size that occupies almost the total body cavity of the host, it engulfs the remaining body of its host (Puttler, 1961; Jowyk and Smilowtiz, 1978; Thompson, 1982). This parasitoid–host interaction provides an excellent opportunity to learn how the uptake of "nutrients" (casein and soy protein) and/or "toxins" (rutin and tomatine) by the host affect the fitness of the parasitoid. We make the assumption that these effects on fitness will be an approximate index of potential incompatibility of HPR and BC in the field.

#### 6.3.4.2. Independent Effects of Nutrients and Toxins on Parasitoid
The approach we have taken to assessing how these natural products affect the parasitoid is to compare the ability of *H. exiguae* to grow, pupate, eclode to an adult, and survive under dietary regimes that inhibit the growth of the two hosts by 50% (the $ED_{50}$ level) when in the unparasitized state. Since we have previ-ously calibrated the response of the unparasitized hosts to rutin, tomatine, and protein, the appropriate levels are defined (Figs. 6.2 and 6.4). For example, with *H. zea* as the host on an optimal protein diet (casein at 2.4% net weight), 3.5 $\mu$mol/g of rutin or 0.9 $\mu$mol/g of tomatine will reduce larval growth by 50% (see

Table 6.2). On the other hand, one can use 0.2 or 6.0% casein, without "toxins" to reduce larval growth by 50%. This approach provides a means to assess independently the effects of "nutrients" and "toxins" upon the development of the parasitoid as mediated by a host relatively less sensitive to these "toxins." Correspondingly, the more sensitive host *S. exigua*, on an optimal diet of 2.4% casein, can have its growth reduced by 50% via the inclusion of 1.5 $\mu$mol/g of rutin, or 0.2 $\mu$mol/g of tomatine in the diet. Similarly, utilizing only casein as a variable, growth can be reduced 50% by rearing the larvae on 0.5 or 5.0% casein (Table 6.2). Also, since we know that the level of casein in the diet has a great hearing upon the toxicity of rutin (altered $ED_{50}$ values; Fig. 6.5), we can assess the effect of the interaction of protein and rutin (or tomatine) upon the parasitoid by varying the level of protein about the above $ED_{50}$ values of the "toxins" (Table 6.2).

Consider the effects of these types of dietary regimes upon *H. exiguae* if the $ED_{50}$ levels for *S. exigua* are used. If one feeds too little (0.5%) or too much casein (5.0%), the larval period of the parasitoid is extended significantly longer in both hosts, but to a greater extent in *S. exigua* (3–4 days), which is generally more sensitive to the quantity and quality of dietary protein. Although these regimens extend the period of larval development, they do not have any overt effects upon any other developmental parameters such as length of pupal period, survivorship of pupae, percent eclosion to adult, adult weight, or adult longevity. When one introduces rutin (at 1.5 $\mu$mol/g), it is only at the high protein regime (4.8%) that any additional antibiotic effect is observed in the parasitoid. The antibiotic effect is a further delay in the larval period ($\sim$5 days), particularly in the more rutin-sensitive host *S. exigua* (+++++ vs. +++), and only in *S. exigua* is the percent eclosion to adults and adult weight reduced (+ vs. 0). These effects are consistent with the ability of high levels of casein to enhance the toxicity of rutin to an unparasitized host (Fig. 6.5). If one adds tomatine (at 0.2 $\mu$mol/g) to the diet instead of rutin, one observes somewhat similar effects upon the parasitoid to that seen for rutin. Tomatine significantly increases the larval period of the parasitoid to a greater degree in the most tomatine-sensitive host, *S. exigua*, and more so at levels of protein that alone would reduce the growth of the unparasitized host (0.6 and 4.8%). Tomatine also, at high and low protein regimes, causes a reduction in the adult weight of the parasitoid in *S. exigua* but not in *H. zea*. When equimolar levels of cholesterol or cholesterol oleate are added to a tomatine-containing diet, the effects of tomatine upon both hosts and the parasitoid are completely eliminated.

As seen above, the effect of these dietary regimes upon the parasitoid are minimal, and hence would likely translate into the compatible use of these chemicals with BC, although the antibiotic effect of these levels of chemicals upon unparasitized *H. zea* would be minimal because of the larval relative insensitivity. However, when we examine the effects of $ED_{50}$ levels of these chemicals for *H. zea*, much more dramatic effects are seen upon the parasitoid. Again, consider the effects of $ED_{50}$ levels of protein alone upon *S. exigua*. At both low and high protein regimes (0.2 or 6.0% protein; Table 6.2), the parasitoid grows even more slowly than it did under the $ED_{50}$ regimes for *S. exigua*. This effect is most prom-

**Table 6.2 Some Effects of Dietary Protein and Various "Nutrient" and "Toxic" Dietary Regimes Upon Certain Fitness Parameters of the Parasitoid *Hyposoter exiguae* Reared in Either *Spodoptera exigua* or *Heliothis zea***

| Dietary Conditions of Host | Larval Period[a] | | Pupal Period[a] | | Adult Period[a] | |
|---|---|---|---|---|---|---|
| | Sp. | Hz. | Sp. | Hz. | Sp. | Hz. |
| ED$_{50}$ values[b] for *S. exigua* | | | | | | |
| Protein low (0.5%) | +++[c] | + | 0[a] | 0 | 0 | 0 |
| Protein high (5.0%) | +++ | ++ | 0 | 0 | 0 | |
| Rutin (1.5 μmol/g wwt) | | | | | | |
| Low protein (0.6%) | +++ | ++ | 0 | 0 | 0 | 0 |
| Medium protein (2.4%) | +++ | ++ | 0 | 0 | 0 | 0 |
| High protein (4.8%) | +++++ | +++ | + | 0 | + | 0 |
| Tomatine (0.2 μmol/g wwt) | | | | | | |
| Low protein (0.6%) | ++++ | + | 0 | 0 | + | 0 |
| Medium protein (2.4%) | +++ | + | 0 | 0 | 0 | 0 |
| High protein (4.8%) | ++++ | ++ | 0 | 0 | + | 0 |
| Cholesterol (eq. molar) | 0 | 0 | 0 | 0 | 0 | 0 |
| ED$_{50}$ values for *H. zea* | | | | | | |
| Protein low (0.2%) | +++++ | +++ | ++ | 0 | + | 0 |
| Protein high (6.0%) | ++++ | +++ | 0 | 0 | + | 0 |
| Rutin (3.5 μmol/g wwt) | | | | | | |
| Low protein (0.6%) | Die[a] | +++ | Dead[a] | 0 | Dead | 0 |
| Medium protein (2.4%) | +++++ | +++ | ++ | 0 | + | 0 |
| High protein (4.8%) | Die | +++++ | Dead | + | Dead | + |
| Tomatine (0.9 μmol/g wwt) | | | | | | |
| Low protein (0.6%) | Die | ++++ | Dead | ++++ | Dead | ++++ |
| Medium protein (2.4%) | Die | +++ | Dead | +++ | Dead | +++ |
| High protein (4.8%) | Die | ++++ | Dead | ++++ | Dead | ++++ |
| Cholesterol (eq. molar) | + | 0 | 0 | 0 | 0 | 0 |

[a] Sp. = *Spodoptera exigua* as host and Hz. = *Heliothis zea* as host; 0 = no effect; Die = parasitoid dies prior to its pupation; Dead = parasitoid dies before reaching that stage. The normal larval period of the parasitoid is about 7 days. The normal pupal period is about 7 days. Normal pupal eclosion is about 87%. Adult weight of parasitoid is about 2.5 mg. Normal adult longevity is about 21 days.

[b] The ED$_{50}$ value specifies the dose of protein (= casein), rutin, and/or tomatine required to reduce the growth of an unparasitized species of larvae by 50% relative to the control (2.4% casein) which supports maximal growth of (un)parasitized host.

[c] The relative scale of + to +++++ indicates an increasing degree of interference with normal fitness in terms of parameters indicated relative to parasitized control. For specific effects of tomatine see Campbell and Duffey (1981, 1979b).

159

inent when *S. exigua* is the host on low protein $(+++++)$; this results from the fact that the growth of even unparasitized larval *S. exigua* on 0.2% casein is minimal (75% inhibition). In comparison, unparasitized larvae of *H. zea* grow better on 0.2% casein (50% reduction in growth) and hence in these dietary conditions serve as a more nutritious host. It is also worth noting that when the parasitoid is grown in *S. exigua* under the low-protein regime a decrease in the number of pupae successfully eclosing to the adult, and a decrease in the adult weight are observed.

### 6.3.4.3. Interactive Effects of Nutrients and Toxins upon the Parasitoid

When the additional stress of an $ED_{50}$ level of rutin (3.5 $\mu$mol/g) is added, it is observed that in the most rutin-sensitive host, *S. exigua*, at both low and high protein regimes, the parasitoids die in the larval stage (host also dies). Death occurs because the combined effects of the nutrient–toxin load and the parasitoid load prevent the larva from ever reaching a critical weight of 25 mg; this critical weight must be reached for the parasitoid to complete its development. In *H. zea*, although the growth rate of the larval parasitoid is reduced, particularly at the high protein regime, little overt effects are observed upon the further development of the parasitoid. Similarly, when an $ED_{50}$ level of tomatine is added, the parasitoids die in the larval stage at all protein regimes for the same reasons given above if the host is the more tomatine-sensitive *S. exigua*. When the host is *H. zea*, the growth of the parasitoid is reduced, fewer pupae eclode to the adult, adult weight is substantially reduced, and longevity is decreased. Developmental abnormalities in the ovipositor and antennae are also found in a significant portion of the parasitoid population arising from such a toxified larval host (Campbell and Duffey, 1981). However, when cholesterol is added at equimolar levels, the detrimental effects of tomatine to the parasitoid in both hosts are alleviated.

How these independent and interdependent effects of "nutrients" and "toxins" upon the parasitoid are mediated is unknown. A more detailed investigation of such phenomenon is very likely to be a fruitful and prominent future area of ecological-physiological research. An in-depth discussion into the possible physiological mechanisms of these chemical–parasitoid–host interactions is beyond the scope of this paper. However, a brief overview of the potential routes and modes of interaction between the host and the parasitoid will serve both to summarize and forecast.

Several modes of toxification and/or nutrification of the parasitoid are possible (Fig. 6.6). The most obvious mode/route of action is what we will specify as a direct effect upon the parasitoid. One can then postulate that rutin and tomatine are rapidly and effectively sequestered by the larval host (see Duffey, 1980) into the hemolymph where the actively feeding parasitoid acquires them by direct ingestion. Hence, as a result of continued feeding upon "poisoned" hemolymph, the parasitoid gradually acquires a toxic dose of either or both chemicals. This acquisition is not dependent upon any metabolic or physiological idiosyncrasy of the host, but put purely upon the nature of the parasitoid. Rather than acquisi-

INDIRECT                                                                            DIRECT

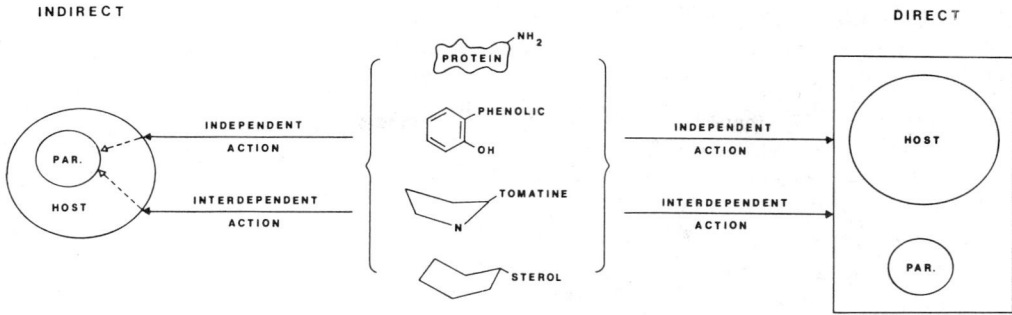

Fig. 6.6 Diagrammatic representation of the effect of various phytochemicals upon host–parasitoid relationship. Direct effects are those in which the action of the chemicals upon the parasitoid (= PARA) are not mediated by the host; indirect effects are those mediated by the host. *Independent* effects of chemicals upon the parasitoid imply that the action of a given chemical is not dependent upon the presence of other plant products; *dependent* effects imply joint or synergistic action of plant products against the parasitoid.

tion by gradual accumulation, it is possible, because the parasitoid engulfs its host before pupation, that a significant portion of the toxic dose of tomatine and/or rutin is acquired at this stage. Likewise, for protein, one can postulate that the parasitoid is able to ingest amino acids from the hemolymph as a result of the host's digestion. The ability of the parasitoid to grow effectively is dependent solely upon the nutritive quality of the protein hydrolysates crossing the gut wall. In all these cases we can also argue that the chemicals are acting *independently* of each other. That is, the effect of rutin, for example, is not dependent upon action of the protein or tomatine upon the host—in other words, if there is multiple sequestration, the effects are additive. On the other hand, it is also possible that direct affects upon the parasitoid can be mediated by *dependent* action—that is a synergistic effect. In these circumstances, the effects of the chemicals upon the host are irrelevant.

It is also possible that the chemicals act upon the parasitoid not in a direct but rather in an indirect but *independent* manner (Fig. 6.6). In this case, tomatine, phenolics, and/or protein exert their effect upon the parasitoid by additively and strictly altering the quality of the host as food for the parasitoid and/or the ability the host to withstand the parasitization. This route of action, therefore, supposes that the chemicals do not directly affect the parasitoid. Such an indirect affect is possible through alteration of the host's metabolism and hormonal status (see Sect. 6.1.3.). This alteration of the physiological status of the host could lead to a reduction, for example, in nutrients and/or hormones in the hemolymph and fat body of the host essential for growth of the parasitoid, and/or even arise from metabolism of "toxins" to forms more toxic to both the host and the parasitoid (Duffey et al., 1985).

Finally, it is possible to envision indirect but *interdependent* effects wherein the chemicals synergistically and strictly alter the physiological status of the host

as described above. In this circumstance the chemicals may be jointly antagoniz-
ing one or several of the host's metabolic-physiological systems, or acting upon
different systems to produce nonadditive effects. The exact biochemical-physio-
logical mechanisms of this synergistic action of these chemicals are unknown.
Our experiments are not designed to probe the exact mechanisms of these complicated potential interactions, nor is it within the realm of this chapter to discuss it in any more detail.

In reality, it is more likely that the chemicals affect the parasitoid by a number of the above routes. Certainly, our results with tomatine–phytosterol, phenolic–protein interactions suggest that both direct and indirect may be occurring
simultaneously in dependent and independent manners. An elucidation of these
mechanisms at the physiological, biochemical, and genetic levels will be important in understanding (1) how hosts and parasitoids are adapted, respectively,
for withstanding or succeeding at parasitism, (2) how resistance to toxic agents
in organisms involved in multitrophic interactions might evolve, and (3) how
durable plant resistance will be when pest insects are subjected to multiple selective pressures such as plant toxins and parasitoids (see Gould, 1985a,b).

### 6.3.4.4. Compatibility of Nutrients and Toxins with BC

On the basis of the above effects, it appears that rutin, at the $ED_{50}$ level of *H. zea*
is a potentially more compatible chemical than is tomatine (assuming no nega-
tion of toxicity by phytosterols). However, as suggested earlier, it might be reasonable to attempt to breed for higher levels of rutin to effect control of insects
such as *H. zea*. If one then uses an $ED_{90}$ dose of rutin ($\sim 6$ $\mu$mol/g at 2.4%
casein in unparasitized state) in the diet, at any protein regime, the stress upon
the host and parasitoid is so great that neither organism survives. Similar results
are obtained with tomatine. Therefore, the compatible level of rutin lies somewhere between the $ED_{50}$ and $ED_{90}$ levels. This level has not been determined.
However, if one is attempting to control both species of noctuids, an $ED_{50}$ level
of rutin (for *H. zea*) at high or low protein (Table 6.2) causes death of both the
host and parasitoid. Somewhere in between these levels represents a compatible
compromise. To decide what the compromise is cannot be deduced from our
experiments because (1) they are not complete in terms of toxicological analyses,
(2) the studies have not been done with tomato plant protein (we assume casein is
a good model), and (3) probably most importantly these types of studies have not
been done in the field taking into account life-table analyses.

It appears then that nutrient–toxin levels in the diet of the host have a major
effect upon the growth, development, and survival of the parasitoid. It is also
obvious that the effects of "nutrients" and "toxins" upon the parasitoid are not
absolute but dependent upon the host, as well as upon the chemical content or
mixture. Thus, if one is concerned about using either rutin and/or tomatine as
bases of resistance against these two pests, one must specify a great deal more
information than just the independent level of rutin or tomatine in the plant.
One must be concerned with accentuating and/or attenuating effects of other
natural products such as protein or phytosterols and how these types of interac-

tions affect a specific host–parasitoid interaction. It should be understood that our results are not a description of reality, what will actually happen in the field, but rather a mechanistic approximation of potential interactions. If such interactions do occur in the field, then the use of HPR in IPM programs is a potentially complicated affair demanding an interdisciplinary approach at both the basic and applied levels. How our findings relate to other crop systems remains to be determined. The specific details will likely not be relatable, but the theoretical principles about the general phenomenon of chemically mediated tritrophic interactions should be extrapolatable to other crops systems.

### 6.3.5. Recommendations about Use of HPR with BC

What recommendations can we make about which natural products might be compatible with the use of the wasp *H. exiguae* as a BC agent in a resistance program to control *H. zea* and *S. exigua* on tomato plants?

If one were only trying to control larval *S. exigua*, an $ED_{50}$ level or higher of rutin (Fig. 6.2) might be functional; the effects of variable levels of dietary protein might be minimal as an interdependent modifier. The use of tomatine may prove to be complicated in that one may have to breed not just for high tomatine but also for low phytosterol content. A limited amount of information is known about the genetic regulation of tomatine in the tomato plant *L. esculentum* (Juvik et al., 1982; Juvik and Stevens, 1982b), and as far as we are aware nothing is known about the genetics controling the production of phytosterols in this plant.

If one is trying to control both pests, our limited information suggests that tomatine will be incompatible or inoperative. On the other hand, rutin (total catecholic phenolics) may offer the possibility of being compatible, providing that one can breed simultaneously for appropriate rutin levels as well as protein levels. Our results do not provide definite answers for creating compatible resistant plants, or implementing them in IPM programs; yet, they do provide mechanistic reasons for caution in using antibiotic natural products unwittingly in HPR programs, unless proven otherwise compatible by definitive laboratory and field experiments.

From a more theoretical and ecological viewpoint, our results support the viewpoint that the realistic development of theories on predator–prey and plant–herbivore interactions, and plant chemical defensive strategies will not be possible until these interactions are analyzed at the minimum as tritrophic systems (Price et al., 1980; Price, 1983). If our results are a reflection of reality, then it seems that a definition of the defensive value of given plant natural product may only be partially possible, because its activity is dependent to such a degree upon context. By context, we mean at the minimum its chemical milieu (its dosage with respect to other "nutrients" and "toxins"), the herbivore, and the parasitoid or predator. A major factor in the context of the chemical milieu seems to be how the interactivity of natural products determines the degree of nutrification or toxification, as evidenced by the rutin–protein and tomatine–phytosterol interactions.

Perhaps a strict distinction between what is a "nutrient" or a "toxin" is more of a semantic convenience, when what appears to be pivotal in determining their biological effects is not just the independent absolute levels of uptake by the host, but rather the absolute balance of uptake of "nutrients" and "toxins." The potential role of plant protein quantity and particularly quality in modulating the toxicity of plant defensive agents to herbivorous insects certainly demands a more critical analysis than is presented here or elsewhere (Mattson, 1980; Price, 1983; Strong et al., 1984). Current theories on plant chemical defensive strategies do not adequately account for the complexity of multitrophic systems (i.e., the apparent complexity of our tritrophic system), and hence can not depict predictively what comprises chemical defense in a multitrophic systems. Consequently, because of lack of adequate theory at both the applied and basic levels, it is not currently possible to be at all predictive in the use of chemically based HPR that is compatible with the use BC agents.

## 6.4. ADDITIONAL ASPECTS OF COMPATIBILITY

A meaningful consideration of implementability and particularly compatibility of chemical bases of resistance in IPM programs extends beyond the realm of the use of parasitic BC agents so far discussed here. The facets of implementability will only receive brief discussion in comparison to compatibility.

For implementation it is assumed that the genetic base of the plant is manipulatable to produce an agronomically efficient plant, possessing the appropriate defensive chemical properties to control the desired pests in given geographic regions. It is assumed that field research has been done to validate that the resistant line(s) are easily included into existent IPM programs [i.e., (1) not facilitating the action of another or new pest, (2) cultural or chemical control do negate resistance], and/or that farmers will use the lines. For long-term resistance, it is assumed that valid sampling and/or monitoring methods exist for assessing continued resistance to both insects and disease. Many other facets exist.

Some of the extended aspects of compatibility deal with the effects of a given chemical bases of resistance against insects upon (1) resistance to plant disease, (2) effect on nontarget organisms, (3) efficacy of synthetic insecticides, (4) effect upon the environment, (5) cultural practices, and (6) human and animal health. These aspects will be briefly discussed sequentially.

### 6.4.1. Compatibility with Plant Pathogens

One primary concern in developing resistance to insects is that compatibility exists with resistance to infection by microorganisms. Is it possible that by breeding for resistance to given insects that resistance to pathogens could be negated or reduced? A moderate body of knowledge strongly implicates both catecholic phenolics (Matta et al., 1969; Hobson and Davies, 1971; Levin, 1971, 1976; Mendez and Brown, 1971; Chadha and Brown, 1974; Pollock and Drysdale, 1976; Carrasco et al., 1978), and tomatine (Allen and Kuc, 1968; Arneson and Durbin, 1968; Roddick, 1974; McCance and Drysdale, 1975; Schonbeck and Schlosser, 1976; Schlosser, 1977; Mahadevan, 1982; Callow, 1983). From this information both natural products superficially seem consistent with the control of both insects and pathogens. However, matters may be not this simple; because these above papers do not, by and large, address cause-and-effect relationships nor invoke mechanisms of action.

As pointed out earlier, the simultaneous use of tomatine to control both insects and pathogens may be difficult because of the necessity for high sterol/tomatine ratios for defense against pathogens (Mahadevan, 1982; Callow, 1983; Defago and Kern, 1983; Defago et al., 1983; Nes et al., 1983), a state that alleviates toxicity of tomatine to some insects (Gestetner et al., 1970; Shany et al., 1970a,b; Campbell and Duffey, 1981). This aspect of dual resistance warrants more detailed study in the tomato plant, but also in crops such as potatoes, alfalfa, and soybeans where saponins are putative constitutive chemical resistance factors against both disease and insects (Horber et al., 1974; Raman et al., 1979; Birk and Peri, 1980; Liener, 1980; Mahadevan, 1982; Callow, 1983; Tingey, 1984).

The use of phenolics (chlorogenic acid and rutin) to combat insect and microorganismal pests may not be without its effects upon nontarget organisms. It is known that some simple phenolics and flavonoids not only have antibacterial (Iizuka et al., 1974; Koike et al., 1979) but also antiviral activity (Verma, 1973; Harborne et al., 1975, Chapter 18). It is suspected that part of this antibacterial–antiviral activity resides in the ability of phenolics, particularly catecholic derivatives, to bind covalently and noncovalently to protein (e.g., cell walls, viral protein coats, and enzymes; Mink and Saksena, 1971; Verma, 1973; van Sumere et al., 1975). This activity raises the possibility that rutin and/or chlorogenic acid might interfer with the concurrent use of *Bacillus thuringiensis* (BT) or nuclear polyhedrosis virus to control larval *H. zea* and *S. exigua*. In fact, on the basis of laboratory experiments, it has been shown that both rutin and chlorogenic acid have the ability, at levels equivalent to that found in the tomato plant, to cause about a 75% *in vivo* inhibition of infectivity of *Heliothis* singly embedded nuclear polyhedrosis virus (HzSNPV) in larval *H. zea*. However, these phenolics do not adversely affect the action of BT (G. W. Felton and S. S. Duffey, unpublished data). We do not know how this type of finding relates to field conditions, but it points to the necessity of understanding the ramifications of the presence of natural chemicals in agricultural systems. If an HPR program is bent upon using phenolics to control both pest insects and disease, then the very properties of this kind of plant defensive system may negate the use of viruses pathogenic to insects.

### 6.4.2. Compatibility with Insect Pathogen

However, still further complications exist in defining whether catecholic phenolics are potentially incompatible with viral control of insect pests. One major factor has not yet been considered, which is also of utmost importance in depicting merely the toxicity of catecholic phenolics to larval *H. zea* and *S. exigua*. This factor is the presence of highly active polyphenoloxidase (PPO) and peroxidase (PO) activity in the leaves and fruits of the tomato plant (Hall et al., 1969; Duffey, 1985; Reuveni and Ferreira, 1985). The consequence of such activity is that catecholic phenolics are oxidized to highly reactive *o*-quinones (Butt, 1980; Pierpoint, 1983) that covalently bind to protein. As discussed above (Sect. 6.4.1.); this binding reduces the utilizability of protein for livestock, and also may be responsible for inactivation of viruses. A further consequence of this oxidative coupling of catechols to protein is that it alleviates the toxicity of chlorogenic acid (but not rutin) to larval *H. zea* and *S. exigua* (Felton and S. S. Duffey, unpublished data). Thus, the effective utilization of catecholic phenolics for control of larval noctuids may be dependent upon the use of rutin rather than chlorogenic acid. At this point, the interrelationships between catecholic phenolics, PPO and PO activity, and resistance to insects and disease is very unclear and warrants more research.

The major IPM tactic for controlling insects is via the use of synthetic insecticides. It cannot be assumed that natural bases of resistance in the plant will not modify the efficacy of pesticides. Ample evidence is accumulating to demonstrate clearly that natural products of plants greatly modify the toxicity of insecticides to insects (Yu, 1982,a,b; Terriere, 1984). Kennedy (1984) demonstrated that the concurrent use of the neurotoxin, 2-tridecanone, against larval *H. zea* with carbaryl primes the larval insect to be more resistant to carbaryl, supposedly by alteration of mixed function oxidase activity. Since catecholic phenolics offer a potential factor for resistance, would their use be compatible with the use of pesticides? On the basis of laboratory studies with carbaryl, we have found (G. W. Felton and S. S. Duffey, unpublished data), that both rutin and chlorogenic acid accentuate the toxicity of carbaryl to larval *H. zea* two to ten times. How these catechols affect the toxicity of other insecticides is unknown. The above results demonstrate the need to consider carefully the effect natural products may have on insecticide-based control of insects in IPM programs.

### 6.4.3. Compatibility with the Environment

Some consideration should be given to the possible effects of the use of tomatine and/or phenolics upon the environment, specifically with reference to the properties and nutrient value of soil to crops. Many phenolics and saponins are potent allelopathic agents (Rice, 1974, 1979). If tomato plants with elevated levels of these chemicals are plowed into the field after each harvest, the potential exists for a buildup of allelopathic agents in the soil and subsequent detrimental effects upon the growth of successive crops. Along similar lines, the effects of a buildup of these chemicals in the soil may alter the nature of the rhizosphere, so that a different fertilization regime is required. Also, there is no guarantee that a "resistant" line will not require different nutrient conditions in the field from that otherwise gained by current cultural practices in order to express resistance. Environmental and soil conditions do affect the production of phenolics in tomato plants (Harborne, 1982, Chapter 18) and the nutrient status of a plant is known to affect its resistance to both disease and insects (Schoeneweiss, 1975; Hare, 1983; Kogan and Paxton, 1983; White, 1984). Additionally, resistant varieties of plants may have slightly different phenologies and/or growth patterns that physically and temporally dispose to be a better host for secondary pests. In tomatoes, vine shape and mulching of soil affects the pest status of the plant (Price and Poe, 1977; Schuster, 1977). Hence, the degree of influence of environment and cultural practices on resistance must be carefully established.

### 6.4.4. Compatibility with Human and Animal Health

One final concern is whether natural products, which serve as bases of HPR of crop plants, are compatible with the health of animals and humans who consume these crops. Tomato foliage and/or fruit are not normally used as animal feed, so that the effect of tomatine and phenolics (as well as proteinase inhibitors) upon animals is not a major consideration. On the other hand, tomato seed protein is sometimes used as cattle feed. A wealth of literature exists demonstrating the detrimental effects of certain phenolics, saponins (e.g., glycoalkaloids), and proteinase inhibitors, as examples, upon the growth of animals (Rosenthal and Janzen, 1979; Liener, 1980; Singleton, 1981). What is of more concern is the potential for natural products like rutin, chlorogenic acid, and tomatine to exert acute and/or chronic effects upon humans arising from the ingestion of fruit. Tomatine does not appear too problematic, since it is not present at significant levels in ripe fruit (Heftmann and Schwimmer, 1972; Roddick, 1974), and is relatively nontoxic to humans (Cayen, 1971; Nishie et al., 1975; Wilson, et al., 1961). In contrast, the presence of phenolics in the fruit could pose problems for human health. Claims are made that chlorogenic acid is a cocarcinogen (Chablis and Bartlett, 1975) and that flavonoids like rutin and quercetin are mutagenic (Ames, 1983; see contradictory information in Singleton, 1981). Also, phenolics such as rutin and chlorogenic acid are known to reduce the nutritional value of protein (Liener, 1980, Chapter 13; Pierpoint, 1983). The assessment of the full risk of an individual eating a few tens of pounds of tomatoes per year (not a staple food), is far from clear, when many other more staple foods and beverages contain potential mutagens and carcinogens (Liener, 1980, Chapter 11; Ames, 1983) of both natural and synthetic origin.

## 6.5. CONCLUSION

Our research has taken a reductionistic (= mechanistic) physiological approach rather than a holistic ecological approach to understanding how HPR can be used compatibly with BC in IPM programs. We fully realize our approach is only an approximation of reality that is only evidenced by validation through field experiments. On the other hand we feel that our mechanistic approach offers, to a degree, proximate explanations as to what specific chemical and biological factors are influencing the efficacy of HPR and BC. These explanations are not usually derivable by more holistic or ecological types of experiments that search for ultimate explanations usually based on correlational evidence rather than cause-and-effect relationships. Both approaches are valid and necessary in learning how to utilize HPR and BC compatibly in IPM programs; however, a more intensive integration of basic and applied research will be required for effective implementation.

Our analysis of the tomato–noctuid–parasitoid interaction should demonstrate the complexities of utilization of natural products in HPR, and likewise point out that the development of current theories about the action of plant natural products in agricultural and natural system may be overly speculative. Certainly, these theories do not permit us to be predictive in our use of specific plant chemicals to effect HPR and BC. However, such predictivity may not be possible, considering how the biological activity of a "toxin" is not just an independent function of its level in the plant, but is also an interdependent function of the quality and quantity of other "nutrients" and "toxins" in the plant. In addition, we have shown that the full effects of these "nutrients" and "toxins" upon a given herbivore cannot be specified unless one also takes into account the influence of the third trophic level (i.e., parasitoids). Current theories do not adequately address the issues of interactivity between plant products nor the extension of their effects through multitrophic systems.

To end on a somewhat pessimistic and philosophical note, resistance of plants to insects is usually thought of as the action of some definable state or entity of the plant upon a given pest. It follows from this line of reasoning that the state or entity can be selectively modified or manipulated by humans to enhance resistance. However, it is very seldom supposed that resistance is not the result of one or few clearly definable characteristics but rather arises as a result of the mosaic or emergent effect of many plant factors. Certainly, the action of 2-tridecanone against the tobacco hornworm *Manduca sexta* (Dimock and Kennedy, 1983; Kennedy and Dimock, 1983) is an example of resistance based on a few clearly definable plant characters. However, the bases of resistance are not always that clearly definable. If, indeed, resistance is in some cases an emergent property (in our case the interdependent and multiple action of "nutrients" and "toxins"), then it may be extremely difficult to define the nature of resistance, let alone use resistance predictively. By analogy, the letter R in Fig. 6.7 emerges. The fact that it is the letter R has nothing to do with the physical–chemical properties of

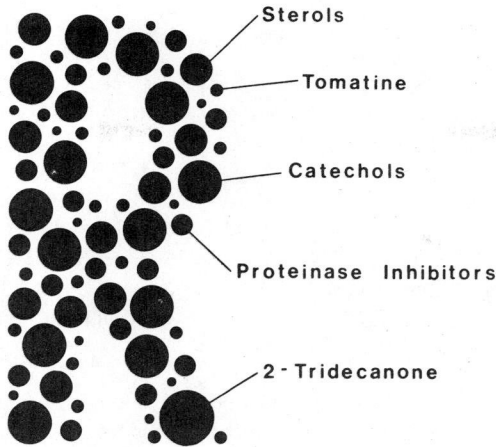

Fig. 6.7 Diagrammatic representation of the emergent nature of resistance.

the paper or of the ink that forms the dots. Imagine that the letter R is diagrammatic of the state of resistance, in which each of the dots represents factors that contribute to varying degrees to resistance. If one of the dots happens to represent phenolics, and if one experimentally only investigates the independent action of phenolics upon resistance (by removing the dot from the letter R), one is no longer looking at the state of resistance but merely at the dot, phenolics in isolation. To further the analogy, the size of the dot expresses the factor's relative importance in contributing to the emergent state of resistance; hence, if one choose a large dot to investigate, correlations with resistance might be correspondingly greater than that achieved from smaller dots. It can also be envisioned, however, that the dots do not remain the same size every year, even though the size of the letter remains constant. This means that degree of resistance remains (emerges) the same, but that the qualitative and quantitative bases behind it have altered. Perhaps some of the lack of success in attempting to identify the chemical bases of resistance in plants arises from the fact that resistance may often be an emergent property (= horizontal resistance?). Such an emergent state would defy breeding for the "magic bullet."

## ACKNOWLEDGMENTS

This research has been supported by U.S. Department of Agriculture competitive grants (82CRCR1-1054 and 5901-041-9-0269-0) to S. S. D. and a National Science Foundation grant to S. S. D. and K. A. B. (DEB-8214555). We thank Gary Felton and Dr. G. G. Kennedy and Dr. F. Gould for critical comments.

## REFERENCES

Adkisson, P. L., and V. A. Dyck. (1980). "Resistant varieties in pest management systems," in F. G. Maxwell and P. R. Jennings, Eds., *Breeding Plants Resistant to Insects*, Wiley, New York, pp. 233-251.

Adkisson, P. L., G. A. Niles, J. K. Walker, L. S. Bird, and H. B. Scott. (1982). Controlling cotton's insect pests: A new system, *Science*, **216**, 19-22.

Ahmad, S., Ed. (1983). *Herbivorous Insects*, Academic, New York, 257 pp.

Allen, E. H. and J. Kuc. (1968). α-Solanine and α-chaconine as fungotoxic compounds in extracts of Irish potatoes, *Phytopathology*, **58**, 776-781.

Al-Zubaidi, F. S. and J. L. Capinera. (1983). Application of different nitrogen levels to the host plant and cannibalistic behavior of beet armyworm, *Spodoptera exigua* (Hübner) (Lepidoptera: Noctuidae), *Environ. Entomol.*, **12**, 1687-1689.

Al-Zubaidi, F. S., and J. L. Capinera. (1984). Utilization of food and nitrogen by the beet armyworm, *Spodoptera exigua* (Hübner) (Lepidoptera: Noctuidae), in relation to food type and dietary nitrogen levels, *Environ. Entomol.*, **13**, 1604-1608.

Ames, B. N. (1983). Dietary carcinogens and anticarcinogens, *Science*, **221**, 1256-1264.

Arneson, P. A., and R. D. Durbin. (1968). Studies on the mode of action of tomatine as a fungitoxic agent, *Plant Physiol.*, **43**, 683-686.

Arthur, A. P. (1962). Influence of host tree on abundance of *Itoplectis conquisitor* (Say) (Hymenoptera:Ichneumonidae), a polyphagous parasite of the European pine shoot moth, *Rhyaconia buoliana* (Schiff) (Lepidoptera:Olethreutidae), *Can. Entomol.*, **94**, 337-347.

Atsatt, P. R., and D. J. O'Dowd. (1976). Plant defense guilds, *Science*, **193**, 24-29.

Barbeau, W. E., and J. E. Kinsella. (1983). Factors affecting the binding of chlorogenic acid to fraction 1 leaf protein, *J. Agric. Food. Chem.*, **31**, 993-998.

Barbosa, P., and J. A. Saunders. (1984). Plant allelochemicals: Lineage between herbivores and their natural enemies, *Phytochem. Soc. N. Amer. Newsletter*, **24**, 23 (abstract).

Barbosa, P., J. A. Saunders, and M. Waldvogel. (1982). "Plant mediated variation in herbivore suitability and parsitoid fitness," in J. H. Visser and A. K. Minks, Eds., *Proc. 5th Int. Symp. Plant-Insect Relationships*, Wageningen, Pudoc, Wageningen, pp. 63-71.

Baronio, P., and F. Sehnal. (1980). Dependence of the parasitoid *Gonia cinerascens* on the hormones of its lepidopterous hosts, *J. Insect Physiol.*, **26**, 619-626.

Beckage, N. E., and Riddiford, L. M. (1978). Developmental interactions between the tobacco hornworm *Manduca sexta* and its braconid parasite *Apanteles congregatus*, *Entomol. Exp. Appl.*, **23**, 139-151.

Beckage, N. E., and L. M. Riddiford. (1982a). Effects of parasitism by *Apanteles congregatus* on the endocrine physiology of the tobacco hornworm *Manduca sexta*, *Gen. Comp. Endocrinol.*, **47**, 308-322.

Beckage, N. E., and L. M. Riddiford. (1982b). Effects of methoprene and juvenile hormone on larval ecdysis, emergence, and metamorphosis of the endoparasitic wasp, *Apanteles congregatus*, *J. Insect Physiol.*, **28**, 329-334.

Beckage, N. E., and L. M. Riddiford. (1983). Growth and development of the endoparasitic wasp *Apanteles congregatus*: Dependence on host nutritional status and parasite load, *Physiol. Entomol.*, **8**, 231-241.

Bell, W. J., and Cardé, R. T., Eds. (1984). *Chemical Ecology of Insects*, Sinauer, Sunderland, MA, 524 pp.

Benedict, J. H., T. F. Leigh, W. Tingey, and A. H. Hyer. (1977). Glandless Acala cotton: more susceptible to insects, *Calif. Agric.*, **31**, 14-15.

Berenbaum, M. (1978a). Toxicity of a furanocoumarin to armyworms: A case of biosynthetic escape from insect herbivores, *Science*, **201**, 532-534.

Berenbaum, M. (1978b). Adaptive significance of midgut pH in larval Lepidoptera, *Am. Nat.*, **112**, 138-146.

Berenbaum, M. and P. Feeny. (1981). Toxicity of angular furanocoumarins to swallowtail butterflies: Escalation in a coevolutionary race? *Science*, **212**, 927-929.

Bergman, J. M. and W. M. Tingey. (1979). Aspects of interaction between plant genotypes and biological control, *Bull. Entomol. Soc. Amer.*, **25**, 275-279.

Bernays, E. A. (1981). Plant tannins and insect herbivores: An appraisal, *Ecol. Entomol.*, **6**, 353-360.

Bernays, E. A., and D. J. Chamberlain. (1980). A study of tolerance of ingested tannin in *Schistocerca gregaria*, *J. Insect Physiol.*, **26**, 415-420.

Birk, Y., and I. Peri. (1980). "Saponins," in I. E. Liener, Ed., *Toxic Constituents of Plant Foodstuffs*, 2nd ed., Academic, New York, pp. 161-182.

Blau, P. A., P. Feeny, L. Contardo, and D. S. Robson. (1978). Allylglucosinolate and herbivorous caterpillars: A contrast in toxicity and tolerance, *Science*, **200**, 1296-1298.

Bouletreau, M., and J. R. David. (1981). Sexually dimorphic response to host habitat toxicity in *Drosophila* parasitic wasps, *Evolution*, **35**, 395-399.

Brewer, F. D., and E. G. King. (1979). Consumption and utilization of soyflour-wheat germ diets by *Heliothis* spp., *Ann. Entomol. Soc. Amer.*, **72**, 415-417.

Brewer, F. D., and E. G. King. (1981). Food consumption and utilization by sugarcane borers parasitized by *Apanteles flavipes*, *J. Georgia Entomol. Soc.*, **16**, 185-192.

Brewer, J. W., J. L. Capinera, R. E. Deshon, Jr., and M. L. Walmsley. (1985). Influence of foliar nitrogen levels on survival, development, and reproduction of western spruce budworm, *Choristoneura occidentalis* (Lepidoptera: Tortricidae), *Can. Entomol.*, **117**, 23-32.

Broadway, R., S. S. Duffey, G. Pearce, and C. A. Ryan. (1986). Plant proteinase inhibitors: A defense against herbivorous insects, *Entomol. Exp. Appl.* **41**, 33-38.

Browning, H. W., and E. R. Oatman. (1981). Effects of different constant temperatures on adult longevity, development time, and progeny production of *Hyposoter exiguae* (Hymenoptera: Ichneumonidae), *Ann. Entomol. Soc. Amer.*, **74**, 79-82.

Buddenhagen, I. W., and M. B. de Ponti. (1983). Crop improvement to minimize future losses to diseases and pests in tropics, *FAO Plant Protec. Bull.*, **31**, 11-30.

Burkett, G. R., J. C. Schneider, and F. M. Davis. (1983). Behavior of the tomato fruitworm, *Heliothis zea* (Boddie) (Lepidoptera: Noctuidae), on tomato, *Environ. Entomol.*, **12**, 905-910.

Burleigh, J. G., and J. H. Farmer. (1978). Dynamics of *Heliothis* spp. larval parasitism in southeast Arkansas, *Envir. Entomol.*, **7**, 692-694.

Burton, R. L., and K. J. Starks. (1977). Control of a primary parasite of the greenbug with a secondary parasite in greenhouse screening for plant resistance, *J. Econ. Entomol.*, **70**, 219-220.

Butt, V. S. (1980). "Direct oxidases and related enzymes," in P. K. Stumpf and E. E. Conn, Eds., *The Biochemistry of Plants. A Comprehensive Treatise*. Vol. 2. *Metabolism and Respiration*, Academic, New York, pp. 81-123.

Callow, J. A., Ed. (1983). *Biochemical Plant Pathology*, Wiley, New York, 484 pp.

Campbell, B. C., and S. S. Duffey. (1979a). Effect of density and instar or *Heliothis zea* on parasitization by *Hyposoter exiguae*, *Environ. Entomol.*, **8**, 127-130.

Campbell, B. C., and S. S. Duffey. (1979b). Tomatine and parasitic wasps: Potential incompatibility of plant-antibiosis with biological control, *Science*, **205**, 700-702.

Campbell, B. C., and S. S. Duffey. (1981). Alleviation of $\alpha$-tomatine-induced toxicity to the parasitoid, *Hyposoter exiguae*, by phytosterols in the diet of the host, *Heliothis zea*, *J. Chem. Ecol.*, **7**, 927-946.

Carrasco, A., A. M. Boudet, and S. Marigo. (1978). Enhanced resistance of tomato plants to *Fusarium* by controlled stimulation of their natural phenolic production, *Physiol. Plant Pathol.*, **12**, 225-232.

Cates, R. G. (1980). Feeding patterns of monophagous, oligophagous, and polyphagous insect herbivores: The effect of resource abundance and plant chemistry, *Oecologia*, **46**, 22-31.

Cates, R. G. (1981). Host plant predictability and the feeding patterns of monophagous, oligophagous, and polyphagous insect herbivores, *Oecologia*, **48**, 319-326.

Cates, R. G., and D. F. Rhoades. (1977). Patterns in the production of antiherbivore chemical defenses in plant communities, *Biochem. System. Ecol.*, **5**, 185-193.

Cayen, M. N. (1971). Effect of dietary tomatine on cholesterol metabolism in the rat, *J. Lipid. Res.*, **12**, 482-490.

Chablis, B. C., and C. D. Bartlett. (1975). Possible cocarcinogenic effects of coffee constituents, *Nature*, **254**, 532-533.

Chadha, K. C., and S. A. Brown. (1974). Biosynthesis of phenolic acids in tomato plants infected with *Agrobacterium tumefasciens*, *Can. J. Bot.*, **52**, 2041-2047.

Chan, B. G., A. C. Waiss, and M. Lukefahr. (1978). Condensed tannin, an antibiotic chemical from *Gossypium hirsutum*, *J. Insect Physiol.*, **24**, 113-118.

Cheng, L. (1970). Timing of attack by *Lypha dubia* Fall. (Diptera:Tachinidae) on the winter moth *Operopthera brumata* (L.) (Lepidoptera:Geometridae) as a factor affecting parasite success, *J. An. Ecol.*, **39**, 313-320.

Cosenza, G. W., and H. B. Green. (1979). Behavior of the tomato fruitworm, *Heliothis zea* (Boddie), on susceptible and resistant lines of tomatoes, *Hort. Sci.*, **14**, 171-173.

Curry, J. P., and D. Pimentel. (1971). Life cycle of the greenhouse whitefly, *Trialeurodes vaporarium*, and population trends of the whitefly and its parasite, *Encarsia formosa*, on two tomato varieties, *Ann. Entomol. Soc. Amer.*, **64**, 1188-1190.

Dahlman, D. L. (1977). Effect of L-canavanine on the consumption and utilization of artificial diet by the tobacco hornworm, *Manduca sexta*, *Entomol. Exp. Appl.*, **22**, 123-131.

Dahlman, D. L., and J. R. Greene, Jr. (1981). Larval hemolymph protein patterns in tobacco hornworms parasitized by *Apanteles congregatus*, *Ann. Entomol. Soc. Amer.*, **74**, 130-133.

Dahlman, D. L., and G. A. Rosenthal. (1975). Non-protein amino acid interactions-I. Growth effects and symptomology of L-canavanine consumption by the tobacco hornworm, *Manduca sexta* (L.), *Comp. Biochem. Physiol.*, **51A**, 33-36.

Dahlman, D. L., and S. B. Vinson. (1976). Trehalose level in haemolymph of *Heliothis virescens* parasitized by *Campoletis sonorensis*, *Ann. Entomol. Soc. Amer.*, **69**, 523-524.

Dahlman, D. L., and S. B. Vinson. (1980). Glycogen content in *Heliothis virescens* parasitized by *Microplitis croceipes*, *Comp. Biochem. Physiol.*, **66A**, 625-630.

Dahms, R. G. (1969). Theoretical effects of antibiosis on insect population dynamics, *USDA Agric. ERD*, 5 pp.

Davies, A. M. C., V. K. Newby, and R. L. M. Synge. (1978). Bound quinic acid as a measure of coupling of leaf and sunflower-seed proteins with chlorogenic acid and congeners: Loss of availability of lysine, *J. Sci. Food Agric.*, **29**, 33-41.

Defago, G., and H. Kern. (1983). Induction of *Fusarium solani* mutants insensitive to tomatine, their pathogenicity and aggressiveness to tomato fruits and pea plants, *Physiol. Plant Pathol.*, **22**, 29-37.

Defago, G., H. Kern, and L. Sedlar. (1983). Genetic analysis of tomatine insensitivity, sterol content and pathogenicity for green fruits in mutants of *Fusarium solani*, *Physiol. Plant Pathol.*, **22**, 39-43.

Denno, R. F., and M. S. McClure, Eds. (1983). *Variable Plants and Herbivores in Natural and Managed Systems*, Academic, New York, 717 pp.

Dimock, M. A., and G. G. Kennedy. (1983). The role of glandular trichomes in the resistance of *Lycopersicon hirsutum* f. *glabratum* to *Heliothis zea*, *Entomol. Exp. Appl.*, **33**, 263–268.

Dimock, M. B., G. G. Kennedy, and W. G. Williams. (1982). Toxicity studies of analogs of 2-tridecanone, a naturally occurring toxicant from a wild tomato, *J. Chem. Ecol.*, **8**, 837–842.

Dryer, D. L., and K. C. Jones. (1981). Feeding deterrency of flavonoids and related phenolics towards *Schizaphis graminum* and *Myzus persicae*: Aphid feeding deterrent in wheat, *Phytochemistry*, **20**, 2489–2493.

Dryer, D. L., R. G. Binder, B. G. Chan, A. C. Waiss, Jr., E. E. Hartwig, and G. L. Beland. (1979). Pinitol, a larval growth inhibitor for *Heliothis zea* in soybeans, *Experientia*, **35**, 1182–1183.

Duffey, S. S. (1980). Sequestration of natural products by insects, *Ann. Rev. Entomol.*, **25**, 447–477.

Duffey, S. S. (1986). "Plant glandular trichomes: Their role in partial defence against insects," in T. R. E. Southwood and B. Juniper, Eds., *Insects and Plant Surfaces*, Edward Arnold, London (in press).

Duffey, S. S., and M. B. Isman. (1981). Inhibition of insect larval growth by phenolics in glandular trichomes of tomato leaves, *Experientia*, **37**, 574–576.

Duffey, S. S., K. A. Bloem, and B. C. Campbell. (1985). "Consequences of sequestration of plant natural products in plant–insect–parasitoid interactions," in W. M. Boethel, and R. D. Eikebary, Eds., *Interactions of Plant Resistance and Parasitoids and Predators of Insects*, Symp. XVII Int. Congr. Entomol., Hamburg, Aug. 20–26, 1984, Ellis Horwood, Chichester, England (in press).

Ehler, L. E. (1976). The relationship between theory and practice in biological control, *Bull. Entomol. Soc. Amer.*, **22**, 319–321.

Ehler, L. E., and L. A. Andres. (1983). "Biological control: Exotic natural enemies to control exotic pests," in C. L. Wilson, and C. L. Graham, Eds., *Exotic Plant Pests and North American Agriculture*, Academic, New York, pp. 395–418.

Eikenbary, R. D., and R. C. Fox. (1968). Responses of Nantucket pine tip moth parasites to tree level, orientation, and host per pine tip, *Ann. Entomol. Soc. Amer.*, **61**, 1380–1384.

El-Shazly, N. Z. (1972a). Der Einfluss aussere Faktoren auf die hamocytare Abwehrreaktion von *Neomyzus circumflexus* (Buck.) (Homoptera: Aphididae), *Z. Angew. Entomol.*, **70**, 414–436.

El-Shazly, N. Z. (1972b). Der Einfluss von Ernahrung und alterbder Muttertieres auf die hamocytare Abwehrreaktion von *Neomyzus circumflexus* (Buck.), *Entomophaga*, **17**, 203–209.

Elliger, C. A., B. C. Chan, and A. C. Waiss. (1980). Flavonoids as larval growth inhibitors, *Naturwissenschaften*, **67**, S. 358–359.

Elliger, C. A., Y. Wong, B. G. Chan, and A. C. Waiss, Jr. (1981). Growth inhibitors in tomato (*Lycopersicon*) to tomato fruitworm (*Heliothis zea*), *J. Chem. Ecol.*, **7**, 753–758.

Elsey, K. D. and J. F. Chaplin. (1978). Resistance of tobacco introduction 1112 to the tobacco budworm and green peach aphid, *J. Econ. Entomol.*, **71**, 723–725.

Faeth, S. H., S. Mopper, and D. Simberloff. (1981). Abundance and diversity of leaf mining insects on three oak host species: Effects of host-plant phenology, nutrition and geographic range, *Oikos*, **37**, 238–251.

Feeny, P. P. (1968). Effect of oak leaf tannins on larval growth of winter moth *Operophthera brumata*, *J. Insect Physiol.*, **14**, 805–817.

Feeny, P. P. (1975). "Biochemical coevolution between plants and their insect herbivores," in L. E. Gilbert and P. H. Raven, Eds., *Coevolution of Animals and Plants*, University of Texas Press, Austin, pp. 3–19.

Feeny, P. P. (1976). Plant apparency and chemical defense, *Rec. Adv. Phytochem.*, **10**, 1–40.

Feeny, P. P., and H. Bostock. (1968). Seasonal changes in tannin content of oak leaves, *Phytochemistry*, **7**, 871–880.

Fox, L. R. (1981). Defense and dynamics in plant–herbivore systems, *Am. Zool.*, **21**, 853–864.

Fox, L. R., and B. J. Macauley. (1977). Insect grazing on *Eucalyptus* in response to variation in leaf tannins and nitrogen, *Oecologia*, **29**, 145–162.

Friend, J. and D. R. Threlfall, Eds. (1976). *Biochemical Aspects of Plant–Parasite Relationships*, Academic, New York, 354 pp.

Fritz, R. S. (1982). Selection for host modification by insect parasitoids, *Evolution*, **36**, 283–288.

Futuyma, D. J., and M. Slatkin, Eds. (1983). *Coevolution*, Sinauer, Sunderland, Massachusetts, 555 pp.

Gestetner, B., S. Shany, T. Tencer, Y. Birk, and A. Bondi. (1970). Lucerne saponins. II.—Purification and fractionation of saponins from lucerne tops and roots and characterisation of isolated fractions, *J. Sci. Food Agric.*, **21**, 502–507.

Gibson, R. W. (1978). Resistance in glandular-haired wild potatoes to flea beetles, *Am. Potato J.*, **55**, 595–599.

Gilbert, L. E., and P. H. Raven, Eds. (1975). *Coevolution of Animals and Plants*, University of Texas Press, Austin, 246 pp.

Gilmore, J. U. (1938a). Observations on the hornworms attacking tobacco in Tennessee and Kentucky, *J. Econ. Entomol.*, **31**, 706–712.

Gilmore, J. U. (1938b). Notes on *Apanteles congregatus* (Say) as a parasite of tobacco hornworms, *J. Econ. Entomol.*, **31**, 712–715.

Gould, F. (1983). "Genetics of plant herbivore systems: Interaction between applied and basic study," in R. F. Denno and M. S. McClure, Eds., *Variable Plants and Herbivores in Natural and Managed Systems*, Academic, New York, pp. 599–653.

Gould, F. (1984). Role of behavior in the evolution of insect adaptation to insecticides and resistant host plants, *Bull. Entomol. Soc. Amer.*, **30**, 34–41.

Gould, F. (1986a). Simulation models for predicting durability of insect resistant germ plasm: A deterministic diploid, two-locus model, *Environ. Entomol.*, **15**, 1–10.

Gould, F. (1986b). Simulation models for predicting durability of insect-resistant germ plasm: Hessian fly (Diptera: Cecidomyiidae)-resistant winter wheat, *Environ. Entomol.*, **15**, 11–23.

Greany, P. D., J. H. Tumlinson, D. L. Chambers, and G. M. Boush. (1977). Chemically mediated host finding by *Biopteres (Opius) longicaudatus*, a parasitoid of tephritid fruit fly larvae, *J. Chem. Ecol.*, **3**, 189–195.

Green, T. R., and C. A. Ryan. (1972). Wound-induced proteinase inhibitor in plant leaves: A possible defense against insects, *Science*, **175**, 776–777.

Greenblatt, J. A., and P. Barbosa. (1981). Effects of host's diet on two pupal parasitoids of the gypsy moth: *Brachymeria intermedia* (Nees) and *Coccygomimus turionellae* (L.), *J. Appl. Ecol.*, **18**, 1–10.

Gregory, P., D. A. Ave, P. Y. Bouthyette, R. L. Plaisted, and W. M. Tingey. (1984). "Research progress: Glandular trichome biochemistry and potato resistance to insects," in *Proceedings of the 27th CIP Planning Conference, Integrated Pest Management for the Potato*, International Potato Center, Lima, Peru, pp. 125–131.

Hagstrum, D. W., and B. J. Smittle. (1977). Host-finding ability of *Bracon hebetor* and its influence upon adult parasite survival and fecundity, *Envir. Entomol.*, **6**, 437–439.

Hall, C. B., F. W. Knapp, and R. E. Stall. (1969). Polyphenoloxidase activity in bacterially induced graywall of tomato fruit, *Phytopathology*, **59**, 267–268.

Harborne, J. B. (1982). *Introduction to Ecological Biochemistry*, Academic, New York, 278 pp.

Harborne, J. B., T. J. Mabry, and H. Mabry, H., Eds. (1975). *The Flavonoids*, Vols. I and II, Academic, New York, 1204 pp.

Hare, J. D. (1983). "Manipulation of host suitability for herbivore pest management," in R. F. Denno and M. S. McClure, Eds., *Variable Plants and Herbivores in Natural and Managed Systems*, Academic, New York, pp. 655–680.

Harrington, E. A., and P. Barbosa. (1978). Host habitat influences on oviposition by *Parasetigena silvestris* (R-D), a larval parasite of the gypsy moth, *Environ. Entomol.*, **7**, 466–468.

Hassel, M. P., and J. K. Waage. (1984). Host-parasitoid population interactions, *Ann. Rev. Entomol.*, **29**, 89–114.

Haukioja, E. (1980). On the role of plant defenses in the fluctuation of herbivore populations, *Oikos*, **35**, 202–213.

Hedin, P. A., Ed. (1983). *Plant Resistance to Insects*, ACS Symposium. Ser., American Chemical Society, Washington, D.C., 375 pp.

Hedin, P. A., F. G. Maxwell, and J. N. Jenkins. (1974). "Insect plant attractants, repellents, deterrents, and other related factors affecting insect behavior," in F. G. Maxwell and F. A. Harris, Eds., *Proceedings of the Summer Institute on Biological Control of Plant Insects and Diseases*, University of Mississippi Press, Jackson, pp. 494–527.

Heftmann, E., and S. Schwimmer. (1972). Degradation of tomatine to 3$\beta$-hydroxy-5$\alpha$-pregn-16-en-20-one by ripe tomatoes, *Phytochemistry*, **11**, 2783–2787.

Hilliard, R. A., and L. L. Keeley. (1984). The effects of dietary nitrogen on reproductive development in the female boll weevil, *Anthonomus grandis*, *Physiol. Entomol.*, **9**, 165–174.

Hobson, G. E., and J. N. Davies. (1971). "*The Tomato*," in A. C. Hulme, Ed., *The Biochemistry of Fruits and Their Products*, Vol. 2, Academic, New York, pp. 437–482.

Holmes, J. C., and W. M. Boethel. (1972). Modification of intermediate host behaviour by parasites, *Zool. J. Linn. Soc.*, **51** (suppl. 1), 123–149.

Horber, E., K. T. Leath, B. Berrang, V. Marcarian, and C. H. Hanson. (1974). Biological activities of saponin components from Dupuits and Lahontan alfalfa, *Entomol. Exp. Appl.*, **17**, 410–424.

Horigome, T., and T. Kandatsu. (1968). Biological value of proteins allowed to react with phenolic compounds in presence of *o*-diphenol oxidase, *Agric. Biol. Chem.*, **32**, 1093–1102.

House, H. L. and J. S. Barlow. (1961). Effects of different diets of a host, *Agria affinis* (Fall.) (Diptera: Sarcophagidae) on the development of a parasitoid, *Aphaereta pallipes* (Say) (Hymenoptera: Braconidae), *Can. Entomol.*, **93**, 1041–1044.

Huebner, L. D., and H. C. Chiang. (1982). Effects of parasitism by *Lixophaga diatraeae* (Diptera: Tachinidae) on food consumption and utilization of European corn borer larvae (Lepidoptera: Pyralidae), *Environ. Entomol.*, **11**, 1053–1057.

Hurrell, R. F., P. A. Finot, and J. L. Cuq. (1982). Protein–polyphenol reactions. 1. Nutritional and metabolic consequences of the reaction between oxidized caffeic acid and the lysine residues of casein, *J. Nutr.*, **47**, 191–211.

Igarashi, K., T. Tsunekuni, and T. Yasui. (1983). Inhibition of proteolytic activity of papain by browning reaction products of quercetin, *J. Nutr. Sci. Vitaminol.*, **29**, 227–232.

Iizuka, T., S. Koike, and J. Mizutani. (1974). Antibacterial substances in feces of silkworm larvae reared on mulberry leaves, *Agric. Biol. Chem.*, **38**, 1549–1550.

Ishii, T., M. Yamaguchi, and M. Kametaka. (1983). Relation between labile body protein and dietary protein metabolism in adult rats, *Nutr. Rep. Int.*, **27**, 993–1004.

Isman, M. B., and S. S. Duffey. (1982a). Toxicity of tomato phenolic compounds to the fruitworm, *Heliothis zea*, *Entomol. Exp. Appl.*, **31**, 370–376.

Isman, M. B., and S. S. Duffey. (1982b). Phenolic compounds in the foliage of commercial tomato cultivars as growth inhibitors to the fruitworm, *Heliothis zea*, *J. Am. Hort. Sci.*, **107**, 167–170.

Isman, M. B., and S. S. Duffey. (1983). Pharmacokinetics of chlorogenic acid and rutin in larvae of *Heliothis zea*, *J. Insect Physiol.*, **29**, 295–300.

Janzen, D. H. (1979). "New horizons in the biology of plant defenses," in G. A. Rosenthal and D. H. Janzen, Eds., *Herbivores: Their Interaction with Secondary Plant Metabolites*, Academic, New York, pp. 331–350.

Jenkins, J. N. (1980). "The use of plants and insect models," in F. G. Maxwell and P. R. Jennings, Eds., *Breeding Plants Resistant to Insects*, Wiley-Interscience, New York, pp. 215-229.

Jones, C. G., and R. D. Firn. (1979a). Resistance of *Pteridium aquilinum* to attack by non-adapted phytophagous insects, *Biochem. System. Ecol.*, **7**, 95-101.

Jones, C. G., and R. D. Firn. (1979b). Some allelochemicals of *Pteridium aquilinum* and their involvement in resistance to *Pieris brassicae*, *Biochem. System. Ecol.*, **7**, 187-192.

Jones, R. L. (1981). "Chemistry of semiochemicals involved in parasitoid-host and predator-prey relationships," in D. A. Nordlund, R. L. Jones, and W. J. Lewis, Eds., *Semiochemicals: Their Role in Pest Control*, Wiley, New York, pp. 239-250.

Jowyk, E. A., and Z. Smilowitz. (1978). A comparison of growth and developmental rates of the parasite *Hyposoter exiguae* reared from two instars of its host, *Trichoplusia ni.*, *Ann. Entomol. Soc. Amer.*, **71**, 467-472.

Juvik, J. A., and M. A. Stevens. (1982a). Physiological mechanisms of host-plant resistance in the genus *Lycopersicon* to *Heliothis zea* and *Spodoptera exigua*, two insect pests of cultivated tomato, *J. Am. Soc. Hort. Sci.*, **107**, 1065-1069.

Juvik, J. A., and M. A. Stevens. (1982b). Inheritance of foliar $\alpha$-tomatine content in tomatoes, *J. Am. Soc. Hort. Sci.*, **107**, 1061-1065.

Juvik, J. A., M. A. Stevens, and C. M. Rick. (1982). Survey of the genus *Lycopersicon* for variability in $\alpha$-tomatine content, *Hort. Sci.*, **17**, 764-766.

Kareiva, P. (1983). "Influence of vegetation texture on herbivore populations: Resource concentration and herbivore movement," in R. F. Denno and M. S. McClure, Eds. *Variable Plants and Herbivores in Natural Systems*, Academic, New York, pp. 259-289.

Kennedy, G. G. (1978). Recent advances in insect resistance of vegetable and fruit crops in North America: 1966-1977, *Bull. Entomol. Soc. Amer.*, **24**, 375-384.

Kennedy, G. G. (1984). 2-Tridecanone, tomatoes and *Heliothis zea*: Potential incompatability of plant antibiosis with insecticidal control, *Entomol. Exp. Appl.*, **35**, 305-311.

Kennedy, G. G., and M. B. Dimock. (1983). "2-Tridecanone: A natural toxicant in a wild tomato responsible for insect resistance," in J. Miyamoto et al., Eds., *Human Welfare and the Environment*, *IUPAC Pesticide Chemistry*, Pergamon, London, pp. 123-128.

Kennedy, G. G., L. R. Romanow, S. F. Jenkins, and D. C. Sanders. (1983). Insects and diseases damaging tomato plants in the coastal plain of North Carolina, *J. Econ. Entomol.*, **76**, 168-173.

Kogan, M. (1975). "Plant resistance in pest management," in R. L. Metcalf and W. H. Luckman, Eds., *Introduction to Insect Pest Management*, Wiley-Interscience, New York, pp. 93-134.

Kogan, M., and J. Paxton. (1983). "Natural inducers of plant resistance to insects," in P. A. Hedin, Ed. *Plant Resistance to Insects*, ACS Symposium Ser. 208, American Chemical Society, Washington, D.C., pp. 153-171.

Koike, S., T. Iizuka, and J. Mizutani. (1979). Determination of caffeic acid in the digestive juice of silkworm larvae and its antibacterial activity against the pathogenic *Streptococcus faecalis* AD-4, *Agric. Biol. Chem.*, **43**, 1727-1731.

Lange, W. H., and L. Bronson. (1981). Insect pests of tomatoes, *Ann. Rev. Entomol.*, **26**, 345-371.

Lange, W. H., and J. S. Kishiyama. (1978). Integrated pest management on vegetable crops in Southern California, *Calif. Agric.*, Feb. 1978, 27-28.

Levin, D. A. (1971). Plant phenolics: an ecological perspective, *Am. Nat.*, **105**, 157-181.

Levin, D. A. (1973). The role of trichomes in plant defense, *Q. Rev. Biol.*, **48**, 3-15.

Levin, D. A. (1976). The chemical defenses of plants to pathogens and herbivores, *Ann. Rev. Ecol. System.*, **7**, 121-159.

Levin, D. B., J. E. Laing, and R. P. Jacques. (1981). Interactions between *Apanteles glomeratus* (L.) (Hymenoptera: Braconidae) and granulosis virus in *Pieris rapae* (L.) (Lepidoptera: Pieridae), *Environ. Entomol.*, **10**, 65-68.

Lewis, J. J., D. A. Nordlund, H. R. Gross, R. L. Jones, and S. L. Jones. (1977). Kairomones and their use for pest management of entomophagous insects. V. Moth scales as a stimulus for predation of *Heliothis zea* (Boddie) eggs by *Chrysopa carnea* Stephens larvae, *J. Chem. Ecol.*, **3**, 483-487.

Liener, I. E., Ed. (1980). *Toxic Constituents of Plant Foodstuffs*, Academic, New York, 500 pp.

Lingren, P. D., and H. J. Lukefahr. (1977). Effects of nectariless cotton on caged populations of *Campoletis sonorensis*, *Environ. Entomol.*, **6**, 586-588.

Loke, W. H., T. R. Ashley, and R. I. Sailer. (1983). Influence of fall armyworm, *Spodoptera frugiperda*, (Lepidoptera: Noctuidae) larvae and corn plant damage on host finding in *Apanteles marginiventris* (Hymenoptera: Braconidae), *Environ. Entomol.*, **12**, 911-915.

MacKenzie, D. R. (1980). "The problem of variable pests," in F. G. Maxwell and P. R. Jennings, Eds., *Breeding Plants Resistant to Insects*, Wiley-Interscience, New York, pp. 183-213.

Mahadevan, A. (1982). *Biochemical Aspects of Plant Disease Resistance*, Today & Tomorrow's Printers and Publishers, New Dehli, India, 381 pp.

Martin, J. S., and M. M. Martin. (1982). Tannin assays in ecological studies: Lack of correlation between phenolics, proanthocyanidins, and protein-precipitating constituents in mature foliage of six oak species, *Oecolgia*, **54**, 205-311.

Matheis, G., and J. R. Whitaker. (1984). Modification of proteins by polyphenol oxidase and peroxidase and their products, *J. Food Biochem.* **8**, 137-162.

Matta, A., I. Gentile, and I. Giai. (1969). Accumulation of phenols in tomato plants infected by different forms of *Fusarium oxysporum*, *Phytopathology*, **59**, 512-513.

Mattson, W. J. (1980). Herbivory in relation to plant nitrogen content, *Ann. Rev. Ecol. Syst.*, **11**, 119-161.

Maxwell, F. G. (1977). Plant resistance to cotton insects, *Bull. Entomol. Soc. Amer.*, **23**, 199-203.

Maxwell, F. G., and Jennings, P. R., Eds. (1980). *Breeding Plants Resistant to Insects*, Wiley-Interscience, New York, 683 pp.

McCance, D. J., and R. B. Drysdale. (1975). Production of tomatine and rishitin in tomato plants inoculated with *Fusarium oxysporum* f. sp. *lycopersici*, *Physiol. Plant Physiol.*, **7**, 221-230.

McClure, M. S. (1983). "Competition between herbivores and increased resource heterogeneity," in R. F. Denno and M. S. McClure, Eds., *Variable Plants and Herbivores in Natural and Managed Systems*, Academic, New York, pp. 135-154.

McCutcheon, G. S., and S. G. Turnipseed. (1981). Parasites of lepidopterous larvae in resistant and susceptible soybeans in South Carolina, *Environ. Entomol.*, **10**, 69-74.

McKey, D. (1974). Adaptive patterns in alkaloid physiology, *Am. Nat.*, **108**, 305-320.

McKey, D., P. G. Waterman, C. N. Mbi, J. S. Gartlan, and T. T. Struhsaker. (1978). Phenolic content of vegetation in two African rain forests: Ecological implications, *Science*, **202**, 61-64.

McNeill, S. and T. R. E. Southwood. (1978). "The role of nitrogen in the development of insect/plant relationships," in J. B. Harborne, Ed., *Biochemical Aspects of Plant and Animal Coevolution*, Academic, New York, pp. 78-98.

Mendez, J., and S. A. Brown. (1971). Changes in the phenolic metabolism of tomato plants infected by *Agrobacterium tumefasciens*, *Can. J. Bot.*, **49**, 2101-2105.

Milliron, H. E. (1940). A study of some factors affecting the efficiency of *Encarsia formosa* Gahan, an Aphelinid parasite of the greenhouse whitefly, *Trialeurodes vaporariorum*, *Mich. State Agric. Exp. Stn. Tech. Bull. 173*, 23 pp.

Mink, G. I., and K. N. Saksena. (1971). Studies on the mechanism of oxidative inactivation of plant virus by *o*-quinones, *Virology*, **45**, 755-763.

Mooney, H. A., and S. L. Gulmon. (1982). Constraints on leaf structure and function in reference to herbivory, *BioScience*, **32**, 198-206.

Morgan, A. C. (1910). Observations recorded at the 236th regular meeting of the Entomological Society of Washington, *Proc. Entomol. Soc. Wash.*, **12**, 72.

Mueller, T. F. (1983). The effect of plants on the host relations of a specialist parasitoid of *Heliothis* larvae, *Entomol. Exp. Appl.*, **34**, 78-84.

Nes, W. D., G. A. Saunders, and E. Heftman. (1983). A reassessment of the role of steroidal alkaloids in the physiology of *Phytophthora*, *Phytochemistry*, **22**, 75-78.

Nickol, B. B., Ed. (1979). *Host–Parasite Interfaces*, Academic, New York, 144 pp.

Nilakhe, S. S., and R. B. Chalfant. (1981). Distribution of insect pest eggs on processing tomato plants, *J. Georgia Entomol. Soc.*, **16**, 445-450.

Niles, G. A. (1980). "Breeding cotton for resistance to insect pests," in F. G. Maxwell and P. R. Jennings, Eds., *Breeding Plants Resistant to Insects*, Wiley-Interscience, New York, pp. 337-369.

Nishie, K., W. P. Norred, and A. P. Swain. (1975). Pharmacology and toxicology of chaconine and tomatine, *Res. Commun. Chem. Pathol. Pharmacol.*, **12**, 657-668.

Nordlund, D. A., and C. E. Sauls. (1981). Kairomones and their use for management of entomorphagous insects. XI. Effect of host plants on kairomonal activity of frass from *Heliothis zea* larvae for the parasitoid *Microplitis croceipes*, *J. Chem. Ecol.*, **7**, 1057-1061.

Nordlund, D. A., R. L. Jones, and W. J. Lewis, Eds. (1981). *Semiochemicals: Their Role in Pest Control*, Wiley, New York, 306 pp.

Norris, D. M., and M. Kogan. (1980). "Biochemical and morphological bases of resistance," in F. G. Maxwell and P. R. Jennings, Eds., *Breeding Plants Resistant to Insects*, Wiley, New York, pp. 23-61.

Oatman, E. R., G. R. Platner, J. A. Wyman, R. A. Van Steenwyk, M. W. Johnson, and H. W. Browning. (1983). Parasitization of lepidopterous pests of fresh market tomatoes in southern California, *J. Econ. Entomol.*, **76**, 452-455.

Obrycki, J. J., M. J. Tauber, and W. M. Tingey. (1983). Predator and parasitoid interaction with aphid-resistant potatoes to reduce aphid densities: A two-year field study, *J. Econ. Entomol.*, **76**, 456-462.

Ortman, E. E. and D. C. Peters. (1980). "Introduction," in F. G. Maxwell and P. R. Jennings, Eds., *Breeding Plants Resistant to Insects*, Wiley-Interscience, New York, pp. 3-13.

Pair, S. D., M. L. Laster, and D. F. Martin. (1982). Parasitoids of *Heliothis* spp. (Lepidoptera: Noctuidae) larvae in Mississippi associated with sesame interplanting in cotton, 1971-74: Implications of host-habitat interaction, *Environ. Entomol.*, **11**, 509-512.

Panda, N. (1979). *Principles of Host-Plant Resistance to Insect Pests*, Allanheld/Universe, New York, 386 pp.

Pathak, M. D., and R. C. Saxena. (1980). "Breeding approaches in rice," in F. G. Maxwell and P. R. Jennings, Eds., *Breeding Plants Resistant to Insects*, Wiley-Interscience, New York, pp. 421-455.

Pencoe, N. L., and R. E. Lynch. (1982). Distribution of *Heliothis zea* eggs and first-instar larvae on peanuts, *Environ. Entomol.*, **11**, 243-245.

Pierpoint, W. S. (1982). A class of blue quinone–protein coupling products: The allagochromes? *Phytochemistry*, **21**, 91-95.

Pierpoint, W. S. (1983). "Reactions of phenolic compounds with proteins, and their relevance to the production of leaf protein," in L. Telek and H. D. Graham, Eds., *Leaf Protein Concentrates*, Avi Publ. Comp., Inc., Wesport, Connecticut, pp. 235-267.

Pimentel, D., and A. G. Wheeler, Jr. (1973). Influence of alfalfa resistance on a pea aphid population and its associated parasites, predators, and competitors, *Environ. Entomol.*, **2**, 1-11.

Pollock, C. J., and R. B. Drysdale. (1976). The role of phenolic compounds in the resistance of tomato cultivars to *Verticillium alboatrum*, *Phytopath. Z.*, **86**, 56-66.

Prestidge, R. A. (1982). Instar duration, adult consumption, oviposition and nitrogen utilization efficiencies of leafhoppers feeding on different quality food (Auchenorrhyncha: Homoptera), *Ecol. Entomol.*, **7**, 91-101.

Price, J. P., and S. L. Poe. (1977). Influence of stoke and mulch on lepidopteran pests of tomatoes, *Fla. Entomol.*, **60**, 173–176.

Price, P. W. (1975a). "Reproductive strategies of parasitoids," in P. W. Price, Ed., *Evolutionary Strategies of Parasitic Insects and Mites*, Plenum, New York, pp. 87–111.

Price, P. W., Ed. (1975b). *Evolutionary Strategies of Parasitic Insects and Mites*, Plenum, New York, 224 pp.

Price, P. W. (1983). "Hypotheses on organization and evolution in herbivorous communities," in R. F. Denno and M. S. McClure, Eds., *Variable Plants and Herbivores in Natural and Managed Systems*, Academic, New York, pp. 559–596.

Price, P. W., C. E. Bouton, P. Gross, B. A. McPheron, J. N. Thompson, and A. E. Weis. (1980). Interactions among three trophic levels: Influence of plants on interactions between insect herbivores and natural enemies, *Ann. Rev. Ecol. and Syst.*, **11**, 41–65.

Prokopy, R. J. (1983). Visual detection of plants by herbivorous insects, *Ann. Rev. Entomol.*, **28**, 337–364.

Prokopy, R. J., and R. P. Webster. (1978). Oviposition-deterring pheromone of *Rhagoletis pomonella*. A kairomone for its parasitoid *Opius lectus*, *J. Chem. Ecol.*, **4**, 481–494.

Puttler, B. (1961). Biology of *Hyposoter exiguae* (Hymenoptera: Ichneumonidae); A parasite of lepidopterous larvae, *Ann. Entomol. Soc. Amer.*, **54**, 25–30.

Rabb, R. L., and J. R. Bradley. (1968). The influence of host plants on parasitism of eggs of the tobacco hornworm, *J. Econ. Entomol.*, **61**, 1249–1252.

Raman, K. V., W. M. Tingey, and P. Gregory. (1979). Potato glycoalkaloids: Effect on survival and feeding behavior of the potato leafhopper, *J. Econ. Entomol.*, **72**, 337–341.

Reese, J. C. (1979). "Interactions of allelochemicals with nutrients in herbivore food," in G. A. Rosenthal and D. H. Janzen, Eds., *Herbivores, Their Interaction with Secondary Plant Metabolites*, Academic, New York, pp. 309–330.

Reese, J. C. (1981). "Insect dietetics: Complexities of plant-insect interactions," in G. Bhaskaran, S. Friedman, and J. G. Rodriguez, Eds., *Current Topics in Insect Endocrinology and Nutrition: A Tribute to Gottfried S. Fraenkel*, Plenum, New York, pp. 317–335.

Reese, J. C., B. G. Chan, and A. C. Waiss, Jr. (1982). Effects of cotton condensed tannin, maysin (corn) and pinitol (soybeans) on *Heliothis zea* growth and development, *J. Chem. Ecol.*, **8**, 1429–1436.

Reuveni, R., and J. F. Ferreira. (1985). The relationship between peroxidase activity and the resistance of tomatoes (*Lycopersicon esculentum*) to *Verticillium dahliae*, *Phytopath. Z.*, **112**, 193–197.

Rhoades, D. F. (1979). "Evolution of plant defense against herbivores," in G. A. Rosenthal and D. H. Janzen, Eds., *Herbivores: Their Interaction with Secondary Plant Metabolites*, Academic, New York, pp. 1–54.

Rhoades, D. F. (1983). "Herbivore population dynamics and plant chemistry," in R. F. Denno and M. S. McClure, Eds., *Variable Plants and Herbivores in Natural and Managed Systems*, Academic, New York, pp. 155–220.

Rice, E. L. (1974). *Allelopathy*, Academic, New York, 353 pp.

Rice, E. L. (1979). Allelopathy—an update, *Bot. Rev.*, **45**, 15–109.

Ritter, K. S., and W. R. Nes. (1981a). The effects of cholesterol on the development of *Heliothis zea*, *J. Insect Physiol.*, **27**, 175–182.

Ritter, K. S., and W. R. Nes. (1981b). The effects of the structure of sterols on the development of *Heliothis zea*, *J. Insect Physiol.*, **27**, 419–424.

Robinson, R. A. (1976). *Plant Pathosystems*, Advanced Series in Agric. Sciences III, Springer-Verlag, New York, 184 pp.

Robinson, R. A. (1980). "The pathosystem concept," in F. G. Maxwell and P. R. Jennings, Eds., *Breeding Plants Resistant to Insects*, Wiley-Interscience, New York, pp. 157–181.

Roddick, J. G. (1974). The steroidal glycoalkaloid, α-tomatine, *Phytochemistry*, **13**, 9–25.

Roddick, J. G. (1979). Complex formation between solanaceous steroidal glycoalkaloids and free sterols in vitro, *Phytochemistry*, **18**, 1467–1470.

Roddick, J. G. (1980). A sterol-binding assay for potato glycoalkaloids, *Phytochemistry*, **19**, 2455–2457.

Rosenthal, G. A. (1982). *Plant Nonprotein Amino and Imino Acids*, Academic, New York, 273 pp.

Rosenthal, G. A., and D. H. Janzen, Eds. (1979). *Herbivores, Their Interaction with Secondary Plant Metabolites*, Academic, New York, 718 pp.

Rosenthal, G. A., D. L. Dahlman, and D. H. Janzen. (1976). A novel means for dealing with L-canavanine, a toxic metabolite, *Science*, **192**, 256–258.

Rosenthal, G. A., D. H. Janzen, and D. L. Dahlman. (1977). Degradation and detoxification of canavanine by a specialized seed predator, *Science*, **196**, 658–660.

Rosenthal, G. A., D. L. Dahlman, and D. H. Janzen. (1978). L-Canaline detoxification: A seed predator's biochemical mechanism, *Science*, **202**, 528–529.

Roth, J. P., E. G. King, and S. D. Hensley. (1982). Plant, host, and parasite interactions in the host selection sequence of the tachinid *Lixophaga diatraeae*, *Environ. Entomol.*, **11**, 273–277.

Schlosser, E. (1977). Role of saponins in antifungal resistance VII. Significance of tomatine in species-specific resistance of tomato fruits against fruit-rotting fungi, *Meded. Fac. Landbouwwet. Rijksuniv. Genet.*, **41**, 499–503.

Schoeneweiss, D. F. (1975). Predisposition, stress, and plant disease, *Ann. Rev. Phytopathol.*, **13**, 193–211.

Schonbeck, F., and E. Schlosser. (1976). "Preformed substances as potential inhibitors," in R. Heitefuss and P. H. Williams, Eds., *Physiological Plant Pathology*, Springer-Verlag, Berlin, pp. 653–678.

Schultz, J. C. (1983a). "Impact of variable plant defensive chemistry on susceptibility of insects to natural enemies," in P. A. Hedin, Ed., *Plant Resistance to Insects*, American Chemical Society, Washington, D.C., pp. 37–54.

Schultz, J. C. (1983b). "Habitat selection and foraging tactics of caterpillars in heterogeneous trees," in R. F. Denno and M. S. McClure, Eds., *Variable Plants and Herbivores in Natural and Managed Systems*, Academic, New York, pp. 61–90.

Schultz, J. C., and I. T. Baldwin. (1982). Oak leaf quality declines in response to defoliation by gypsy moth larvae, *Science*, **217**, 149–151.

Schuster, D. J. 1977. Effect of tomato cultivars on insect damage and chemical control, *Fla. Entomol.*, **60**, 227–232.

Schuster, D. J., and K. J. Starks. (1975). Preference of *Lysiphlebus testacipes* for greenbug resistant and susceptible small grain species, *Environ. Entomol.*, **4**, 887–888.

Scott, D. R., and L. E. O'Keeffe, Eds. (1976). *Lygus Bug: Host Plant Interactions*, Proceedings of a workshop at the XV Int. Cong. Entomol., University Press of Idaho, Moscow, 38 pp.

Scriber, J. M. (1978). The effects of larval feeding specialization and plant growth form on the consumption and utilization of plant biomass and nitrogen: An ecological consideration, *Entomol. Exp. App.*, **24**, 494–510.

Scriber, J. M. (1979a). Post-ingestive utilization of plant biomass and nitrogen by Lepidoptera: Legume feeding by the southern armyworm, *N.Y. Entomol. Soc.*, **87**, 141–153.

Scriber, J. M. (1979b). Effects of leaf-water supplementation upon post-ingestive nutritional indeces of forb-, shrub-, vine-, and tree-feeding lepidoptera, *Entomol. Exp. Appl.*, **25**, 240–252.

Shany, S., Y. Birk, B. Gestetner, and A. Bondi. (1970a). Preparation, characterisation and some properties of saponins from lucerne tops and roots, *J. Sci. Food Agric.*, **21**, 131–135.

Shany, S., B. Gestetner, Y. Birk, and A. Bondi. (1970b). Lucerne saponins III—Effect of lucerne saponins on larval growth and their detoxication by various sterols, *J. Sci. Food Agric.*, **21**, 508–510.

Shapiro, V. A. (1956). The influence of the nutritional regimes of the host on the growth of certain insect parasites (in Russian), *Zhurnal Obshcher Biology*, **17**, 218–227.

Shukle, R. H., and L. L. Murdock. (1983). Lipoxygenase, trypsin inhibitor, and lectin from soybeans: Effects on larval growth of *Manduca sexta* (Lepidoptera: Sphingidae), *Environ. Entomol.*, **12**, 787–791.

Sinden, S. L., J. M. Schalk, and A. K. Stoner. (1978). Effects of daylength and maturity of tomato plants on tomatine content and resistance to the Colorado potato beetle, *J. Am. Soc. Hort.*, **103**, 596–599.

Singleton, V. L. (1981). Naturally occurring food toxicants: Phenolic substances of plant origin common in foods, *Adv. Food Res.*, **27**, 149–242.

Slansky, F., Jr. (1978). Utilization of energy and nitrogen by larvae of the imported cabbageworm, *Pieris rapae*, as affected by parasitism by *Apanteles glomeratus*, *Environ. Entomol.*, **7**, 179–185.

Slansky, F., Jr. (1982). Insect nutrition: An adaptionist's perspective, *Fla. Entomol.*, **65**, 45–71.

Slansky, F. Jr., and P. Feeny. (1977). Stabilization of the rate of nitrogen accumulation by the larvae of the cabbage butterfly on wild and cultivated food plants, *Ecol. Monogr.*, **47**, 209–228.

Smilowitz, Z., and C. L. Smith. (1977). Haemolymph proteins of developing *Pieris rapae* larvae parasitized by *Apanteles glomeratus*, *Ann. Entomol. Soc. Amer.*, **70**, 447–454.

Smith, J. M. (1957). Effects of the food plant of California red scale, *Aonidiella aurantii* (Mask.) on reproduction of its hymenopterous parasites, *Can. Entomol.*, **89**, 219–230.

Snodderly, L. J., and P. L. Lambdin. (1982). Oviposition and feeding sites of *Heliothis zea* on tomato, *Environ. Entomol.*, **11**, 513–515.

Starks, K. J., and R. L. Burton. (1977). Greenbugs: a comparison of mobility on resistant and susceptible varieties of four small grains, *Environ. Entomol.*, **6**, 331–332.

Starks, K. J., R. Muniappan, and R. D. Eikenbary. (1972). Interaction between plant resistance and parasitism against greenbug on barley and sorghum, *Ann. Entomol. Soc. Amer.*, **65**, 650–655.

Stiling, P. D., B. V. Brodbeck, and D. R. Strong. (1982). Foliar nitrogen and larval parasitism as determinants of leafminer distribution patterns on *Spartina alterniflora*, *Ecol. Entomol.*, **7**, 447–452.

Stipanovic, R. D. (1983). "Function and chemistry of plant trichomes and glands in insect resistance," in P. A. Hedin, Ed., *Plant Resistance to Insects*, ACS Symposium Series 208, American Chemical Society, Washington, D.C., pp. 70–100.

Streams, F. A., M. Shahjahan, and H. G. LeMasurier. (1968). Influence of the plant on the parasitization of the tarnished plant bug by *Leiophron pallipes*, *J. Econ. Entomol.*, **61**, 996–999.

Strong, D. R., J. H. Lawton, and R. Southwood, Sir. (1984). *Insects on Plants. Community Patterns and Mechanisms*, Harvard University Press, Cambridge, Massachusetts, 313 pp.

Teetes, G. L. (1980). "Breeding sorghums resistant to insects," in F. G. Maxwell and P. R. Jennings, Eds., *Breeding Plants Resistant to Insects*, Wiley-Interscience, New York, pp. 421–485.

Teetes, G. L., J. R. Schaefer, and J. W. Johnson. (1974). Resistance in sorghums to the greenbug: laboratory determinations of mechanisms of resistance, *J. Econ. Entomol.*, **67**, 393–396.

Telek, L., and H. D. Graham, Eds. (1983). *Leaf Protein Concentrates*, AVI Publ., Wesport, CN, 844 pp.

Terriere, L. C. (1984). Induction of detoxication enzymes in insects, *Ann. Rev. Entomol.*, **29**, 71–88.

Thompson, S. N. (1977). Lipid nutrition during larval development of the parasitic wasp, *Exeristes*, *J. Insect Physiol.*, **23**, 579–583.

Thompson, S. N. (1982). Effects of parasitization by the insect parasite *Hyposoter exiguae* on the growth, development and physiology of its host *Trichoplusia ni*, *Parasitology*, **84**, 491–510.

Thurston, R., and P. M. Fox. (1972). Inhibition by nicotine of emergence of *Apanteles congregatus* from its host, the tobacco hornworm, *Ann. Entomol. Soc. Amer.*, **65**, 547–550.

Tingey, W. M. (1984). Glycoalkaloids as pest resistance factors, *Am. Potato J.*, **61**, 157–167.

Tingey, W. M., and R. W. Gibson. (1978). Feeding and mobility of the potato leafhopper impaired by glandular trichomes of *Solanum berthaultii* and *S. polyadenium*, *J. Econ. Entomol.*, **71**, 857–858.

Tingey, W. M., and S. L. Sinden. (1982). Glandular pubescence, glycoalkaloid composition, and resistance to the green peach aphid, potato leafhopper, and potato fleabeetle in *Solanum berthaultii*, *Am. Potato J.*, **59**, 95–106.

Tingey, W. M., R. L. Plaisted, J. E. Laubengayer, and A. Mehlenbacher. (1982). Green peach aphid resistance by glandular trichomes in *Solanum tubersosum* × *S. berthaultii* hybrids, *Am. Potato J.*, **59**, 241–251.

Tingey, W. M., P. Gregory, R. L. Plaisted, and M. J. Tauber. (1984). "Research progress: Potato glandular trichomes and steroid glycoalkaloids," in *Proceedings of the 27th CIP Planning Conference, Integrated Pest Management for the Potato*, International Potato Center, Lima, Peru, pp. 115–124.

Todd, G. W., A. Getahun, and D. C. Cress. (1971). Resistance in barley to the greenbug, *Schizaphis graminum*. 1. Toxicity of phenolic and flavonoid substances, *Ann. Entomol. Soc. Amer.*, **64**, 718–722.

van Emden, H. S. (1969). Plant resistance to *Myzus persicae* induced by a plant regulator and measured by aphid relative growth rate, *Entomol. Exp. Appl.*, **12**, 125–131.

van Emden, H. S., and M. A. Bashford. (1969). A comparison of the reproduction of *Brevicoryne brassicae* and *Myzus persicae* in relation to soluble nitrogen concentration and leaf age (leaf position in Brussels sprout plant), *Entomol. Exp. Appl.*, **12**, 351–364.

van Emden, H. F., and C. H. Wearing. (1965). The role of the aphid host plant in delaying economic damage levels in crops, *Ann. Appl. Biol.*, **56**, 323–324.

van Lenteren, J. C., P. M. J. Ramakers, and J. Woets. (1980). "Integrated control of vegetable pests in greenhouses," in A. K. Minks and P. Gruys, Eds., *Integrated Control of Insect Pests in the Netherlands*, Centre for Agricultural Publishing and Documentation, Wageningen, The Netherlands, pp. 109–120.

van Sumere, C. F., J. Albrecht, A. Dedonder, H. de Pooter, and I. Pe. (1975). "Plant proteins and phenolics," in J. B. Harborne and C. F. van Sumere, Eds., *The Chemistry and Biochemistry of Plant Proteins*, Academic, New York, pp. 211–264.

Verma, V. S. (1973). Effect of flavonoids on the infectivity of potato X virus, *Zentralbl. Bakteriol., Parasentik., Infektionskr. Hyg., Abt. 2*, **128**, 467–472.

Versoi, P. L., and W. G. Yendol. (1982). Discrimination by the parasite, *Apanteles melanoscelus*, between healthy and virus-infected gypsy moth larvae, *Environ. Entomol.*, **11**, 42–45.

Vinson, S. B. (1984). "Parasitoid–host relationship," in W. J. Bell, and R. T. Cardé, Eds., *Chemical Ecology of Insects*, Sinauer, Sunderland, MA, pp. 205–233.

Vinson, S. B. (1976). Host selection by insect parasitoids, *Ann. Rev. Entomol.*, **21**, 109–133.

Vinson, S. B. (1975). "Biochemical coevolution between parasitoids and their hosts," in P. W. Price, Ed., *Evolutionary Strategies of Parasitic Insects and Mites*, Plenum, New York, pp. 14–48.

Vinson, S. B., and G. F. Iwantsch. (1980). Host regulation by insect parasitoids, *Q. Rev. Biol.*, **55**, 143–165.

Waage, J. K. (1979). Foraging for patchily-distributed hosts by the parasitoid, *Nemeritis canescens*, *J. An. Ecol.*, **48**, 353–371.

Waage, J. K., and M. P. Hassel. (1982). Parasitoids as biological control agents—A fundamental approach, *Parasitology*, **84**, 241–268.

Waiss, A. C., B. G. Chan, C. A. Elliger, D. L., Dryer, R. G. Binder, and R. C. Guledner. (1981). Insect growth inhibitors in crop plants, *Bull. Entomol. Soc. Amer.*, **27,** 217–221.

Way, M. J. (1973). "Objectives, methods and scope of integrated control," in P. W. Geier, L. R. Clark, D. J. Anderson, and H. A. Nix, Eds., *Insects: Studies on Population Management*, Ecolog. Soc. Australia, Memoirs 1, pp. 1–17.

Weseloh, R. M. (1981). "Host location by parasitoids," in D. A. Nordlund, R. L. Jones, and W. J. Lewis, Eds. *Semiochemicals: Their Role in Pest Control*, Wiley, New York, pp. 79–95.

Weseloh, R. M., T. G. Andreadis, R. E. Moore, and J. F. Anderson. (1983). Field confirmation of a mechanism causing synergism between *Bacillus thuringiensis* and the gypsy moth parasitoid, *Apanteles melanoscelus*, *J. Invert. Pathol.*, **41,** 99–103.

White, T. C. R. (1984). The abundance of invertebrate herbivores in relation to the availability of nitrogen in stressed food plants, *Oecologia*, **63,** 90–105.

White, T. C. R. (1976). Weather, food, and plagues of locusts, *Oecologia*, **22,** 119–134.

White, T. C. R. (1978). The importance of relative shortage of food in animal ecology, *Oecologia*, **33,** 71–86.

Whitehead, D. L., and W. S. Bowers, Eds. (1983). *Natural Products for Innovative Pest Management*, Pergamon, New York, 586 pp.

Whitham, T. G. (1983). "Host manipulation of parasites: Within-plant variation as a defense against rapidly evolving pests," in R. F. Denno and M. S. McClure, Eds., *Variable Plants and Herbivores in Natural and Managed Systems*, Academic, New York, pp. 15–41.

Wilson, R. H., G. W. Poley, and F. Deeds. (1961). Some pharmacological and toxicological properties of tomatine and its derivatives, *Toxicol. Appl. Pharmacol.*, **3,** 39–48.

Wyatt, I. J. (1970). The distribution of *Myzus persicae* (Sulz.) on year-round chrysanthemums, *Ann. Appl. Biol.*, **65,** 31–41.

Wyman, J. A., E. R. Oatman, and M. W. Johnson. (1979). Key to efficient tomato production— Insect pest management, *Calif. Fresh Mkt. Tomato Adv. Brd. Info. Bull. 22*, 6 pp.

Yu, S. J. (1982a). Induction of microsomal oxidases by host plants in the fall armyworm, *Spodoptera frugiperda* (J. E. Smith), *Pestic. Biochem. Physiol.*, **17,** 59–67.

Yu, S. J. (1982b). Host-plant induction of glutathione-*S*-transferase in the fall armyworm, *Pestic. Biochem. Physiol.*, **18,** 101–106.

Zhody, N. (1976). On the effect of the food of *Myzus persicae* Sulz. on the hymenopterous parasite *Aphelinus asychis* Walker, *Oecologia*, **26,** 185–191.

Zucker, W. V. (1983). Tannins: Does structure determine function? An ecological perspective, *Am. Nat.*, **121,** 335–365.

# 7

---

# ECOLOGY OF INSECT–PATHOGEN INTERACTIONS AND SOME POSSIBLE APPLICATIONS

*Robert M. May*

## 7.1. INTRODUCTION

Viral, bacterial, protozoan, fungal, and other pathogens can kill insects. The possibility of using such pathogens—either alone or in combination with other agents—to control insect pests is attracting increasing attention.

At the empirical level, there are now several case studies of the use of pathogens in efforts to control particular insect pests; the use of the bacterium *Bacillus thuringiensis* against the gypsy moth [*Lymantria dispar* (L.)] in North America is probably the best known example. These empirical studies are well reviewed by Tinsley and Entwhistle (1974), Tinsley (1979), Falcon (1982), and others. At the theoretical level, Anderson and May (1981) have given a monographical analysis of the population dynamics of pathogens and their invertebrate hosts, with the aim of bringing together two separate literatures (one dealing with the ecology of animal populations and the other with invertebrate pathology) to understand how infections persist within, and may regulate or even eradicate, populations of invertebrates. Remaining conspicuous by their absence are studies, either empirical or theoretical, that consider the interplay between pathogens and other regulatory agents in the integrated control of insect pest populations: It would be nice to see, for example, studies that combined theory and experiment to understand the possible effects of pathogens together with natural enemies, or pathogens together with pesticides, upon populations of insect pests.

Against this broad background, this chapter is very narrowly focused. It aims to present no more than a sketchy outline of the theory of invertebrate host-pathogen interactions, as developed by Anderson and May (1981), McNamee, et al. (1981), Getz and Pickering (1983), Ewald (1985), and others. To this end, Section 7.2 presents a basic mathematical model, emphasizing the biological

185

ingredients of the model and the biological conclusions that can be drawn from it. A variety of realistic refinements are then grafted onto this model, and the biological effects of these modifications are discussed (and summarized in Table 7.2). Section 7.3 briefly reviews some of the more complicated dynamical behavior—including regular or irregular cycles, or apparently random population fluctuations—that can emerge from such simple and deterministic population models. Section 7.3 indicates that host–pathogen dynamics can be complex and unsteady, even before environmental stochasticity is accounted for. Section 7.4 deals with control of a pest by a pathogen that, once introduced, is left to maintain itself within the pest population. In contrast, Section 7.5 examines the rate at which a pathogen must be released, year after year, to exterminate a pest population. The concluding Section 7.6 draws attention to some of the shortcomings of the work presented here, and suggests some problems that need more study.

## 7.2. BASIC MODEL AND SOME REFINEMENTS

### 7.2.1. A Basic Model

Some basic ideas can be developed by considering the idealized case of a population that undergoes pure exponential growth, except insofar as it is affected by an infectious disease. Most invertebrate populations will consist of several distinct but overlapping age classes, or even discrete and nonoverlapping age classes. It is, however, often a useful approximation to treat the total population of adults or larvae as being a continuous function $N(t)$ of the time $t$ (so that the dynamics of the population is described by differential equations rather than by some set of difference equations).

In the basic model, the invertebrate host population is assumed to have a per capita birth rate $a$ and a per capita death rate $b$ from all causes other than the pathogen, both of which are density independent. The total population $N(t)$ is partitioned into uninfected and infected hosts, with population densities $X(t)$ and $Y(t)$, respectively ($X + Y = N$). Although invertebrate species are usually able to mount cellular or humoral responses to infection, current evidence suggests they are not able to develop acquired immunity to agents of infectious disease (see the discussion in Anderson and May, 1981). Thus, the basic model is lacking the class of recovered-and-immune hosts found in conventional epidemiological models for vertebrates.

The basic model is further restricted to infections that are transmitted directly, as opposed to those with indirect transmission (involving one or more species of intermediate hosts). The simplest assumption is that the rate at which infections are acquired by such direct transmission is proportional to the number of encounters between susceptible and infected hosts. That is, the net rate of transmission of the infection is $\beta XY$, where $\beta$ is called the transmission parameter. This description of the infection process is reasonably realistic in some cases, as reviewed by Anderson and May (1981). Infected hosts either recover (at a rate $\gamma$, so that the characteristic recovery time is $1/\gamma$) or are killed by the disease (at a rate $\alpha$).

When all this is put together, the dynamics of the model population is given by the pair of equations

$$dX/dt = a(X + Y) - bX - \beta XY + \gamma Y \tag{1}$$

$$dY/dt = \beta XY - (\alpha + b + \gamma)Y \tag{2}$$

New susceptibles appear by birth (from either infected or uninfected individuals) or by recovery from the infected state; infectious individuals appear at the rate $\beta XY$, and remain infectious for an average time of $1/(\alpha + b + \gamma)$ before they die of the disease or of other causes, or recover. These dynamical processes are illustrated schematically in Fig. 7.1.

In reality, a greater or lesser degree of environmental and demographic stochasticity will be superimposed on this deterministic description of the dynamics of the population. In some crude sense, the population density, $N(t)$, and the demographic and epidemiological parameters, such as $a$, $b$, $\alpha$, $\gamma$, and $\beta$, are to be thought of as rough averages, around which there may be substantial fluctuations.

### 7.2.2. Biological Properties of the Basic Model

In the absence of infection ($Y = 0$, $N = X$), the model population will obviously grow exponentially, with a per capita population growth rate $r = a - b$. If a small number of infectious individuals are introduced into such a disease-free population (corresponding to $X \simeq N$, $Y \ll N$), the infection will spread ($Y$ will increase; $dY/dt > 0$) provided the right-hand side of Eq. (2) is positive (which

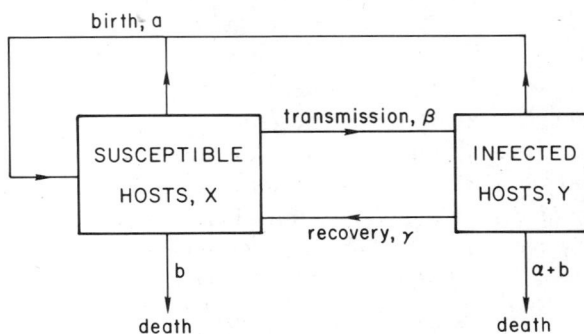

FIG. 7.1 Schematic representation of the assumptions embodied in the basic model defined by Eqs. (1) and (2). The "flow rates" of hosts into and out of the susceptible and infected classes are governed by the demographic and epidemiological parameters $a$, $b$, $\alpha$, $\gamma$, and $\beta$, as defined in the text.

will be so if $\beta N > \alpha + b + \gamma$). That is, the infection can establish itself provided the host population exceeds some threshold density, $N_T$, defined by

$$N_T \equiv (\alpha + b + \gamma)/\beta \qquad (3)$$

The requirement $N > N_T$ can equivalently be thought of as the criterion for the basic reproductive rate of the parasite within the host population to be greater than unity ($R_0 > 1$; see Anderson and May, 1981). Given the exponential growth of this host population in the absence of disease, the threshold will always be attained eventually, and thus the infection can always establish itself sooner or later.

Once established, the virus, bacterium, or other infectious agent can regulate the host population density provided it is sufficiently pathogenic:

$$\alpha > r \qquad (4)$$

This criterion, Eq. (4), can be derived from Eqs. (1) and (2). The criterion can, however, also be appreciated in an intuitive way: In the absence of infection, the host population grows at the per capita rate $r = a - b$; for this overall growth to be halted, the per capita death rate of infected hosts must outrun their birth rate, $\alpha + b > a$, whence Eq. (4). The criterion of Eq. (4) is significantly simpler than the corresponding result for most vertebrate hosts, where the presence of a large class of recovered-and-immune hosts adds complexity (Anderson and May, 1979).

If the regulatory criterion of Eq. (4) is satisfied, then in this simplest model the host population is regulated to a constant equilibrium value $N^*$ given by

$$N^* = [\alpha/(\alpha - r)]N_T \qquad (5)$$

Here $N_T$ is defined by Eq. (3). The fraction of this equilibrium host population that is infected at any one time is simply $y^* \equiv Y^*/N^* = r/\alpha$; the more virulent the infection (the larger $\alpha$), the smaller the equilibrium fraction infected. On the other hand, if Eq. (4) is not satisfied, the host population continues to grow exponentially at a diminished per capita rate, $r' = r - \alpha$, until other kinds of regulatory effects can no longer be ignored.

Table 7.1 lists some estimates of per capita mortality rates induced by pathogens $\alpha$ and from other natural causes $b$ for a miscellany of invertebrate hosts and viral, bacterial, protozoan, and fungal infections. Table 7.1 lacks corresponding estimates of the per capita birth rates $a$ and consequently lacks comparisons between $\alpha$ and $r = a - b$; there are, unfortunately, very few field studies that give estimates of all three parameters ($a$, $b$, $\alpha$). Thus, it is hard to say how often $\alpha$ exceeds $r$ in the field. The data compiled in Table 7.1 do no more than indicate that many disease agents of invertebrates are highly pathogenic, with the most pronounced pathogenicity tending to arise among viral and bacterial agents. Overall, the disease-induced death rates $\alpha$ in Table 7.1 are typically an order of magnitude greater than the corresponding natural death rates $b$. The incompleteness of the information in Table 7.1 is compounded by the fact that laboratory studies tend to underestimate natural mortality rates (because many natural causes of death are not present in the laboratory), and by the biases introduced by invertebrate pathologists' interest in the more pathogenic disease agents by virtue of their possible use in biological control programs.

For a fuller account of the dynamical properties of Eqs. (1) and (2), and of the inadequacies and interpretations of Table 7.1, see Anderson and May (1981).

### 7.2.3. Refinements to the Basic Model

The following list of complications that must be added to the basic model is a summary of Sections 6–11 in the monograph by Anderson and May (1981). Here I indicate the effects such complications have upon the ability of the pathogen to maintain itself within, and to regulate, the host population; the detailed dynamics and such date as are available are presented in the above-mentioned monograph.

Many pathogens of invertebrates *decrease or entirely eliminate the reproductive capacity* of infected hosts. Suppose this results in the per capita birth rate of infected hosts being $(1 - f)a$, rather than $a$, where $1 \geq f \geq 0$. In the discussion of Eq. (4), we saw the intuitive basis of the criterion for the pathogen to be capable of regulating the host population is that the death rate outrun the birth rate for infected hosts. Applied to the more general circumstance where infected hosts have a depressed birth rate, this control criterion becomes

$$\alpha > a(1 - f) - b \tag{6}$$

Indeed, such an infection can regulate the host population even if it induces no mortality ($\alpha = 0$), provided $f > 1 - (b/a)$; this condition is obviously satisfied if there is no reproduction from infected hosts ($f = 1$). If Eq. (6) is satisfied, the host population is regulated to a stable equilibrium level that is lower than it would be for $f = 0$. A decrease in the reproductive capacity of infected hosts has, however, no effect on the threshold host density needed to maintain the infection.

*Vertical transmission* occurs when the infection is transmitted directly to some or all of the offspring of infected parents. One kind of vertical transmission may occur when the pathogen gains entry to the egg or embryo within the host, via infection of the reproductive organs of a parent; this "transovarial" transmission can come from male or female parent. A second kind of vertical transmission arises from the pathogen being absorbed on, or contaminating, the exterior of the egg during the birth process; such "transovum" transmission comes only from the female parent. Neither kind of vertical transmission has any effect on the ability of the pathogen to regulate its host population, so that the regulatory criterion continues to be given by Eq. (4), regardless of the fraction $q$ of offspring of infected hosts that are born infected. The threshold host density for

**Table 7.1 Natural and Pathogen-Induced Mortality Rates for Insect Hosts of Some Viral, Bacterial, Protozoan and Fungal Infections.**[a]

| Pathogen | Host | Natural Mortality Rate $b$ (per week) | Pathogen-Induced Mortality Rate $\alpha$ (per week) | Ratio $\alpha/b$ (one-figure accuracy) |
|---|---|---|---|---|
| Viruses | | | | |
| Sack brood virus | *Apis mellifera* | 0.17 | 1.2 | 7 |
| Nuclear-Polyhedrosis virus | *Cadra cautella* | 0.061 | 0.54 | 9 |
| Nuclear-polyhedrosis virus | *Hyphantria cunea* | 0.003 | 0.80 | 300 |
| A.B.P. virus | *Apis mellifera* | 0.25 | 1.9 | 8 |
| Nuclear-polyhedrosis virus | *Lymantria dispar* | 0.060 | 0.63 | 10 |
| Nuclear-polyhedrosis virus | *Malacosoma americanum* | 0.070 | 0.37 | 5 |
| Bacteria | | | | |
| Bacillus thuringiensis | *Simulium vittatum* | 0.035 | 2.4 | 70 |
| Bacillus thuringiensis | *Choristoneura fumiferana* | 0.001 | 4.0 | 4000 |
| Aeromonas punctata | *Anopheles annulipes* | 0.36 | 2.9 | 8 |
| Erwinia spp. | *Colladonus montanus* | 0.031 | 0.17 | 5 |

## Protozoa

| | | | | |
|---|---|---|---|---|
| Nosema stegomyiae | Anopheles albimanus | 0.23 | 0.41 | 2 |
| Pleistrophora schubergi | Hyphantria cunea | 0.003 | 0.036 | 10 |
| Herpetomonas muscarum | Hippelates pusio | 0.17 | 0.43 | 3 |
| Tetrahymena pyriformis | Culex tarsalis | 0.26 | 0.66 | 3 |

## Fungi

| | | | | |
|---|---|---|---|---|
| Beauveria tenella | Aedes siemensis | 0.026 | 0.50 | 20 |
| Beauveria tenella | Culex tarsalis | 0.11 | 0.84 | 8 |
| Beauveria bassiana | Musca domestica | 0.27 | 0.74 | 3 |
| Beauveria bassiana | Delia antiqua | 0.30 | 0.55 | 2 |
| Beauveria bassiana | Phormia regina | 0.24 | 0.56 | 2 |
| Metarrhizium anisopliae | Musca domestica | 0.27 | 0.38 | 1 |
| Metarrhizium anisopliae | Delia antiqua | 0.30 | 0.48 | 2 |
| Metarrhizium anisopliae | Phormia regina | 0.24 | 0.42 | 2 |
| Aspergillus flavus | Culex peus | 0.020 | 0.17 | 9 |
| Aspergillus flavus | Culex tarsalis | 0.061 | 0.22 | 4 |
| Fusarium oxysporum | Culex pipiens | 0.027 | 0.62 | 20 |

maintenance of the pathogen is, however, lowered by the effects of vertical transmission, with Eq. (3) being replaced by

$$N_T = (\alpha + b + \gamma - aq)/\beta \tag{7}$$

If $aq$ is low enough ($aq > \alpha + b + \gamma$), the pathogen can maintain itself in an
arbitrarily small host population ($N_T \to 0$). The equilibrium host density, $N^*$, is
still given by Eq. (5), but with $N_T$ given by Eq. (7) it follows that $N^*$ is lower than
in the absence of vertical transmission.

   Many pathogens undergo an *incubation or latent period* within the host before beginning to produce transmission stages (for horizontal transmission) or to
contaminate or infect unborn progeny of the host (for vertical transmission). As
one might expect, this has the effect of raising the threshold host density and of
making it harder for the pathogen to regulate the host. Specifically, the threshold host density $N_T$ of Eq. (3) is increased by a factor $(b + \sigma)/b$, and the regulatory criterion of Eq. (4) is replaced by

$$\alpha > r[1 + (\alpha + b + \gamma)/\sigma] \tag{8}$$

Here $\sigma$ is the rate at which infected but latent hosts move into the infectious
stage; in other words, the typical time spent in the latent phase is $1/\sigma$. These
results pertain to infections that are transmitted directly and horizontally, and
the presence of vertical transmission makes for additional complexities. In addition to the above effects, a latent period of significant duration can result in the
regulated state being one of sustained cyclic oscillations in host population density, rather than a constant equilibrium point.

   A certain amount of evidence suggests that the pathogenicity $\alpha$ may depend
on the degree to which the host population is under "stress" from prevailing
environmental conditions, such as food shortages, overcrowding, or extremes of
temperature or humidity. One crude way of incorporating such stress-related
pathogenicity into models for the host–pathogen dynamics is to let the parame-
ter $\alpha$ depend on the host population density [$\alpha \to \alpha(N)$, with $\alpha(N)$ some increasing function of $N$]. The ensuing complications are not easily summarized, but
they tend to have the effect that the disease can always regulate the host population, provided its intrinsic transmission efficiency is high enough.

   More generally there will always be density-dependent effects, produced by
resource limitation, predation, and so on, that ultimately can regulate the host
population in the absence of the pathogen. In brief, suppose such effects tend to
keep the host population around some "carrying capacity" $K$ when disease is
absent. Such carrying capacity effects will now complicate the definition of the
threshold density $N_T$ but an appropriately generalized version of Eq. (3) can be
given (Anderson and May, 1981). When this is done, it is found that the disease
cannot maintain itself within the host population if $N_T > K$. Conversely, if $N_T <
K$, the disease can be established and will depress the host population density

below the level $K$ found in the absence of disease. We will return to this particular complication in Section 7.4, below.

Up to this point, it has been assumed that transmission occurs by direct contact with infectious individuals. Many directly transmitted infections, however, possess long-lived transmission stages. Thus, the spores of many bacteria, protozoans, and fungi often possess tough proteinaceous coats, which enhance the ability of the infective agent to survive outside the host. Alternatively, relatively unspecialized stages, such as the capsules, polyhedra, or free particles of viruses, may be produced. As reviewed by Anderson and May (1981), these free-living transmission stages can survive for long time, sometimes up to many years or even decades, in the external environment. In such cases, it can become necessary to expand the basic model to include a dynamical description of these populations of free-living infective stages of the pathogen. When this is done, it is found that the pathogen may regulate its host population [essentially if Eq. (4) is satisfied], or that it may slow the exponential growth of its host population, or (a new possibility, not found in the basic model above) that the pathogen may simply be incapable of ever maintaining itself within the host population. If the pathogen does indeed regulate its host, it may do so at a stable level, or in a stably cyclic fashion (producing marked oscillations in host density, with periods up to 40 years or more). These dynamical complications are pursued in the next section.

Table 7.2 summarizes the above discussion, cataloging the ways the various realistic refinements affect the threshold and regulatory criteria. Background data for actual host–pathogen systems are given in Tables 2–8 in Anderson and May (1981).

## 7.3. HOST–PATHOGEN DYNAMICS CAN BE COMPLEX

The time lags introduced into the system by the presence of a "seedbank" of long-lived transmission stages can induce oscillations in the density of the disease-regulated host population, and in the prevalence of infection within it. The oscillations can be understood in an intuitive way that may be more familiar from prey–predator associations: If the host population is driven to low density, relatively few transmission stages of the pathogen are produced, whence the host population is relatively disease-free and grows exponentially; eventually the host population climbs to levels high enough to build up the pool of transmission stages, which at last attain levels high enough for the depredations of disease to cause the host population to crash to low levels; and thus the cycle continues. These lags in the feedback loop that lead to cycles rather than to regulation about a constant equilibrium level will tend to be marked if the average lifetime $\tau$ of the free-living transmission stages is long, if the host population growth rate is low (that is, if the characteristic population growth time in the absence of disease, $1/r$, is also long), and if the disease is highly pathogenic ($\alpha \gg r$).

**Table 7.2  Ways in Which Realistic Refinements Affect Threshold and Regulatory Criteria**[a]

| Complicating Factor | Effect on Regulatory Capacity of Pathogen [relative to basic model defined by Eqs. (1) and (2)] | Effect on Threshold Host Density to Maintain Pathogen (relative to basic model) |
|---|---|---|
| Diminution in reproductive capacity of infected hosts | Regulation is easier; regulated host population density is lower | No effect on threshold criterion |
| Vertical transmission | No effect on regulatory condition; regulated host population density is lower | Threshold host density lower (can be arbitrarily small) |
| Latent period of infection | Regulation is harder; regulated host population density is higher | Threshold host density is higher |
| Pathogenicity is stress related | Can always regulate if pathogenicity is sufficiently strongly enhanced by stress | Pathogen can persist provided transmission efficiency is high enough |
| Other density-dependent constraints on host population growth | Disease-related depression of host population density maximized for an intermediate degree of pathogenicity | Pathogen cannot persist if threshold host density is too high (if $N_T > K$) |
| Free-living infective stages of pathogen | Regulatory criterion as for basic model; regulated state may be a stable point or cyclic oscillations | Easier to maintain pathogen, especially if infective stages are long lived |

[a]This table summarizes the way in which some realistic complications can affect the ability of a pathogen to regulate its host population, and to persist within a population of hosts that either is of low density or fluctuates widely in abundance.

This circumstance arises for many polyhedrosis and granular virus and many microsporidian protozoan infections of univoltine forest insects. Anderson and May have argued that such interactions may be responsible, wholly or in part, for many of the long-term (3–40 year) cycles observed in forest insect pests. Only for the larch budmoth, *Zeiraphera diniana*, and an associated granulosis virus could Anderson and May find sufficient data to estimate all parameters in the relevant model, but in this one instance they did find an encouraging fit between the observed and theoretically estimated period of the oscillations in host density, and the cyclic patterns of prevalence of infection (the agreement between the observed and theoretically estimated amplitudes of the cycles in host density were less encouraging, with the observed amplitude being an order of magnitude larger than the theoretical estimate (Anderson and May, 1981; May, 1983). McNamee et al., (1981) have refined the theoretical analysis, but there remains much scope for further research, particularly in field studies of forest insects and their pathogens.

Importantly, there is need for theoretical studies that go beyond treating host population growth as continuous (differential equations), acknowledging the reality of discrete age classes that may, or may not, overlap.

A hint of the further complications that such studies are likely to find is given by May's (1985) recent analysis of a model for a host population with discrete, nonoverlapping generations that is regulated by a lethal pathogen which spreads in epidemic fashion throughout each generation before reproductive age is attained. This system is "completely chaotic," in the sense defined by May (1976). It possesses no stable points and no stable cycles; although the equations describing the system are fully deterministic, the dynamical trajectories are best described by a probability density function (May, 1985; Rogers et al., 1986).

This simple example is interesting in that straightforward and natural biological assumptions lead to a form of deterministic randomness. Borrowing from the terminology used elsewhere in this volume (Strong, Chapter 3), we could call this phenomenon "deterministic density vagueness." In other words, nonlinearities of the kind that enter naturally into host–pathogen and other interactions can lead to apparently random fluctuations and effectively probabilistic dynamics, even within the framework of a purely deterministic model. Deterministic models do not necessarily result in stable equilibrium states. In short, there is no crisp distinction between intrinsically deterministic and intrinsically stochastic models; the nonlinearities found in host–pathogen associations and elsewhere mean that even deterministic models can behave in very complex ways, producing stable equilibrium points, or stable cycles, or apparent chaos.

## 7.4. OPTIMAL PATHOGENICITY FOR SELF-MAINTAINING REGULATION

Suppose a pathogen is to be introduced as an agent to help control an insect pest, in circumstances where it is hoped that the pathogen, once introduced, will be able to maintain itself indefinitely within the pest population.

Notice first that such a one-time-only introduction will not in general be capable of eradicating the target pest species; once the host population is driven to sufficiently low densities, it will be below threshold for maintenance of the pathogen. Exceptions can arise by virtue of some of complications discussed in Section 7.2: Eradication could, for instance, be made possible by extreme cases of vertical transmission. And demographic or environmental stochasticity can result in local or widespread extinction of a pest that is reduced to low densities—either stably or at the trough of cyclic oscillations—by a pathogen.

Given that a self-sustaining pathogen is unable to eradicate the pest, the aim is to reduce the host population significantly below its disease-free average level $K$ (around which the pristine population may have exhibited a greater or lesser degree of density-independent fluctuation). The question thus arises, what is the degree of pathogenicity $\alpha$ that produces maximum depression of the pest population?

At first sight, it may seem that the greater the pathogenicity the better. Certainly, if the pathogen has $\alpha \rightarrow 0$ (and has no effect on the reproductive capacity of infected hosts) it has little effect on host population growth. But, at the other extreme, too large an $\alpha$ value leads to too large a value of $N_T$ (essentially because the disease kills hosts so fast that a very large host population is needed to perpetuate the infection), and thence to the regulated population having high density. The optimum self-sustaining control, in the sense of lowest equilibrium levels of pest population density, will usually be attained from an infectious disease agent with intermediate pathogenicity.

We can see this directly in the unrealistically simple "basic model" defined in Section 7.2. Here the disease-regulated host population density (attained if $\alpha > r$) is given as an explicit function of $\alpha$ from Eqs. (3) and (4) as

$$N^* = \frac{\alpha(\alpha + \gamma + b)}{(\alpha - r)\beta} \qquad (9)$$

In this simple, illustrative model, $N^*$ becomes infinite if $\alpha$ is too big or too small ($\alpha \rightarrow \infty$ or $\alpha \rightarrow r$, respectively), and has its minimum value for the intermediate $\alpha$-value of $\alpha_{min} = r + [r(a + \gamma)]^{1/2}$.

A more realistic, though still grossly oversimplified, model will acknowledge the existence of other density-dependent constraints upon the host population, which regulate it around some carrying capacity $K$ in the absence of the disease. For an extremely pathogenic agent, the high $\alpha$ value will result in a large value of $N_T$, and thus (as discussed in Sect. 7.2) in the disease being unable to establish itself (because $N_T > K$). Conversely, if $\alpha$ is very small, the pathogen will have little impact on the pristine pest level $K$. Again, in this more realistic model the maximum impact of the pathogen upon the pest population is attained for an intermediate $\alpha$ value. This point is explored in detail in Anderson and May (1981, Sect. 11; see also the more general discussion for vertebrate hosts in Anderson, 1979).

## 7.5.  RATE OF CONTINUOUS RELEASE OF A PATHOGEN TO ERADICATE AN INSECT PEST

For the reasons outlined in the previous section, eradication of a pest species will usually require continuous introduction of transmission stages of the pathogen. This section summarizes results for the critical rate, above which such transmission stages must be introduced, in order to achieve extinction of the pest. The results themselves are derived in Anderson and May (1981).

We have established in Section 7.2 that for the viral, bacterial, protozoan, or fungal pathogen to be capable of regulating or eradicating the insect pest population we require $\alpha > r$ (or some appropriately modified version of this relation if the infection decreases reproductive capacity, or if pathogenicity is stress-related, or if other complications enter). If this criterion is obeyed, then the host population will be extinguished if free-living transmission stages of the pathogen are introduced at a rate $A$ in excess of a critical rate $A_c$:

$$A_c = \frac{\mu r(\alpha + b + \gamma)}{\nu(\alpha - r)} \tag{10}$$

Here the parameters $\alpha$, $b$, $\gamma$, and $r$ have their usual meanings; $\mu$ is the death rate of free-living infective stages (whence their life expectancy is $1/\mu$); and $\nu$ is the rate at which infective stages successfully infect hosts (that is, $\nu$ is the transmission coefficient of free-living infective stages). Eradication efforts of this kind are more likely to be successful if they require relatively low rates of introduction of the pathogen. Thus, the criterion $A > A_c$, with $A_c$ defined by Eq. (10), is more likely to be met for infectious agents that are highly pathogenic ($\alpha$ large), have long-lived transmission stages ($\mu$ small), and have high transmission efficiency ($\nu$ large). In accord with common sense, both Eq. (10) and the overriding constraint $\alpha > r$ state that pest species with high population growth rates (large $r$) are relatively difficult to control.

With the exception of $\nu$, all the parameters in Eq. (10) can be measured, although there are very few host–pathogen systems for which all such measurements have been made. Direct assessment of the transmission parameter $\nu$ is, however, exceedingly difficult. It is therefore helpful to note that the critical introduction rate $A_c$ can be alternatively expressed as

$$A_c = \lambda Y_0^* \tag{11}$$

Here $Y_0^*$ is the equilibrium population of infected hosts, and $\lambda Y_0^*$ is the equilibrium net rate of production of infective stages by infected hosts, in a natural system when no pathogens are being artifically introduced ($A = 0$). Thus, if we are using a pathogen found in natural systems and believed to have $\alpha > r$, we have only to introduce infective stages at a net rate in excess of the rate at which they are produced by the pristine host–pathogen association, to eradicate the host population. It seems plausible that this critical rate can, for example, be estimated (and attained in practice) for some of the known baculoviruses of forest insect pests.


## 7.6. CONCLUSION

The foregoing discussion has dealt exclusively with the dynamics of host–pathogen associations. Ultimately, however, the dynamics of populations are always entwined with their genetics. The studies of the relations among the demographic and epidemiological parameters determining the dynamics of a host–pathogen association need also to take account of the evolutionary pressures acting on both host and pathogen. Particularly in cases where an "exotic" pathogen is introduced in the hope of reducing the density, or of eradicating, a particular insect pest, it is likely that coevolution will shift the average genotypic constitution of host and/or pathogen populations in a relatively short time. As has been the experience with insecticides and antibiotics, control measures will almost always be aimed at a moving target. These complications are explored elsewhere, with emphasis on the example of myxoma virus among Australian rabbits (May and Anderson, 1983, and references therein; see also May and Dobson, 1985).

Focusing only on the dynamics, the present review has sought to outline the criteria that govern whether a pathogen is capable of maintaining itself within a host population, and of regulating that population to relatively low levels. Table 7.2 summarizes these threshold and regulatory conditions, indicating how they are affected by a variety of particular circumstances (diminished reproduction of infected hosts, vertical transmission, latent periods, stress-related pathogenicity, other carrying capacity constraints, and long-lived infective stages). It cannot be too strongly emphasized that the extreme case of deterministic density-dependent regulation of a host population by a pathogen does not necessarily imply an unvarying equilibrium state: The nonlinearities inherent in host–pathogen associations can easily produce sustained cycles or irregular ("chaotic") fluctuations, quite independently of any externally imposed stochastic effects. Such dynamical complexities should be reckoned with when thinking about particular control projects.

As discussed in Section 7.4, a self-sustaining pathogen is unlikely to eradicate its host species. The optimum such self-maintaining pathogen—in the sense of the pathogen giving the maximum reduction of host population density—is likely to be one with some intermediate level of pathogenicity (rather than the most pathogenic agent discoverable). As discussed in Section 7.5, eradication of an insect pest species by a pathogen will usually entail continued introduction of the pathogen, at a rate $A$, in excess of a critical value $A_c$; this critical value can possibly be estimated from field studies of the parameters characterizing the natural, unperturbed host–pathogen system [Eq. (11)]. Eradication will usually require maintaining $A > A_c$ for many years, and even then constant surveillance and reintroduction of infective stages are likely to be needed to prevent resurgence of the pest species.

   This review has largely been a synopsis of Anderson and May (1981). There is, however, considerable scope for further theoretical and empirical work, exploring the effects of pathogens in conjunction with other regulatory agents, in the integrated management of insect pests. Hassell and May (1985a) have begun a study of the combined effects of a pathogen and a parasitoid upon a host species with discrete, nonoverlapping generations; the dynamics of such systems can exhibit features not easily understood by considering the host–pathogen or the host–parasitoid system separately. Parallel work on the combined effects of pathogens and insecticides, or of pathogens and natural enemies other than parasitoids, is also needed.

   Another grievous shortcoming in the work summarized here is that the models make no mention of spatial heterogeneity. It is increasingly apparent that the effects of nonuniform distributions of hosts in a spatially heterogeneous environment are crucial for understanding the dynamics of many insect prey-

**Table 7.3  Glossary of Symbols Used in This Chapter**

Population Variables

| | |
|---|---|
| $N$ | Total number of hosts |
| $X$ | Number of susceptible hosts |
| $Y$ | Number of infectious hosts |
| $K$ | Carrying capacity for host population |
| $N_T$ | Threshold host density |

Rate and Other Parameters

| | |
|---|---|
| $a$ | Host birth rate (per individual) |
| $b$ | Natural mortality rate of hosts |
| $r$ | Intrinsic growth rate of host population (per individual), $a - b$ |
| $\alpha$ | Disease-induced mortality rate |
| $\gamma$ | Rate of host recovery from infection |
| $\beta$ | Transmission coefficient (in models where infection is transmitted from infecteds to susceptibles) |
| $f$ | Fractional decrease in birth rate of infected hosts (birth rate of infected hosts is $a(1 - f)$). |
| $q$ | Proportion of offspring of infected hosts acquiring infection by vertical transmission |
| $\sigma$ | Rate at which infected but latent hosts move into the infectious stage |
| $\mu$ | Mortality rate of free-living infective stages |
| $\lambda$ | Rate of production of infective stages, per infected host |
| $\nu$ | Rate at which infective stages successfully infect hosts (transmission coefficient of free-living infective stages) |
| $A$ | Rate of introduction of infective stages (in a biological control program) |
| $R_0$ | Basic reproductive rate of the pathogen |

predator and competitive associations [see, for example, the review by Hassell and May (1985b) and references therein]. Studies of the use of pathogens in pest management, or as a component in integrated pest management, should in general incorporate spatial heterogeneity. By the same token, the spatially uniform character of many laboratory studies should make us cautious about extrapolating their conclusions into the field. In particular, the analysis of the possibility of eradication has ignored immigration of pests into the population from other areas.

Although theoretical work on insect host–pathogen associations is in a relatively undeveloped state, it looks like an ocean liner beside the skiff representing studies that integrate theory with field observations. It would be nice to see more applied control programs deliberately designed as experiments, so that basic understanding could be gained at the same time as practical needs are responded to.

Many of the issues raised in this chapter also appear, *mutatis mutandis*, when one considers how accidental introductions of pathogens and parasites can affect native flora and fauna (Elton, 1958; for a recent review, see Dobson and May, 1985). The depredations wrought on native populations by invading pathogens, or the competitive advantage sometimes enjoyed by an invading species that does not bring its natural complement of pathogens and parasites into a new environment, provide many case studies that could widen the data base pertaining to the use of pathogens in the integrated management of insect pests.


## ACKNOWLEDGMENTS

This survey has benefitted from discussions with R. M. Anderson, A. P. Dobson, and M. P. Hassell. The work was supported in part by the National Science Foundation under grant BSR83-03772.


## REFERENCES

Anderson, R. M. (1979). Parasite pathogenicity and the depression of host population equilibria, *Nature*, **279**, 150–152.

Anderson, R. M., and R. M. May. (1979). Population biology of infectious diseases, *Nature*, **280**, 361–367; 455–461.

Anderson, R. M., and R. M. May. (1981). The population dynamics of microparasites and their invertebrate hosts, *Phil. Trans. R. Soc. B*, **291**, 451–524.

Dobson, A. P., and R. M. May. (1985). "Patterns of invasions by pathogens and parasites," in H. A. Mooney, Ed., *Ecology of Biological Invasions of North America and Hawaii*, Springer-Verlag, New York, (in press).

Elton, C. S. (1958). *The Ecology of Invasions by Animals and Plants*, Methuen, London, 181 pp.

Ewald, P. W. (1985). "Pathogen induced cycling of outbreak insect populations," in P. Barbosa and J. C. Schultz, Eds., *Insect Outbreaks: Ecological and Evolutionary Processes*, Academic, New York, (in press).

Falcon, L. A. (1982). "Use of pathogenic viruses as agents for the biological control of insect pests," in R. M. Anderson and R. M. May, Eds., *Population Biology of Infectious Diseases*, Springer-Verlag, New York, pp. 191–210.

Getz, W. M., and J. Pickering. (1983). Epidemiological models: Thresholds and population regulation, *Am. Nat.*, **121**, 892–898.

Hassell, M. P., and R. M. May. (1985a). Generalist and specialist natural enemies in predator-prey interactions, *J. Anim. Ecol.* (in press).

Hassell, M. P., and R. M. May. (1985b). "From individual behaviour to population dynamics," in R. Sibly and R. Smith, Eds., *Behavioural Ecology*, Blackwell, Oxford, pp. 3–32.

McNamee, P. J., J. M. McLeod and C. S. Holling. (1981). *The Structure and Behavior of Defoliating Insect/Forest Systems*, University of British Columbia, Institute of Resource Ecology Research Publication, R-25.

May, R. M. (1976). Simple mathematical models with very complicated dynamics, *Nature*, **261**, 459–467.

May, R. M. (1983). Parasitic infections as regulators of animal populations, *Am. Sci.*, **71**, 36–45.

May, R. M. (1985). Regulation of populations with non-overlapping generations by microparasites: A purely chaotic system, *Am. Nat.*, **125**, 573–584.

May, R. M., and R. M. Anderson. (1983). Epidemiology and genetics in the coevolution of parasites and hosts, *Proc. R. Soc. B*, **219**, 281–313.

May, R. M., and A. P. Dobson. (1985). "Population dynamics and the rate of evolution of pesticide resistance," in *Pesticide Resistance Management*, National Academy of Sciences, Washington, D.C., (in press).

Rogers, T. D., Z.-C. Yang, and L.-W. Yip. (1986). Complete chaos in a simple epidemiological model, *J. Math. Biol.*, **23**, 263–268.

Tinsley, T. W. (1979). The potential of insect pathogenic viruses as pesticidal agents, *Annu. Rev. Entomol.*, **24**, 63–87.

Tinsley, T. W., and P. F. Entwistle. (1974). "The use of pathogens in the control of insect pests," in D. Price-Jones and M. E. Solomon, Eds., *Biology in Pest and Disease Control*, Blackwell, Oxford, pp. 115–129.

# 8

---

# PLANT-PLANT-PATHOGEN-INSECT INTERACTIONS[1]

*George G. Kennedy*

The topic ecological theory and integrated pest management (IPM) practice concerning plant–plant-pathogen–insect interactions is very broad and cannot be covered completely within the context of a single chapter. Thus, this chapter is restricted in scope to a discussion of aphid- and leafhopper-transmitted plant pathogens. However, the information presented should apply as well to other arthropod vectors and any nonvirus plant pathogens they transmit. I have focused on two bodies of ecological theory, namely life history theory and coevolutionary theory, and have attempted to illustrate how they help to explain why certain types of vector-dependent plant pathogens are particularly severe agricultural problems and why some of the approaches used to manage the diseases they cause are effective.

## 8.1. ARTHROPOD TRANSMISSION OF PLANT VIRUSES

The transmission of plant viruses by arthropods is a complex phenomenon involving interactions between plant and virus, plant and vector, virus and vector, and plant and virus and vector (see Harris and Maramorosch, 1977, 1980; Maramorosch and Harris, 1979, 1981). Arthropod vectors can transmit viruses in a nonpersistent, persistent, or semipersistent manner (Table 8.1). Even the simplest of these relationships, nonpersistent transmission, is very complex and the mechanisms of such transmission are but poorly understood (Harris 1977; Pirone and Harris 1977). The type of transmission greatly influences the patterns of virus spread that occur and the tactics that are effective in managing virus spread in crop plants.

[1]Paper No. 9959 of the Journal Series of the North Carolina Agricultural Research Service, Raleigh, NC 27695-7630.

Table 8.1  Characteristics of Nonpersistent, Persistent, and Semipersistent
Transmission of Aphid- and Leafhopper-Borne Plant Viruses

| Category | Characteristics |
|----------|-----------------|
| Nonpersistent | Virus acquired in seconds<br>No latent period in vector<br>Virus inoculated in seconds<br>Retained for minutes to hours<br>Noncirculative in the vector |
| Persistent | Virus acquired in minutes<br>Latent period in vector<br>Virus inoculated in minutes<br>Retained through molt<br>Circulative<br>Some replicate in vector; most do not |
| Semipersistent | Virus acquired in minutes<br>No latent period in vector<br>Inoculated in minutes<br>Retained for hours to days<br>Noncirculative in vector |

In practice, the focus of IPM relative to arthropod-transmitted plant viruses is on minimizing the impact of disease on the crop by limiting or delaying the incidence of disease or by increasing the plant's ability to tolerate infection by the pathogen. The occurrence of diseased plants involves complex interactions over time between the pathogen, the plant community, the vectors and the environment. The minimum requirements for spread of arthropod-borne plant pathogens to occur are (1) a source of inoculum, (2) vectors, (3) a susceptible host for the pathogen, and (4) a suitable environment. Thus, the incidence of disease is the result of interactions involving these populations and the environment. In the case of the vector, not only is the number of individuals important, but also their degree of activity and the efficiency with which they transmit the virus among plants. In addition, the temporal pattern of vector abundance and activity relative to the incidence of pathogen inoculum and susceptible host plants is important. For the pathogen, the distribution and abundance of infected plants is important, as is the abundance of inoculum within infected plants and its availability to potential vectors. For the plant, the number and distribution of susceptible plants is important.

In natural terrestrial ecosystems (as opposed to agroecosystems), pathogens may alter the population dynamics of plants in a variety of ways. These include (Harper 1977; Burdon and Shattock, 1980) (1) reduced reproductive output (lower seed yield), (2) reduced competitive ability, (3) selection for resistant genotypes with subsequent (or coincident) counterselection for more virulent pathotypes, (4) decreased host density, and (5) increased community diversity.

The basic defenses of plants against arthropod-transmitted pathogens in natural ecosystems appear to involve escape in space and time (nonapparency), resistance to the pathogen and/or the vector, and tolerance to the pathogen such that the infection is symptomless or the symptoms are extremely mild. The pathogens are adapted to persist in natural ecosystems in which their hosts use these defensive strategies.

In modern agriculture, through breeding and production practices, we have effectively stripped many of our crop plants of some or all of their natural defenses and thereby created situations highly conducive to disease epidemics (Day, 1974; Harper 1977; Thresh 1982). Before examining this in detail, I would like to discuss some of the ecological theory that helps to explain why certain types of virus diseases are severe agricultural problems while others are not and that provides a basis for understanding why certain control measures are effective.

## 8.2. LIFE HISTORY THEORY, PATTERNS OF DISEASE SPREAD, AND IPM

Life history theory states that life histories are complex phenotypic characters that function together and covary under selection (Denno and Dingle, 1981). The body of theory seeks to explain associations of life history traits and the way in which selective forces mold them. The main selective forces are habitat variability and the various mortality factors that affect a species. Principal life history traits for animals are considered to be fecundity, size of young, age to first reproduction, interaction between adult mortality and reproduction, the ability to track host resources in space and time, and variation among these traits among an individual's progeny (Stearns, 1977; Denno and Dingle, 1981). Since survival of arthropod-transmitted plant viruses requires a continuous sequence of infection, the principal life history traits for these viruses are ability to persist at a site (perennation), cycling time in host (i.e., time between infection and the ability to be transmitted), and propensity to spread. The latter includes vector specificity, host range, and whether the virus is transmitted nonpersistently, semipersistently, or persistently, and if persistently transmitted whether the virus replicates in its vector (propagative) or not (nonpropagative).

The association of life history traits that occurs in a particular species or population, regardless of the selective forces that molded them [$r/K$ selection, bethedging, etc. (see Stearns, 1977; Denno and Dingle, 1981)], reflect an evolutionary compromise that enables the species to persist in its habitat. Because of design constraints (Stearns, 1977), that compromise may fall short of what we, in our limited wisdom, perceive to be ideal. For viruses, design constraints are likely to be extreme because of the very limited viral genome [e.g., Cauliflower mosaic virus has only six genes (Gardner et al., 1981)]. Thus, for viruses, life history options are limited and trade-offs necessary.

Harrison (1981) hypothesized that the ability of a virus to persist at a particu-

lar site (perennation) varies inversely with the ability to spread to new sites. He provided examples that suggest that there are trade-offs between perennation ability and ability to spread within and among sites or host patches (Table 8.2). Barley yellow dwarf virus, beet curly top virus, lettuce mosaic virus, and alfalfa mosaic virus are illustrative of the types of viruses that are important as problems in agriculture. They have life histories characterized by a limited perennation capacity but a high dispersal capacity mediated by aphids or leafhoppers. They are well-adapted to diverse plant communities characterized by a high degree of spatial and temporal heterogeneity and in which escape is a major defensive strategy of their hosts. In addition, they are characterized by rapid multiplication in their host (short cycling time) and the ability to recover from drastic decreases in the abundance of infected plants (i.e., resilience) (Thresh, 1983a,b).

The major aphid and leafhopper vectors of agriculturally important plant viruses are well-adapted to exploiting spatially and temporally heterogeneous host plants. Aphids are extremely important as vectors of plant viruses (Eastop, 1977; Harris, 1980). They reproduce rapidly and disperse in large numbers (Irwin and Goodman, 1981) and over wide areas (Taylor, 1979). In addition, aphid life histories, even for a particular species, are plastic (Kennedy and Stroyan, 1959; Swenson, 1968; Dixon 1985) and provide for the simultaneous occurrence of both holocyclic (host alternating) and anholocyclic (nonhost alternating) populations. For example, Tamaki et al. (1979) have documented that *Myzus persicae* (Sulzer), a major vector of a number of plant viruses, overwinters in the Yakima Valley of Washington both as eggs on peach trees (holocyclic) and as viviparae on weeds in deep irrigation ditches (anholocyclic). The abundance of the latter is reduced by severe winters; but when winters are mild, the

**Table 8.2  Life History Characteristics for Several Plant Viruses**

| Virus | Vector | Perennation | Spread At Site | Spread to New Sites | Host Range |
|---|---|---|---|---|---|
| Soil-borne wheat mosaic virus | Fungus | High (spores) | Moderate | Poor | Narrow |
| Raspberry-ring spot virus | Nematode | High (plants/seed) | Moderate | Poor | Wide |
| Barley yellow dwarf virus | Aphid (persistent) | Moderate (plant) | High | High | Intermediate |
| Curly top virus | Leafhopper (persistent) | Poor | High | Very high | Wide |
| Lettuce mosaic virus | Aphid (nonpersistent) | Poor | High | Moderate (aphid/seed) | Narrow |
| Alfalfa-mosaic virus | Aphid (nonpersistent) | Moderate | High | High | Wide |

Modified from Harrison (1981).

aphids increase rapidly in spring and disperse in large numbers before the holo-cyclic populations disperse from peach trees. Severe problems with beet western yellows virus in sugar beet are related to the early spread of the virus by anholo-cyclic aphids from weeds. Similar phenomena have been well documented for other viruses (see Thresh, 1983b).

Aphids and leafhoppers are capable of spreading viruses over long distances, but the distance a particular virus is spread from a source is related to both the dispersal of the vector and the persistence of the virus in the vector. Although aphids can disperse over large distances (Taylor, 1979) on wind systems, nonper-sistently transmitted viruses are retained by the vector for only a limited period of time (hours) and, thus, the distance over which they are spread from particu-lar foci is generally limited (Thresh 1976; but see Adlerz, 1981; Thresh, 1983a for exceptions). Typically, for nonpersistently transmitted viruses the spatial gradients of infection from a source are rather steep even when the distance units are meters (Fig. 8.1). Factors in addition to retention time by the vector also appear to be involved in the steepness of the gradient, since under some condi-tions the aphids can move farther than the gradients suggest during the time they remain infective. Since nonpersistent viruses are soon lost when the aphid probes a nonhost (Cockbain and Heathcote, 1964; Thresh, 1983a), alates that are infective when they take flight are unlikely to transmit virus at the end of their flight unless they alight first on a susceptible host. Thus, the probability of successful transmission is low unless the virus has a wide host range or the sus-ceptible host is widely grown in monoculture (Thresh, 1983a).

In contrast, persistently transmitted viruses are retained by their vectors for long periods, often for the life of the vector. They can be successfully carried as far as the vector disperses. Infective individuals can also inoculate large numbers

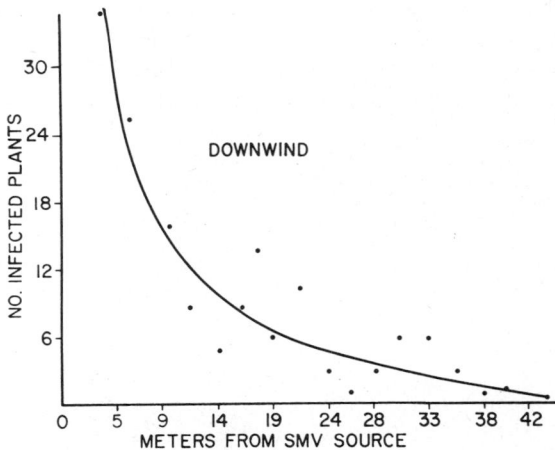

Fig. 8.1 Gradient of downwind spread of a nonpersistent aphid-borne virus, soybean mosaic virus, during a single growing season (source Irwin and Goodman, 1981).

of plants. This compensates, at least in part, for the lower reproductive rate and generally lower populations of leafhopper vectors as compared with aphids. Spatial gradients of infection over distance from a virus source tend to be rather shallow for persistently transmitted viruses and can span distances of hundreds of kilometers, as in the case of curly top virus of sugar beets (Fig. 8.2) (Thresh, 1976, 1983a).

By virtue of the processes involved in nonpersistent transmission (Harris, 1977; Pirone and Harris, 1977) and the manner in which aphids select their host plants (brief probes following random landing on hosts and nonhosts alike), the nonpersistent viruses are not restricted in their host range to plants colonized by their vectors (Swenson, 1968; Irwin and Goodman, 1981). Because persistently transmitted viruses require prolonged probes for acquisition and inoculation, they are generally restricted in host range to plants within the host range of their vector [see Rochow (1977) for discussion of ways in which such viruses can expand their host range].

## 8.3. COEVOLUTION AND PLANT APPARENCY

While escape in space and time provides a mechanism of defense against arthropod-borne pathogens, plants in natural communities also rely on the more direct defenses of resistance to the pathogen or its vector and tolerance of infection. Indeed, epidemics require a high degree of genetic uniformity among the host population vis-à-vis the pathogen (Day, 1974). Epidemics occur most commonly when either the pathogen or the host have been introduced into a new area and have not had the opportunity to coevolve (Harper, 1977). In natural communities competition among plants is usually severe and diseased plants that cannot compete effectively die, whereas resistant plants that avoid infection or tolerant plants that become infected but remain asymptomatic (or manifest very mild symptoms) can compete effectively. Because infected, tolerant plants can serve as a source for spread of pathogens and persist longer than infected, susceptible plants, there is also likely to be selection for mild strains of the pathogen that

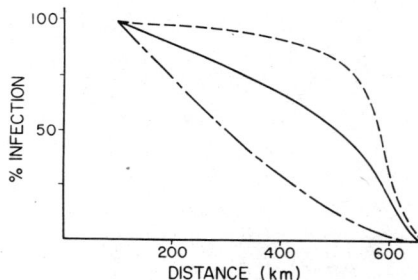

Fig. 8.2 Gradient of spread of curly top virus in sugar beet by the beet leafhopper during each of three growing seasons (source Romney, 1939).

infect tolerant hosts but allow them to serve as effective sources for subsequent spread of the pathogen (Matthews, 1981).

The coadaptation of plants and their pathogens is illustrated by the data of Tomlinson and Walker (1973) which show that *Stellaria media* (L.) plants from Great Britain, North America, and Australia were least affected by the strains of cucumber mosaic virus from their country of origin and showed more severe reactions when infected by alien strains (Table 8.3).

Coevolutionary theory (Feeny, 1976; Rhoades and Cates, 1976) would predict that apparent plant species, which cannot rely heavily on escape in space and time as a defensive strategy, would have higher levels of resistance or tolerance or both to viruses than would nonapparent species. Thresh (1982; 1983b) has indicated that serious virus epidemics occur far less frequently and progress less rapidly among woody perennials (apparent) in agriculture than among annuals (nonapparent), despite the large catchment area they provide for incoming vectors. Resistance and tolerance appear largely responsible for this (Thresh, 1983b). Woody perennials are usually much more difficult to infect than are herbaceous annuals. Once they are infected, woody perennials have a long infection cycle. For example, a 410-day interval was observed between the first detection of citrus Tristeza virus in one citrus tree and the stage when all 44 samples taken from that tree were positive (Burnett, 1961). In addition, woody plants tend to be tolerant of infection and frequently develop only low virus titers. Thus, they show few if any symptoms of infection and may often be characterized by limited availability of virus to vectors. Perennial herbaceous plants would be expected to differ similarly from annuals in their response to viruses.

The role of resistance to virus vectors in protecting plants from virus infection in natural communities is unknown. It might be expected to play a greater role in protecting plants from infection by persistently transmitted viruses than nonpersistently transmitted viruses because the acquisition or inoculation of a persistently transmitted virus generally requires prolonged feeding by the vector. A plant highly resistant to a vector might only rarely be fed on long enough for a persistently transmitted virus to be acquired or inoculated. In contrast, it is unlikely that most mechanisms of vector resistance would operate fast enough to

**Table 8.3 Degree of Symptom Expression by *Stellaria media* Plants from Several Locations in Response to Infection by Isolates of Cucumber Mosaic Virus from Those Locations**[a]

| Plant from | CMV Isolate from | | |
| --- | --- | --- | --- |
| | Great Britain | North America | Australia |
| Great Britain | None | Mild | Mild |
| North America | Severe | Mild | Severe |
| Australia | Severe | Severe | Mild |

[a]Based on results of Tomlinson and Walker (1973).

prevent the very brief probes that result in transmission of nonpersistent viruses [see Kennedy (1976) for a discussion of relevant interactions]. A great deal more research is needed in this area.

## 8.4. ARTHROPOD-BORNE PLANT PATHOGENS IN AGROECOSYSTEMS

Most of our annual crops are early successional species that, in their natural habitats, are nonapparent (see also Kogan, Chapter 5). Modern agricultural practices have stripped them of that defense by making them apparent in space and in time by virtue of large-scale monoculture and extended cropping seasons via sequential cropping, overlapping production seasons, and the use of irrigation (see also Herzog and Funderburk, Chapter 10). In South Africa, where a number of potato viruses and their aphid vectors are common, local potato varieties planted in gardens quickly succumb to virus but volunteers growing as widely separated plants in fields or along roadsides rarely become infected (Van der Plank, 1948). Similarly, rice Tungro virus, spread by several species of leafhoppers (Ling et al., 1983) has been known from southeast Asia and the Pacific for over 100 years. However, epidemics occurred following the introduction of modern high-yielding rice varieties. These are grown in dense stands with irrigation to extend the growing season and allow successive cropping and overlapping crop sequences. Such practices permit the development and continuous presence of large vector populations and abundant virus inoculum. In Malaysia, heavy losses to rice Tungro virus occurred even in traditional varieties following an early and heavy influx of leafhoppers from earlier plantings of new short-season varieties (Ling, 1972; Thresh, 1980). There are numerous other examples where continuous cropping and overlapping crop sequences have resulted in severe virus problems (Zitter and Simons, 1980; Kiritani, 1983; Reddy et al., 1983).

The coevolved plant defenses of tolerance and resistance are also frequently weakened or destroyed by modern agricultural practices. The main production areas of many, if not most, crops are geographically distant from the native home of the crop species. This enables the crop to evade many of the arthropods and pathogens that would attack it in its native habitat when grown in pure stands. This is a two-edged sword, however, in that this practice also exposes the crop to organisms with which it has had no history and against which its defenses may not be effective. Cacao, for example, is native to the Amazonian region of South America, but the main production area is in western Africa. There, the cacao crop suffers devastating losses from the mealybug-transmitted cocoa swollen shoot virus which also occurs in native trees (Owusu, 1983; Thresh, 1980). Similarly, the viruses causing cassava mosaic and maize streak in Africa obviously did not originate in cassava or maize since they do not occur in the Americas where these crops evolved (Buddenhagen, 1983).

The extensive movement of crop propagules around the world with only minimal attention to whether or not they are infected with viruses has undoubtedly resulted in numerous new contacts between viruses and crop plants (Plumb and Thresh, 1983). To illustrate, potato virus Y (PVY), a nonpersistent, aphid-transmitted virus, apparently has been spread around the world in infected tubers. In Florida, infections of PVY in tomato and pepper during the 1950s were found to be restricted to areas where potatoes had been produced commercially. The virus was found to have spread from the potato crop and become established in weeds and pepper (Simons et al., 1956; Simons, 1959).

Modern crop varieties typically have been selected for uniformity, quality, yield, and resistance to a relatively few major pests and diseases. In the breeding process, much of the natural resistance and tolerance to many other diseases has been lost. For example, the virus disease leaf curl of peppers in Sri Lanka is prevalent in modern high-yielding varieties but not among the traditional varieties that are highly resistant (Shivanathan, 1981).

Modern agricultural practices clearly have eliminated or weakened many of the natural defenses of our crop plants. These practices have been adopted in an effort to improve yields and maximize profits. In many agricultural systems and for many pest problems, we have had the luxury of being able to use additional production inputs to replace the natural plant defenses and to produce crops profitably. In other instances we have, in effect, recreated at least a portion of the plant's natural defenses. In the case of arthropod-borne virus diseases, a brief examination of some of the control strategies used will illustrate the point.

## 8.5. APPROACHES TO MANAGING ARTHROPOD-TRANSMITTED PLANT VIRUSES

A general strategy commonly used to minimize the spread of plant viruses involves a reduction in plant apparency. Reducing apparency in time generally involves disrupting the synchrony between the crop and the pathogen and its vector. By implementing a crop-free period for celery in California and Florida, for example, the continuous occurrence of both crop plant and high inoculum levels of celery mosaic virus was eliminated and the incidence of disease was reduced to acceptable levels (Zitter, 1977). For a number of crops, the incidence of virus is controlled in most years by selecting planting dates to avoid peak flight activity of the vectors. In New Zealand, barley yellow dwarf virus is controlled in wheat by delaying planting of wheat until mid-June after the peak flight of aphid vectors has occurred. A similar approach is used to control carrot motley dwarf virus in carrot (Lowe, 1973).

There are numerous agricultural and economic constraints that limit the flexibility of using this approach in many crops. In the Mediterranean countries, for example, losses to a number of virus diseases of vegetable crops can be reduced by delaying planting until after the peak flights of aphids from drying vegetation at the end of the wet season. However, delaying planting not only interferes with established profitable crop sequences, but also causes the growers to miss the high early season prices (Thresh, 1982).

A number of practices designed to reduce the spread of plant virus diseases in effect make the crop plants less apparent in space. A key component of the strategy used in producing certified plant materials involves growing them in isolation from known sources of virus and vectors. Thus, seed potatoes are grown a long distance away from the commercial potato production areas. Isolation *per se* can be achieved directly by growing the crop a safe distance away from known sources of virus and/or vector, as in the case of seed potato production (Thresh, 1976) or it can be achieved by eliminating nearby sources of virus or vector (Broadbent, 1964; Zitter and Simons, 1980; Thresh, 1982). The distances required for effective isolation vary with virus and vector, depending on the type of transmission (persistent, nonpersistent, semipersistent; see Figs. 8.1 nd 8.2) as well as the abundance and mobility of the vector.

With some crops, changes in plant spacing within a field can alter the incidence of infection (Broadbent, 1964). Virus incidence in peanuts and sugar beet, for example, is reduced in very dense stands. This reduction in the number of plants infected apparently results because widely spaced plants are landed upon and are colonized more frequently by certain vectors than plants presenting a closed canopy (Thresh, 1982). Changes in plant spacing alter plant apparency to the vector by changing the way the plants are perceived by the vectors. The ameliorating effect of increasing plant density on virus spread might in some crops, be more than offset by increases in herbivory (see Simberloff, Chapter 2).

The use of protection (i.e., barrier and cover) crops, which are not susceptible to either the vector or the virus, is effective in reducing the incidence of nonpersistent viruses. For example, sugar beet plants grown for seed production can be protected from yellows viruses by overseeding barley [12% vs. 100% yellows with and without barley cover crop (Hull, 1952, cited in Broadbent, 1964)]. The protection crop reduces the apparency of the main crop. This approach is effective when the protection crop is intermixed with the main crop as well as when it is planted in rows between the main crop. It has been used successfully in a number of crops (Broadbent, 1964; Toba et al., 1977).

Genetically based tolerance and resistance to plant viruses have been used to reduce losses to disease in a number of crops. This approach seeks to use directly the plant's natural defenses. If the virus and the crop plant coevolved, then the greatest likelihood of finding genes conferring tolerance or resistance is in the center of origin of the crop, since selection has presumably occurred for such genes (Leppik, 1970; Buddenhagen, 1983). Where the plant and the virus represent recent encounters and coevolution has not occurred, resistance genes are likely to be found only by a thorough search of available germ plasm since any genes conferring resistance would likely not have been selected by the pathogen. That such resistance genes do occur is illustrated by the fact that genes conferring resistance in rice to the temperate rice virus diseases of Japan came from an introduced tropical (indica) subspecies that is genetically remote from the temperate (japonica) cultivars (Buddenhagen, 1983).

Although it may be argued that coevolved (single) genes for resistance to a virus are not likely to be durable (see Harris, 1975; Robinson, 1980), there is little evidence to date that resistance-breaking virus strains are a major problem. The use of virus resistance genes has been quite successful. The apparent infrequency of resistance-breaking virus strains may be an artifact of the relatively limited use of virus-resistant (as opposed to tolerant) varieties. However, it may also be due to the very limited size of the virus genome relative to the genomes of other types of pathogens such as the pathogenic fungi (Buddenhagen, 1983) and to the total dependence of viruses on their vectors for spread to new hosts.

Virus tolerance enables the production of a crop despite infection. However, tolerant varieties may become widely infected and provide a source of inoculum for infection of other nontolerant crops.

Resistance to the vector offers promise of limiting the spread of plant viruses, especially in the case of those that are persistently transmitted. It can suppress virus spread by reducing the vector population or by reducing the frequency of vector–plant contacts. The situation is complex, however, and depends upon the type of resistance, the type of transmission (nonpersistent, persistent, semipersistent), and the relative importance of resident and transient vectors (see Kennedy, 1976).

There are various other procedures for reducing the spread of arthropod-transmitted plant viruses. These include, for example, the use of insecticides to reduce vector populations, reflective mulches to repel aphids, oil sprays to interfere with inoculation of plants by aphids, and repellent sprays (Broadbent, 1964; Zitter, 1977; Zitter and Simons, 1980; Gibson et al., 1982). These have been developed largely on the basis of an understanding of the virus–vector–plant relationship, the virus transmission process, and the epidemiology of arthropod-borne plant viruses. That understanding has emerged from empirical studies of specific virus–vector–crop plant interactions and of vector behavior and population dynamics. It did not emerge from any particular body of theory. Rather, those studies have contributed to the development of a sound conceptual basis for approaching the management of arthropod transmitted plant viruses.

## 8.6. CONCLUSION

Although it would be nice to be able to say that ecological theory has provided guidance in the development of strategies for managing arthropod-transmitted virus diseases in agricultural crops, it would be dishonest to do so. Nonetheless, many of the management approaches that have been proven effective are consistent with those that emerge from a consideration of the relevant ecological theory.

# REFERENCES

Adlerz, W. C. (1981). "Weed hosts of aphid-borne viruses of vegetable crops in Florida," in J. M. Thresh, Ed., *Pests, Pathogens and Vegetation*, Pitman, London, pp. 467–478.

Broadbent, L. (1964). "Control of plant virus diseases," in M. K. Corbett and H. D. Sisler, Eds., *Plant Virology*, University of Florida Press, Gainesville, pp. 330–364.

Buddenhagen, I. W. (1983). "Crop Improvement in relation to virus diseases and their epidemiology," in R. T. Plumb and J. M. Thresh, Eds., *Plant Virus Epidemiology*, Blackwell, Oxford, pp. 7–23.

Burdon, J. J., and R. C. Shattock. (1980). "Disease in plant communities," in T. H. Coaker, Ed., *Applied Biology*, Vol. 5, Academic, New York, pp. 145–219.

Burnett, H. C. (1961). Systemic spread of tristeza in one Valencia orange tree, *Plant Dis. Reptr.,* **45,** 697.

Cockbain, A. J., and G. D. Heathcote. (1964). Transmission of sugar beet viruses in relation to the feeding, probing and flight activity of alate aphids, *Proc. XII Int. Cong., Entomol.*, London, p. 521.

Day, P. R. (1974). *Genetics of Host–Parasite Interactions*, Freeman, San Francisco, 238 pp.

Denno, R. F., and H. Dingle. (1981). "Considerations for the development of a more general life history theory," in R. F. Denno and H. Dingle, Eds., *Insect Life History Patterns, Habitat and Geographic Variation*, Springer-Verlag, New York, pp. 1–6.

Dixon, A. F. G. (1985). Structure of aphid populations, *Ann. Rev. Entomol.*, **30,** 155–174.

Eastop, V. F. (1977). "Worldwide importance of aphids as virus vectors," in K. F. Harris and K. Maramorosch, Eds., *Aphids as Virus Vectors*, Academic, New York, pp. 3–62.

Feeny, P. P. (1976). "Plant apparency and chemical defense," in J. Wallace and R. Mansell, Eds., *Biochemical Interactions Between Plants and Insects. Recent Adv. Phytochem*, 10, Plenum, New York, pp. 1–40.

Gardner, R. C., A. J. Howarth, P. Hahn, M. Leudi-Brown, R. J. Shepherd, and J. Messing. (1981). The complete nucleotide sequence of an infectious clone of cauliflower mosaic virus by M13 mp7 shotgun sequencing, *Nucleic Acids Res.*, **9,** 2871–2888.

Gibson, R. W., A. D. Rice, M. C. Smith, and R. M. Sawicki. (1982). The effects of the repellents dodecanoic acid and polygodial on the acquisition of non-, semi-, and persistent plant viruses by the aphid *Myzus persicae*, *Ann. Appl. Biol.*, **100,** 55–59.

Harper, J. L. (1977). *Population Biology of Plants*, Academic, New York, 892 pp.

Harris, K. F. (1977). "An ingestion-egestion hypothesis of noncirculative virus transmission," in K. F. Harris and K. Maramorosch, Eds., *Aphids as Virus Vectors*, Academic, New York, pp. 166–220.

Harris, K. F. (1980). "Aphids, leafhoppers and planthoppers," in K. F. Harris and K. Maramorosch, Eds., *Vectors of Plant Pathogens*, Academic, New York, pp. 1–13.

Harris, K. F. and K. Maramorosch, Eds. (1977). *Aphids as Virus Vectors*, Academic, New York, 559 pp.

Harris, K. F. and K. Maramorosch, Eds. (1980). *Vectors of Plant Pathogens*, Academic, New York, 467 pp.

Harris, M. K. (1975). Allopatric resistance: searching for sources of insect resistance for use in agriculture, *Environ. Entomol.*, **4,** 661–669.

Harrison, B. D. (1981). Plant virus ecology: ingredients, interactions and environmental influences, *Ann. Appl. Biol.*, **99,** 195–209.

Irwin, M. E., and R. M. Goodman. (1981). "Ecology and control of soybean mosaic virus," in K. Maramorosch and K. F. Harris, Eds., *Plant Diseases and Vectors: Ecology and Epidemiology*, Academic, New York, pp. 182–220.

Kennedy, G. G. (1976). Host plant resistance and the spread of plant viruses, *Environ. Entomol.*, **5,** 827-831.

Kennedy, J. S., and H. L. G. Stroyan. (1959). Biology of aphids, *Ann. Rev. Entomol.*, **4,** 139-160.

Kiritani, K. (1983). "Changes in cropping practices and the incidence of hopper-borne diseases of rice in Japan," in R. T. Plumb and J. M. Thresh, Eds., *Plant Virus Epidemiology*, Blackwell, Oxford, pp. 239-247.

Leppik, E. E. (1970). Gene centers of plants as sources of disease resistance, *Ann. Rev. Phytopathol.*, **8,** 323-344.

Ling, K. C. (1972). *Rice Virus Diseases*, International Rice Research Institute, Los Baños, Philippines, 134 pp.

Ling, K. C., E. R. Tiongco, and Z. M. Flores. (1983). "Epidemiological studies of rice tungro," in R. T. Plumb and J. M. Thresh, Eds., *Plant Virus Epidemiology*, Blackwell, Oxford, pp. 249-257.

Lowe, A. D. (1973). "Aphid biology in New Zealand," in A. D. Lowe, Ed., *Perspectives in Aphid Biology*, Entomol. Soc. New Zealand Bull, No. 2, pp. 7-19.

Maramorosch, K. and K. F. Harris, Eds. (1979). *Leafhopper Vectors and Plant Disease Agents*, Academic, New York, 654 pp.

Maramorosch, K., and K. F. Harris, Eds. (1981). *Plant Diseases and Vectors: Ecology and Epidemiology*, Academic, New York, 368 pp.

Matthews, R. E. F. (1981). *Plant Virology*, 2nd Ed., Academic, New York, 897 pp.

Owusu, G. K. (1983). "The cocoa swollen shoot disease problem in Ghana," in R. T. Plumb and J. M. Thresh, Eds., *Plant Virus Epidemiology*, Blackwell pp. 73-83.

Pirone, T. P., and K. F. Harris. (1977). Nonpersistent transmission of plant viruses by aphids, *Ann. Rev. Phytopathol.*, **15,** 55-73.

Plumb, R. T., and J. M. Thresh, Eds. (1983). *Plant Virus Epidemiology*, Blackwell, Oxford, 377 pp.

Reddy, D. V. R., P. W. Amin, D. McDonald, and A. M. Ghanekar. (1983). "Epidemiology and control of groundnut bud necrosis and other diseases of legume crops in India caused by tomato spotted wilt virus," in R. T. Plumb and J. M. Thresh, Eds., *Plant Virus Epidemiology*, Blackwell, Oxford, pp. 93-102.

Rhoades, D. F., and R. G. Cates. (1976). "Toward a general theory of plant antiherbivore chemistry," in J. Wallace and R. Mansell, Eds., *Biochemical Interactions Between Plants and Insects*, *Recent Adv. Phytochem.*, 10, Plenum, New York, pp. 168-213.

Robinson, R. A. (1980). New concepts in breeding for plant disease resistance, *Ann. Rev. Phytopathol.*, **18,** 189-210.

Rochow, W. F. (1977). "Dependent virus transmission from mixed infections," in K. F. Harris and K. Maramorosch, Eds., *Aphids as Virus Vectors*, Academic, New York, pp. 253-273.

Romney, V. E. (1939). Breeding areas and economic distribution of the beet leafhopper in New Mexico, southern Colorado and western Texas, USDA Cir. No. 518, Washington, D.C., 31 pp.

Shivanathan, P. (1981). Transmission efficiency of whiteflies in the epidemiology of three diseases, *Plant Virus Dis. Epidemiol. Conf.*, Oxford, p. 48, (abstract).

Simons, J. N. (1959). Potato virus Y appears in additional areas of pepper and tomato production in south Florida *Plant Dis. Rep.*, **43,** 710-711.

Simons, J. N., R. A. Conover, and J. M. Walter. (1956). Correlation of occurrence of potato virus Y with areas of potato production in Florida, *Plant Dis. Rep.*, **40,** 531-533.

Stearns, S. C. (1977). The evolution of life history traits: a critique of the theory and a review of the data, *Ann. Rev. Ecol. Syst.*, **8,** 145-171.

Swenson, K. G. (1968). Role of aphids in the ecology of plant viruses, *Ann. Rev. Phytopathol.*, **6,** 351-374.

Tamaki, G., L. Fox, and B. A. Butt. (1979). Ecology of the green peach aphid as a vector of beet western yellows virus of sugar beets, USDA Tech. Bull., No. 1599, 16 pp.

Taylor, L. R. (1979). "The Rothamsted Insect Survey—an approach to the theory and practice of synoptic pest forecasting in agriculture," in R. L. Rabb and G. G. Kennedy, Eds., *Movement of Highly Mobile Insects: Concepts and Methodology in Research*, University Graphics, North Carolina State Univ., Raleigh, NC, pp. 148–185.

Thresh, J. M. (1976). Gradients of plant virus diseases, *Ann. Appl. Biol.*, **82**, 381–406.

Thresh, J. M. (1980). "The origin and epidemiology of some important plant virus diseases," in T. H. Coaker, Ed., *Applied Biology*, Vol. 5, Academic, New York, pp. 2–65.

Thresh, J. M. (1982). Cropping practices and virus spread, *Ann. Rev. Phytopathol.*, **20**, 193–218.

Thresh, J. M. (1983a). The long-range dispersal of plant viruses by arthropod vectors, *Phil. Trans. R. Soc. Lond. B.*, **302**, 497–528.

Thresh, J. M. (1983b). Progress curves of plant virus diseases, *Adv. Appl. Biol.*, **8**, 1–85.

Toba, H. H., A. N. Kishaba, G. W. Bohn, and H. Hield. (1977). Protecting muskmelons against aphid-borne viruses, *Phytopathology*, **67**, 1418–1423.

Tomlinson, J. A., and V. M. Walker. (1973). Further studies on seed transmission in the ecology of some aphid-transmitted viruses, *Ann. Appl. Biol.*, **73**, 293–298.

Van der Plank, J. E. (1948). The relation between the size of fields and the spread of plant disease into them. I. Crowd diseases, *Emp. J. Expt. Agric.*, **16**, 134–142.

Zitter, T. A. (1977). "Epidemiology of aphid-borne viruses," in K. F. Harris and K. Maramorosch, Eds., *Aphids as Virus Vectors*, Academic, New York, pp. 385–412.

Zitter, T. A., and J. N. Simons. (1980). Management of viruses by alteration of vector efficiency and by cultural practices, *Ann. Rev. Phytopathol.*, **18**, 289–310.

# 9

## ECOLOGICAL BASES FOR HABITAT MANAGEMENT AND PEST CULTURAL CONTROL

*Donald C. Herzog and Joseph E. Funderburk*

### 9.1. INTRODUCTION

An ecosystem has been described by Huffaker et al. (1984b) as a subdivision of the global environment containing a restricted number of living species that are adapted to survive within well-defined abiotic and biotic conditions. Natural resource ecosystems have been defined by Spurr (1969) as integrated ecological systems the product of which is directly or indirectly useful to humans. That product may be either biological (i.e., forest, range, agricultural products, fish, wildlife) or physical (i.e., water, air, soil) or both. Our food and fiber crops are produced in intensively managed, simplified ecosystems that have come to be known as agroecosystems. Agroecosystems have been characterized as domesticated ecosystems that are intermediate between natural (e.g., grasslands and forests) and fabricated (e.g., cities) ecosystems (Odum, 1984; see also Risser, Chapter 12).

In our struggle to maximize agricultural production for an ever-increasing human population, we encounter other organisms competing for our crops. Historically, we have attempted to defend our agricultural communities against the instabilities to which they are susceptible. Through trial and error (see Newsom, 1975; Botrell and Adkisson, 1977), we have determined that pests are not managed most efficiently by utilizing a single pest control tactic, especially if that tactic is chemical control. Hence, the current emphasis is on integrated pest control (management). The basis of pest management has been described as the *planned manipulation* of the processes that relate to potential economic loss due to pest(s) (Southwood and Way, 1970). As we enter an era of potential food and energy shortages, the challenge to the pest management research scientist or practitioner (the applied ecologist) is to assist in the development of a technology

217

that will permit higher-yielding cropping patterns or the maintenance of current production levels without creating conditions that will favor increased pest population pressure (Litsinger and Moody 1976). Furthermore, management programs implemented should be economically feasible and should strive for minimal depletion of both renewable and nonrenewable resources.

Many pest problems result from the creation or evolution of environmental conditions that favor pest population development. Historically, efforts to increase agricultural productivity have inadvertently altered the within-crop environment to the benefit of some pests and thus produced intensified crop protection problems. According to Rudd (1971) and Corbet (1976), an organism becomes a pest only after humans require something that it needs and are not prepared to share with it; humans often make that organism a more severe pest by creating an environment that favors its increase and survival. Apple and Smith (1976) enumerated eight factors that have enhanced agroecosystem vulnerability to pests:

1. Reliance on "monocultures" of major agricultural crops;
2. Increased use of fertilizers;
3. Improved water management to increase crop yields;
4. Multiple cropping;
5. Stringent plant breeding selection programs that have narrowed the genetic base of our principal crops;
6. Reduced tillage systems for "energy conservation";
7. Introduction of pests (pathogens, nematodes, insects, and weeds) into new areas;
8. Dependence on broadly toxic chemical pesticides and other single-tactic approaches to crop protection.

Five of these related directly to habitat modification and direct effects of cultural practices.

## 9.2. THE DIVERSITY–STABILITY QUESTION

Natural ecosystems are destroyed through clearing of land to plant crops. Modern agricultural practices tend to reduce diversity by promoting monocultures of single plant species that are genetically and phenologically uniform in both space and time (Southwood and Way, 1970; Dempster and Coaker, 1974; Flint and van den Bosch, 1981). For some time, ecologists have maintained that agricultural systems, by nature of their lack of diversity, are inherently unstable and prone to upsets in the form of pest outbreaks (Wilhelm, 1976). In fact, as Ehrlich et al. (1977) indicate, "That there is a causal connection between complexity (or diversity) and stability . . . in biological communities [has become] part of the folk wisdom of ecology." May (1973), however, has shown convincingly that system stability does not increase as a simple mathematical consequence of an increase in species diversity [see Goodman (1975) for additional review of these concepts]. Indeed, van Emden and Williams (1974) and Rabb (1978), perhaps heretically, maintain that the label of agroecosystem instability has arisen only because agricultural producers and their regulatory activities have been ignored as major interactive components of such systems. They imply that if the agriculturalist is considered as an interacting organism in the (agro)ecosystem, then through regulatory actions a modicum of system stability may perhaps be achieved. If the activities involved in agricultural production promote system instability then alteration of cultural (production) practices (manipulation of the system environment) should assist in achieving an amelioration of pest problems associated with crop production. Lowrance et al. (1984b) described various aspects of agricultural ecosystems and attempted to describe mechanisms involved in the development of crop production and protection problems, and to describe alterations of agricultural systems that would lead to system stability.

That there is no reason to believe that diversity produces stability in natural ecosystems was concluded by Goodman (1975) and Murdoch (1975). Murdoch (1975) maintains that the instability of agroecosystems results from the frequency of disruption of crops and from lack of coevolutionary links between the interacting species (crops and pests). Ehrlich et al. (1977) tentatively conclude that "merely adding more species to agricultural ecosystems will not necessarily stabilize them. Indeed, it may have exactly the opposite effect." An interesting support for that conclusion has been provided by O'Neil (1984), who found that the level of predation on a herbivore is closely linked to diversity of the predator complex. He found that as diversity of the predator complex increased, rate of predation declined.

## 9.3. CULTURAL PRACTICES AND CULTURAL CONTROLS

"Cultural practices" refers to that broad set of management techniques or options which may be manipulated by agricultural producers to achieve their crop production goals (Kennedy et al., 1975). Another definition, proposed by Thompson (1975) is "the manipulation of the environment to improve crop production. . . ." "Cultural control," on the other hand, is the deliberate alteration of the production system, either the cropping system itself or specific crop production practices, to reduce pest populations or avoid pest injury to crops (Ashdown, 1977). Cultural methods may not of themselves reduce pest populations below acceptable levels (established economic thresholds), but may be of benefit in reducing losses due to pests (Glass, 1975) through delaying pest outbreaks or reducing the magnitude or frequency of such outbreaks. Such practices are frequently the first line of defense against pest populations. Most result in little or no added production cost, because they are merely variations in the timing or manner of performing operations that are normally necessary to the production of the crop (Metcalf et al., 1962). Cultural control practices are un-usually free of undesirable ecological consequences. Both the cultural practices used to produce the crop and the cultural methods designed to manage pest populations are essentially habitat management: The former is directed at en-hancement of crop growth and yield, while the latter is directed at reducing pest population development, injury to the crop, and potential loss of productivity. Cultural control practices are usually compatible, both among themselves and with other pest control tactics and strategies, and thus are highly desirable for the management of pests.

Many diverse cultural practices are employed to plant, maintain, and harvest crops in all crop production systems. These manipulations alter the within-crop environment and thus influence the ecology of the crop production (agroeco)sys-tem. Even slight modifications in crop production practices can substantially affect the population dynamics of pests and their natural enemies, as alterations of such cultural activities alter the microenvironment of the crop (Barfield and Gerber, 1979; Arkin and Taylor, 1981; Hatfield and Thomason, 1982). Conse-quently, any change in cropping practices or in crop diversity may lead either directly or indirectly to reduced populations of certain pests and to an increase of others, with corresponding increases or decreases in crop stress from those pests (Glass, 1975; Hanson, 1975). Certainly, it seems reasonable that agricultural producers should maximize the use of any ecologically and economically feasible cultural control tactics and minimize the use of other, less desirable tactics. Huf-faker (1982) stressed the need for ecological studies of the interrelationships of pests and other elements in crop ecosystems, and called for detailed studies to quantify those relationships closely. Therefore, we will follow the rationale of Rolston and McCoy (1966) when considering cultural practices and habitat ma-nipulations and their effects on insect (pest) populations. They state:

> As we attempt to manage insect [pest] populations to further our own ends, we need (1) to recognize the needs of the insects [pests]; (2) to know how they normally realize their needs; and (3) to know what modifications of the environment will either simplify or further aggravate their problems.

Thus, in the discussions that follow, both the beneficial and detrimental ef-fects of cropping practices (habitat manipulation) will be discussed. Before re-viewing the various ways in which cropping practices may be modified to provide cultural control of pests, it is desirable to discuss relevant aspects of ecological theory.

## 9.4. HABITAT AND NICHE

The aggregate of plant and animal communities that occur together in a habitat collectively comprise the *biotic community*. The biotic community, together with its habitat, comprises the *ecosystem* (Tansley, 1935). The term agroecosystem results from the application of ecosystem concepts to agriculture (Lowrance et al., 1984a).

The *habitat* is often simplistically described as "the place in which an organism lives." More precisely, the habitat is the specific set of physical conditions (e.g., space, substratum, climate) that surrounds a single species, a group of species, or a large community (Clements and Shelford, 1939). The habitat thus consists of factors both intrinsic and extrinsic to the growth, development, reproduction, and survival of resident organisms and/or populations of those organisms.

Grinnell (1917) coined the word niche to describe the place of an organism in the environment. In the classroom, the ecological niche is often simply defined as "the sum total of everything that the organism does." The ecological niche has been defined by Kendeigh (1961) as "the particular position in a community and habitat occupied by an animal as the result of its peculiar structural adaptations, its physiological adjustments, and the special behavior patterns that have evolved to make best use of these potentialities." The niche concept was redefined by MacFayden (1957) to include "the ecological conditions that permit a species to effectively exploit a source of energy and thus be able to reproduce and colonize further such sets of conditions." Hutchinson (1957) viewed the niche as an *n*-dimensional hypervolume in the environment, the perimeter of which circumscribes the space in which a species can reproduce indefinitely. It is considered to be the ultimate spatial unit occupied by a single species, and to which that species is held by structural limitations such as climatic factors, quality and quantity of food, shelter, and perhaps other factors. A distinction is often made between the potential (or fundamental) and actual (or realized) niche (Price, 1984). The former is defined as that which could be utilized by an organism in the absence of competitors and other enemies; the latter as that which is actually utilized by an organism.

The habitat of an organism is easily defined in spatial terms. The niche of an organism is intangible and can be described only through in-depth studies of the population dynamics of that organism. A proscribed habitat, therefore, may contain an infinite number of niches, which may be either inhabited or uninhabited. Modification of the habitat, often easily described in descriptive terms, may effectively create niches for, or alter or destroy existing niches inhabited by the organisms of interest. The creation, alteration, or destruction of niches within a habitat are cryptic events. Often it is difficult, perhaps even technically impossible, to ascertain by what mechanisms habitat modification results in either extinction or outbreaks of species within a habitat (Herzog and Funderburk, 1985).

The explanations for such phenomena would probably be couched in terms of thresholds and tolerances (Kendeigh, 1961). Population regulation may result from a single factor or a combination of factors acting concurrently or in sequence [in the sense of Liebig (1840) and Shelford (1913)]. Huffaker et al. (1984a) enumerate the following regulatory factors involved in the natural control of insect populations: (1) temporal and spatial relationships, (2) climate and weather, (3) a place to live, (4) food, (5) other organisms of the same kind, and (6) organisms of other kinds. These regulatory factors often do not function independently (Andrewartha and Birch, 1954), but collectively to comprise the realized niche. Suboptimal availability of (1)–(4) or an overabundance of (5) or (6) may alter the niche or make it uninhabitable by an organism.

### 9.4.1. Functional Mechanisms

Three primary mechanisms function through manipulation or modification of cultural practices to reduce either pest populations on a crop or injury to the crop following and despite infestation. These, in brief, are: (1) impediments to pest colonization of the crop, (2) the creation of adverse biotic or abiotic conditions that reduce survival of individuals or populations of the pest, and (3) modifications of the crop in such a way that pest infestation results in reduced injury to the crop. Here, injury is defined as damage to the crop that results in decreased marketable yield. In the following sections, we will present a general discussion of these mechanisms, and then describe specific examples whereby habitat manipulations have resulted in either reduction or intensification of pest populations and subsequent crop injury.

The applied ecologist is concerned with the mechanisms by which the manipulation of the habitat through alteration of cropping practices alters the microenvironment affecting the organism, and ultimately its niche. The availability of this knowledge then allows the agricultural producer to manipulate the cropping system such that crop production becomes more efficient and, therefore, more profitable.

### 9.5. PLANT (CROP) APPARENCY TO PESTS

The dependency of consumers on the dispersion of their food resources plays a central role in studies of plant–insect interactions. For example, the argument that plant apparency shapes the evolution of antiherbivore defenses of plants rests upon the assumption that plant dispersion determines the frequency of encounter between herbivores and their food plants (Feeny, 1975; Rhoades, 1979; Fox, 1981; Kogan, Chapter 5). There is increasing evidence that pest problems in crops are aggravated by the concentration in time and space of resources associated with modern agricultural practices (Cromartie, 1981), and several authors have noted the potential for controlling insect pests by manipulating cropping patterns (Perrin, 1977; Perrin and Phillips, 1978; Coaker, 1980; Finch, 1980; Cromartie, 1981; Hare, 1983; but see Kareiva, Chapter 4).

### 9.6. RESOURCE HETEROGENEITY

Herbivores must locate and select appropriate species of host plants from among a mosaic of less suitable or unacceptable ones; at the same time, they must contend with differences in nutritional quality that may exist among individuals of the appropriate plant species. Therefore, differences in the abundance and dispersion of host plants in the field, phenological and spatial variation in quality among individuals in a plant population, the nature of the surrounding or intermingled vegetation, and the range of variation in these factors relative to the mobility of the herbivore can all affect the ability of the herbivore to find and remain on plant(s) and, subsequently, the fitness of either individuals or populations of the herbivore (Denno and McClure, 1983a). Different spatial arrangements or combinations of plant species are known to influence patterns of herbivore abundance. However, there have been few attempts to elucidate the relationship among herbivory, plant dispersion, and plant competition (Kareiva, 1983; see also Chapter 4).

### 9.6.1. Resource Concentration (Vegetation Texture)

We usually expect to find consumers concentrated and thriving where their resources are abundant and easy to find. This idea is a prominent component of several ecological theories. For instance, the aggregation of predators in regions of high prey density has received much attention as a potentially stabilizing influence in predator–prey interactions (Hassell, 1978). The concentration of consumers where resources are abundant is also a feature of optimal foraging theory; if searching for resources is costly, then an optimal forager should focus its efforts in regions of high resource density (Kareiva, 1983).

These ideas indicate that "texture" of an ecosystem (the pattern arising out of characteristic plant densities and diversities) should influence patterns of herbivore abundance (Kareiva, 1983). Root (1973) formalized this interaction and emphasized the relationship between host plant dispersion and herbivore community structure in cole crops in proposing the resource-concentration hypothesis: Specialized herbivores are more likely to find and remain in areas where host plants are concentrated, that is, where they grow in pure, large, and/or dense stands such as monocultures of agricultural crops.

### 9.6.2. Host Nutrition (Vigor)

There is an extensive body of literature indicating that insect population growth can be dramatically affected by host vigor (Mattson and Addy, 1975). Fertilization can alter host plants in ways that make them more or less susceptible to herbivore attack (Rodriguez, 1951, 1972; Mattson and Addy, 1975; Jones, 1976; Slansky and Feeny, 1977; McNeill and Southwood, 1978; Mattson, 1980; McClure, 1980; Tingey and Singh, 1980; Scriber, 1982). Pruning of trees and shrubs results in new foliage and alters host physiology such that herbivore populations increase (Owen, 1971; Hall, 1977).

### 9.6.3. Synchrony of Plant and Herbivore Phenologies

Because of the profound, potentially negative effects of plant variability on herbivore abundance, it is not surprising that the reduction of interplant variability by agriculturists and silviculturists elevates some herbivores to pest status (Feeny, 1976; Edmunds and Alstad, 1978; Hare, 1983). However, agriculturalists may be able to turn the tables on herbivores by manipulating or exploiting host plant variability to reduce pest density. Aside from trying to control or manipulate plant heterogeneity, simply understanding how herbivores are affected by plant variability may promote the development of more efficient, finely tuned pest management programs (Denno and McClure, 1983a). Producers of annual crops can manipulate planting and harvesting dates to some extent to increase asynchrony between host and herbivore. In conjunction with other cultural practices to hasten crop maturity, early planting of short-season varieties serves as the basis for a contemporary program of insect pest management (Adkisson et al., 1982). Similarly, in conjunction with the use of sprout inhibitors, timely destruction of tobacco stalks reduces the quantity of young, succulent foliage available to the tobacco hornworm and the subsequent size of the overwintering population (Tappan, 1965; Rabb, 1969). There are, however, definite limits over which plant phenology can be shifted before crop production is affected, and reliance on manipulation of planting and harvesting dates as a primary means of insect control would require extensive knowledge of the relationships between plant growth and insect development and daily and seasonal temperatures (Hare, 1983).

## 9.7. FACTORS AFFECTING CROP GROWTH

Light, soil moisture and fertility, and temperature determine crop growth. In turn, crop growth, along with all other factors, combine to determine potentials for growth and reproduction of crop pests, innocuous species, and natural enemies in the crop ecosystem. Thus, it is essential to understand well the influences of such biometeorological factors as temperature, light, and moisture on the main biological processes (Huffaker, 1982).

Modification of the environment to hinder the pest and/or to favor natural enemies may take many forms: (1) either destruction or provision of breeding or overwintering refuges, (2) either destruction or provision of alternate hosts or volunteer plants, (3) crop rotation or maintenance of a host-free season, (4) tillage, (5) timing of planting or harvest, (6) trap crops, (7) habitat diversification, (8) manipulations of the plant itself, such as pruning or sucker removal, or (9) water or nutrient management (adapted from Rabb et al., 1984).

## 9.8. CULTURAL CONTROL BY MODIFYING VEGETATION IN AGROECOSYSTEMS

The vegetation located in all habitats of agroecosystems influences the diversity and numbers of resident arthropod species, including both pest and innocuous herbivores and their natural enemies. Therefore, specific vegetative species in both crop and noncrop habitats may potentially be established, manipulated, or destroyed for the deliberate purpose of managing pestiferous herbivore populations. Such modifications may involve either crop or noncrop (wild) vegetation.

## 9.8.1. Noncrop Vegetation

Many natural enemy species require food sources in the form of pollen, nectar, or innocuous arthropods that are not present in particular crop habitats. These food requirements may be provided to support natural enemy populations by encouraging or deliberate development of certain wild vegetation habitats near plantings of the crop (see Altieri et al., 1977; Altieri and Whitcomb, 1979 for reviews). Alternative food requirements of predators and parasitoids have been reviewed by Altieri et al. (1977) and Altieri and Whitcomb (1979).

Native perennial stinging nettle *Urtica dioica* L. serves in England as a source of aphid natural enemies without supporting economically important pests (Perrin, 1975). Populations of these natural enemies increase in the spring on aphid and psyllid species inhabiting the nettle, and may disperse to crops with the onset of summer. Cutting of stands of the nettle in mid-June is one potential cultural method of enhancing biological control in nearby crops.

Natural biological control of the grape leafhopper *Erythroneura elegantula* Osborn, the most important pest of grapes in the San Joaquin Valley of California, can be achieved by an egg parasitoid, *Anagrus epos* Girault; however *Anagrus* is an effective control agent only in vineyards located within 5.6 km of streams and rivers (Doutt et al., 1966). *A. epos* overwinters on populations of the blackberry leafhopper *Dikrella californica* (Lawson), present on blackberry stands in stream and river bottoms, but does not successfully overwinter in grape leafhopper populations. Limited dispersal capabilities of the egg parasitoid from these overwintering refuges prevents biological control of grape leafhopper in the more distant vineyards until midsummer or later. Doutt and Nakata (1965, 1973) describe in detail the biology of *A. epos* in the agroecosystem. Recently, blackberry plantings have been established near vineyards that are greater than 5.6 km from streams and rivers in an effort to provide overwintering refuges for *A. epos* (see Flaherty et al., 1985). These practices, however, have not led to successful biological control of grape leafhopper in the nearby vineyards, since the habitat of the established refuges has not provided adequate shade and moisture conditions necessary to support substantial blackberry leafhopper populations. Therefore *A. epos* has failed to overwinter in substantial numbers.

Establishment and/or manipulation of wild vegetation (i.e., weeds) within or near crop habitats may also reduce crop injury from arthropod pests by increasing natural enemy populations and/or by decreasing the colonization of crop pest populations into fields. For example, weeds present within Brussels sprout fields in England greatly reduce damage by phytophagous species, particularly the cabbage aphid *Brevicoryne brassicae* (L.). Many of the pests, including *B. brassicae*, are more attracted to Brussels sprout fields without weeds than to fields with weeds (Smith, 1969, 1976a). Increased natural enemy activity has been shown to be partially responsible for reduced pest abundance in the weedy Brussels sprout crop (Smith, 1976b).

Soybean fields in Georgia infested with sicklepod *Cassia obtusifolia* L. support greater predator populations than comparable but weed-free fields (Altieri, 1981). These predators may reduce soybean injury from pests, notably the velvetbean caterpillar *Anticarsia gemmatalis* Hubner and the southern green stinkbug *Nezara viridula* (L.). Similarly, Shelton and Edwards (1983) noted that both pest and predator populations in Indiana soybean fields are affected by the quantity and species of weeds present (Table 9.1). The predator *Coleomegilla maculata* (deGeer) was more abundant in soybean fields with grassy and/or broadleaf weeds than in fields with no weeds. Other predators including *Orius insidiosus* (Say), *Nabis* spp., and *Harpalus* spp. preferred fields with grassy weeds or mixtures of grassy and broadleaf weeds. Conversely, weed-free fields harbored greater numbers of the Mexican bean beetle *Epilachna varivestis* Mulsant than weedy fields. This increase in pest population was attributed to (1) reduced mortality from natural enemies (e.g., predators) in weed-free fields, (2) stronger olfactory or visual cues to host plants in weed-free fields, and (3) more favorable microclimatic conditions in weed-free fields.

Despite the promising results obtained in these and other studies, there is reason for caution in undertaking manipulations of the system that may potentially alter the predator complex (Herzog et al., 1984). Studies on the velvetbean caterpillar in Florida soybean systems have shown that (1) alteration of species diversity of the predator complex resulted in reduced rates of predation, (2) rates of predation are highly sensitive to changes in the composition of the predator complex, and (3) as density of indigenous generalist predators increased, rate of predation also increased (O'Neil, 1984).

Cultural control practices involving establishment and maintenance of wild vegetation (including the above mentioned examples) are rarely utilized in crop

**Table 9.1  Mean Number of *E. varivestis* and Predators in Sweep-Net Samples and Harpalus spp. in Pitfall Samples in Four Soybean Habitats**[a]

| Soybean Habitat | Number/Sweepnet Sample | | | | Number/ Pitfall Trap/Week |
| | *Epilachna varivestis* | *Coleomegilla maculata* | *Orius insidiosus* | *Nabis* spp. | *Harpalus* spp. |
| --- | --- | --- | --- | --- | --- |
| Weed free | 8.0 | 0.1 | 0.7 | 0.3 | 9.3 |
| Broadleaf weeds | 1.9 | 0.4 | 1.0 | 0.3 | 11.9 |
| Grass weeds | 1.8 | 0.5 | 4.7 | 0.6 | 33.6 |
| Mixed weeds | 1.0 | 0.7 | 3.4 | 0.2 | 25.5 |

[a]Means represent the average of two seasons data. Adapted from Shelton and Edwards (1983).

production and probably will not be acceptable pest control strategies to most crop producers and agricultural scientists. They are usually ecologically desirable, but are usually not economically feasible when compared with other types of pest control in commercial-scale agriculture (e.g., insecticidal control, resistant varieties). In fact, growth of wild vegetation generally is discouraged both actively and passively in modern, intensively managed agroecosystems, where much of the available habitat consists of monocultures (i.e., large fields of one crop produced with clean cultivation).

Certain technical problems must also be addressed before wild vegetation management techniques can be more widely accepted. Methods must be developed to establish and maintain wild vegetation habitats and to manipulate weeds so that they do not compete unacceptably with crops. Altieri and Whitcomb (1979) discuss and review such problems.

In addition to serving as reservoirs for beneficial arthropods, wild vegetation may serve as suitable habitat for certain arthropod pests. Therefore, the destruction of such wild plant species may reduce the populations of these pests and subsequent movement to crop plants. This practice can be especially beneficial if the pest utilizes the wild vegetation for overwintering or spring population increases. A review of the literature reveals that this cultural control method is sometimes used in integrated pest management programs.

The clover seed chalcid *Bruchophagus roddi* (Gussakovsky), a major pest of alfalfa seed production in the western United States, thrives on bur clover and volunteer alfalfa growing along roads and ditch banks, and emigration of the pest from these noncrop habitats results in damage to alfalfa seed fields. Destruction of bur clover and volunteer alfalfa has been shown to reduce damage greatly in nearby alfalfa fields (Sorenson, 1930; Bacon et al., 1959). Through use of this practice, average crop losses in the San Joaquin Valley of California have been reduced from 12 to 14% (Bacon et al., 1959) to less that 1% (Anonymous, 1978).

The wheat curl mite *Eriophyes tulipae* Keifer, a pest in Kansas of winter wheat and a vector of wheat streak mosaic virus (Sylkhuis, 1953), survives the summer months on any suitable green or living plant hosts, primarily volunteer wheat, volunteer barley, volunteer oats, or grasses. Successful field control measures involve the use of two cultural control practices and planting of tolerant or resistant varieties (Somsen and Sill, 1970). These cultural control measures are destruction of volunteer wheat (and other small grains) and grasses in and around fields at least 2 weeks before wheat is planted and planting wheat as late in the fall as agronomically acceptable.

The beet leafhopper *Circulifer tenellus* (Baker) in southern Idaho utilizes graminaceous hosts located in desert and range areas of the Snake River basin for overwintering and spring survival. A large-scale program involving destruction of these grass hosts and replacement with broadleaved annual plants has successfully controlled beet leafhopper and curly top disease (transmitted by the leafhopper) on several commercially important crops (Gibson and Fallini, 1963). The program drastically reduced crop losses from 1965–1971. But, support was not continued when the pest diminished in importance, and problems have increased on some commercial crops (Knipling 1979).

As early as the beginning of this century, destruction of overwintering habitats was suggested as a management strategy for the cotton boll weevil *Anthonomus grandis* Boheman (Hunter, 1909; Isely, 1930). Such methods as burning and raking of leaf litter were shown to be effective. These practices were never widely adapted, probably because the control proved to be only temporary (Slosser et al. 1985). More recently, however, a new boll weevil management strategy has been developed for cotton in the Rolling Plains of Texas. Destruction of stands of sand shinnery oak *Quercus havardii* Rydb., a range shrub, effectively eliminates the deposition of leaf litter that normally serves as overwintering habitat for the weevil, and has been shown to reduce overwintering survival by 80% (Slosser and Boring, 1980; Slosser et al., 1984, 1985).

A few other examples of cultural control through destruction of wild vegetation are worthy of mention. Destruction of perennial weeds, such as field bindweed *Convolvulus arvensis* L. and Canadian thistle *Cirsium arvanse* (L.) have been shown to drastically reduce populations of the redbacked cutworm, *Euxoa ochrogaster* (Guenee), in Washington asparagus fields (Tamaki et al., 1975). Elimination of wild host plants in and around fields is an important method of controlling the sorghum midge *Contarinia sorghicola* (Coquillet) on sorghum (Randolph and Montoya, 1965; Bowden, 1965; Bottrell, 1971; Huddleston et al., 1972).

Although pest populations may often be suppressed by destruction of their alternate hosts, the result may not always be desirable. For example, preliminary reports from Mississippi (H. N. Pitre, Mississippi State Univ., personal communication) indicate that *Heliothis* spp. damage to soybean is reduced in fields infested with teaweed (prickly sida) (*Sida spinosa* L.). This weedy host apparently attracts ovipositing moths and holds the developing larvae. Destruction of this weed through an agronomically sound practice may serve also to intensify seed and pod damage by the corn earworm and/or the tobacco budworm. A possible implication is that development and implementation of economic thresholds for teaweed may allow the acceptance of infestation levels capable of reducing *Heliothis* injury.

On the other hand, maintenance of clean field borders, ditch banks, and fence rows often serves to reduce populations of generalist herbivores, such as grasshoppers (*Melanoplus* spp.) and blister beetles (*Epicauta* spp.). Specifically, the Japanese beetle *Popillia japonica* Newman is attracted to elder (*Sambucus canadensis* L.) and smartweed (*Polygonum* spp.) in field borders. These hosts, in areas infested by the beetle, serve as a reservoir for continuous crop infestation (Anonymous, 1969).

Weedy field borders and wind breaks in the midwestern states serve as reservoirs for aphid vectors of soybean mosaic virus: such species as the green peach aphid [*Myzus persicae* (Sulzer)] and the cowpea aphid (*Aphis craccivora* Koch). Similarly, vectors of tobacco ringspot virus (bud blight)—for example, grasshoppers (*Melanoplus* spp.), the onion thrips (*Thrips tabaci* Lindeman), and spider mites (*Tetranychus* spp.)—are assumed to be responsible for movement of the virus into soybean fields from surrounding weed reservoirs (Newsom et al., 1980). Maintenance of clean field borders may serve to limit the spread of these important plant pathogens into and through a crop by limiting populations of their arthropod vectors.

### 9.8.2. Crop Vegetation

The crops grown in an agroecosystem also influence the diversity and abundance of pest and natural enemy populations. Consequently, pest stress to crops can frequently be reduced by altering the crops grown in the agroecosystem. Such cultural modifications may include crop diversification, or rotation, or planting different crops in the same or adjacent fields (i.e., intercropping or trap cropping). In agroecosystems where only one or a few crops are grown routinely, the use of these practices may require that additional crops be produced.

Rotating susceptible crops with nonhost crops may either greatly reduce or prevent damage from some major arthropod pests. Furthermore, the practice has only rare economic or ecological drawbacks, and agricultural scientists and crop producers tend to regard crop rotation as a prudent management practice even when arthropod pest damage to crops is anticipated to be minimal. Rotation apparently is most successful against arthropod pest species with long generation cycles and with limited dispersal capabilities. Numerous species of major soil pests are successfully controlled by crop rotation. Rotation of host and nonhost crops often provides an effective and economical means of managing insects that become established during one cropping season, overwinter in a flightless stage, and are again capable of damaging the crop the following cropping season. Most pests controlled by rotation have a subterranean larval stage that produces most of the damage. In soybean, an excellent example is the whitefringed beetle complex, composed of *Graphognathus leucoloma* (Boheman) and *G. peregrinus* (Buchanan). Adults of these species lay many eggs when fed on soybean. However, the grass crops, including corn, are in some way nutritionally deficient for support of feeding, and do not suffer heavy damage by larvae, nor do adults fed on them produce many eggs (Ottens and Todd, 1979, 1980). Therefore, rotation between soybean and corn or grain sorghum is beneficial.

Crop rotation interrupts the life cycle of the northern (*Diabrotica longicornis* Smith and Lawrence) and western (*D. virgifera* LeConte) corn rootworm, and both species can be effectively controlled by rotation throughout the corn belt of the United States (see Chiang, 1973). Recent trends of continuous corn production have resulted in increased rootworm damage (Ortman and Fitzgerald, 1964; Chiang and Flaskerd, 1969) and the need for increased application of insecticidal controls. Specific rotation of crops that are favorable to wireworm populations (i.e., clovers, vegetables, and field crops) with crops that are detrimental to wireworm populations (i.e., alfalfa and pasture sod) greatly reduces damage to the vegetable and field crops in the Pacific Northwest (Lane, 1941; Landis and Onsager, 1966).

Populations of some above-ground pests also can be reduced by crop rotation. The migratory grasshopper *Melanoplus sanguinipes* (F.) and the differential grasshopper *M. differentialis* (Thomas) are greatly reduced in Arizona alfalfa fields when that crop is rotated with cotton or other crops (Barnes, 1959). Routine practice of rotation among crops, however, is not always wise. While the soybean–corn rotation described above may be beneficial in the southeast, the same rotation scheme in the midwest has been shown to increase populations of white grubs (*Phyllophaga* spp.) and the black cutworm [*Agrotis ipsilon* (Hufnagel)] (Metcalf et al., 1962).

The diversity of crops in an agroecosystem affects populations of some pests and their natural enemies. The pest status of polyphagous, highly mobile species may be influenced by crop diversity throughout the ecosystem. In Louisiana, outbreaks of the soybean looper *Pseudoplusia includens* (Walker) are believed to be caused by adult feeding on cotton nectar which increases oviposition, mating frequency, and longevity (Jensen et al., 1974). Outbreaks of soybean looper are common in agroecosystems containing soybean and cotton, but rare when soybean in planted after rice and cleared forest areas (Burleigh, 1972; Wuensche, 1976; R. M. Beach and J. W. Todd, University of Georgia, Coastal Plain Experiment Station, personal communication). Obviously, discontinuation of the planting of cotton would reduce soybean looper damage in some areas, but is not utilized as a cultural control practice due to the economic importance of cotton as a crop. Also in Louisiana, expansion of sweet sorghum production as a biomass energy crop has resulted in a large and rapid buildups of populations the sugarcane borer *Diatraea saccharalis* (F.). Such buildups pose a potential threat to the production of sugarcane, which is the primary host of this species (Reagan and Flynn, 1985).

The planting of a different variety or crop species in the same field as the main crop is known as interplanting. Arthropod pest damage in the main crop may be reduced if the interplanted habitat is more attractive to pests (i.e., trap crop), interferes with pest colonization of the main crop, and/or provides a reservoir for natural enemies. Intercropping systems may be an effective cultural control practice, and are most practical when the main crop is an intensively managed, high-value commodity and when the interplanted crop can withstand large pest populations with only limited economic losses. Obvious ecological advantages could hasten acceptance of interplanting programs in the future.

Strips of alfalfa interspersed in cotton fields in the San Joaquin Valley of California attract and act as an excellent trap crop for *Lygus* spp. Resulting economic losses to alfalfa are lower than the cost of alternative controls in the cotton (Stern et al., 1969). Alfalfa is also excellent habitat in this agroecosystem for important natural enemy species of *Lygus* spp., spider mites, *Heliothis zea*, and other major pests (i.e., nabids, chrysopids, anthocorids, coccinellids, and geocorids). These natural enemies can reproduce rapidly in alfalfa habitats, and adults readily disperse to nearby crops. The practice of interplanting of alfalfa in occasional strips further increases benefits of biological control to adjacent crops such as cotton (Stern, 1969).

Despite the economic advantages of interplanting alfalfa within fields of cotton in the San Joaquin Valley, producers have not adopted the practice (see Stern, 1981). The primary reason apparently is that numerous production problems have not been adequately overcome. Regulations require that certified alfalfa seed cannot be grown unless the field has been free of any alfalfa crop for 3 years, and interplanting disrupts the typical cotton–alfalfa seed rotation programs. Alfalfa also requires more frequent irrigation than cotton to remain attractive to *Lygus* spp. populations. When using the interplanting program, water management problems result, many alfalfa strips dry out, and cotton fields sometimes require treatment to suppress dispersing *Lygus* spp. populations.

Alfalfa also has disadvantages for strip cropping in other cotton-growing areas of the United States where the alfalfa weevil *Hypera postica* (Gyllenhal) is a serious alfalfa pest. Insecticides applied in the early spring to eliminate the weevil also eliminate its natural enemies (Laster, 1974). However, damage from such pests as bollworms (*Heliothis* spp.) can be reduced in cotton grown in these areas by interplantings of sorghum (Robinson et al., 1972) or sesame (Laster and Furr, 1972).

Injury to crucifers produced by the cabbage maggot *Delia radicum* (L.) and cabbage aphid is greatly reduced by between-row interplantings of beans, grass, clover, or spinach (Coaker, 1980). Cabbage aphid injury declines because of reduced immigration of populations onto crucifers and increased predation by ground-dwelling arthropods. Cabbage maggot populations are reduced because of reduced oviposition on the crucifer crop. Herbivore response to intercroppings or interplantings, however, are not always so clear-cut. For example, in studies of intercropping pinto beans and sweet corn in Colorado, results were variable, with positive, negative, and neutral effects observed, depending both on the crop and on the insect under study (Capinera et al., 1985).

Early planted, early maturing soybean varieties planted near main soybean plantings attract the bean leaf beetle *Cerotoma trifurcata* (Forster) and southern green stinkbug (Newsom and Herzog, 1977). Appropriately timed insecticide applications to the trap crop control the target pests in the trap reservoir and prevent spread of the pest to the main crop, thereby conserving natural enemy populations throughout the remainder of the ecosystem. Similarly, spread of bean pod mottle virus, which is vectored by the bean leaf beetle, is retarded. Border rows of snap beans planted prior to soybean have been used to attract and function as a trap crop for the Mexican bean beetle (Stevens et al., 1975; Rust, 1977). Release of a parasite, *Pediobius foveolatus* (Crawford), into such trap rows creates a reservoir or "nurse crop" from which parasites disperse and effectively suppress populations of the Mexican bean beetle in the soybean crop.

Historically, it has been hypothesized that pest populations reach outbreak levels more frequently in monocultures of a crop than in the same crop grown in a more diverse habitat. This apparently is not always true for soybean. A seasonal succession of hosts favorable for oviposition has been shown to stimulate outbreaks of the corn earworm on soybean (Johnson et al., 1975; Bradley and Van Duyn, 1980). Similarly, a dry summer, which causes all plantings of corn to senesce over a short time interval, will almost insure a heavy outbreak of this species on soybean.

## 9.9. CULTURAL CONTROL BY MODIFYING PRACTICES USED TO PRODUCE CROPS

Many practices are employed to plant, maintain, and harvest a crop. These practices influence the level of injury to the crop due to arthropod pests, and purposeful modification may help to alleviate crop losses. Pest injury may be avoided by producing a crop at times when pest populations are in nondamaging stages or at low population levels. Modification of certain management practices can reduce colonization of pest populations into crops. Modification of cultural practices often elicit altered microclimatic conditions in crop habitats in such a way that survival or development of pest populations is reduced. Such types of cultural control are common in most cropping systems. In many cases, producers and agricultural scientists are unaware of the biological mechanism(s) by which pest injury is reduced. They simply know that crop yields are improved. In other cases, the effect of a production practice on pest biology is understood, and production is deliberately modified in such a way that crop stress due to infestation is reduced. Production practices that can sometimes be modified to control pests culturally include planting date, harvest date, irrigation, fertilization, row spacing or orientation, seeding rate, and tillage. The following discussion will illustrate how these production practices can be altered to reduce damage from some major arthropod pests.

### 9.9.1. Planting and Harvest Dates

Alterations in planting date and harvest date can frequently result in escape from damaging pest infestation. Manipulation of planting date and harvest date is an important management strategy in cotton. Planting early, growing short-season varieties, and early defoliation enhance maturity of the crop in some regions before boll weevil and bollworm populations reach injurious levels, whereas in other regions planting late to allow for maximum mortality of over-wintered populations of key pests, primarily boll weevil, is the better strategy. Frisbie and Walker (1981) review literature and explain the role of planting date, harvest date, and variety selection on integrated pest management programs for cotton.

As early as 1930, Watson observed that early planting of soybean provided a means of evading peak velvetbean caterpillar populations in central Florida. These observations were later supported by model predictions (Menke, 1973) and experimentally under field conditions (Herzog, 1979). Choice of early-maturing soybean varieties and/or early planting of the crop allows the majority of reproductive plant growth to take place before peak pest populations occur. In this case, pest populations are not reduced, rather the phenology of the crop has advanced to the point that defoliation resulting from infestation poses a decreasing threat to production (Herzog et al., 1987). There are several reasons that this practice has not been universally adopted by producers. Primary among these is the fact that no early-maturing, high-yielding soybean variety has been developed that will fit this type of production scheme. However, breeding efforts are underway to satisfy this need.

Sorghum midge populations increase on johnsongrass and a few other grassy hosts in spring and early summer, then migrate to sorghum fields as the crop begins to flower (Thomas, 1969; Summers et al., 1976). Early-planted sorghum escapes economic infestations of the pest, but injury intensifies with later planting dates (Wiseman and McMillian, 1967; Summers et al., 1976).

Infestation of corn by the Japanese beetle may be reduced by planting the crop either before or after the peak oviposition period (Woodside, 1954; Cory and Langford, 1955). Early planting of soybean has been suggested as a means whereby the crop can escape injury by the southern green stinkbug (Thomas et al., 1974; Newsom and Herzog, 1977). Delayed fall planting of winter wheat is an effective and widely used method of escaping injury by the Hessian fly *Mayetiola destructor* (Say) (Walton and Packard, 1936).

Alterations of planting or harvest date may influence colonization of a crop by potential herbivorous pests. Movements of adult meadow spittlebugs [*Philaenus spumarius* (L.)] into red clover or alfalfa fields in Ohio are conditioned by availability of succulent foliage. The cutting or grazing of such crops for hay or forage encourages new growth during the fall meadow spittlebug ovipositional period. These practices should be avoided to greatly decrease damaging nymphal populations the following spring (King and Weaver, 1953). Adult plant bugs, *Lygus* spp., *Adelphocoris* spp., and the potato leafhopper *Empoasca fabae* (Harris) migrate into soybean from alfalfa fields when harvested (Poston and Pedigo, 1975). Movement of populations into soybean are particularly damaging when that crop is small. In situations where the soybean is located in close proximity to alfalfa fields, soybean planting and alfalfa harvest may be timed so that the first alfalfa cutting occurs before soybean emergence (Pedigo et al., 1981).

Several species of very important stored product pests infest beans, cowpeas, soybean, corn, rice, and wheat in the field before seed is harvested. Examples with *Acanthoscelides obtectus* (Say), *Callosobruchus maculatus* (F.), *Sitophilus zeamais* Motschulsky, and *Sitotroga cerealella* (Olivier) are reviewed by Hagstrum (1985). Either timing of planting such that grain maturity occurs out of synchrony with pest population phenology or proper timing of harvest will provide effective cultural control of these and other pests in storage.

Larval populations of the alfalfa weevil and its parasites are reduced when alfalfa is cut for harvesting (Hamlin et al., 1949). However, survival of both the weevil and its parasites is influenced by date of first cutting (Casagrande and Stehr, 1973). Crop losses are negligible when the first alfalfa cutting is made at 507° degree-days (base 48°F), but cutting either earlier or later results in increased crop injury.

Planting and harvest dates are known to affect the success of biological control in numerous cases. In the San Joaquin Valley of California, harvesting of alfalfa by solid cutting produces conditions unfavorable for the continued survival of numerous natural enemy species (Stern et al., 1964; van den Bosch and Stern, 1969). However, these unfavorable conditions can be avoided by harvesting the alfalfa fields by strip cutting (i.e., harvesting in alternated strips so that two different-aged hay growths occur in the field at the same time). The strip-cutting practice produces a reservoir of natural enemies and increases biological control in the alfalfa fields and also in nearby crops. Natural enemies of phytophagous lepidopteran species in North Carolina soybean fields are affected by planting date (Sprenkel et al., 1979). The affected natural enemies include the predators *Geocoris punctipes* (Say), *Nabis* spp., and spiders and the entomopathogen *Nomuraea rileyi* (Farlow) Samson. Benefits from these natural enemies apparently are substantial when soybean is planted early.

### 9.9.2. Plant Density and Row Width

Alterations in plant density and row width frequently can affect the suitability of a crop for colonization by pest populations. Planting of wheat at high densities and in narrow rows decreases moisture in stems, stem diameter, and plant height. The wheat stem sawfly *Cephus cinctus* Norton prefers larger, more succulent plants for oviposition, and damage to wheat decreases as seeding density increases and row spacing decreases (Luginbill and McNeal, 1958). Conversely, damage to corn by larvae of the southwestern corn borer, *Diatraea grandiosella* (Dyar), is decreased by factors that enhance general stalk vigor. Therefore, damage to corn decreases as plant density increases (Zepp and Keaster, 1977).

Plant density and row width of a crop may be altered to allow escape from damaging pest populations. High plant densities of cotton in narrow rows increases the number of older bolls earlier in the fruiting period, and the practice enchances escape from boll weevil damage in short-season production schemes (Walker et al., 1977). In South Texas, use of this practice increases cotton earliness, crop yield, and producer profit (Walker et al., 1976).

Seeding rate and row width influence the microclimate (e.g., temperature and humidity) within and under a crop canopy, and these factors can greatly affect mortality of pest populations by biological control agents. Narrow rows and high seeding rates increase biological control of lepidopterous pests in North Carolina soybean fields, but not as much as does planting date (Sprenkel et al., 1979). In North Carolina, biological control is maximized in double-cropped soybean (planted late) following wheat by planting in narrow rows at high seeding rates. Planting soybean in narrow rows in Mississippi has been shown to enchance the activity of certain biological control agents (Bushman et al., 1984). Infection of *Heliothis* spp. by *N. rileyi* is increased by planting cotton in narrow rows (Burleigh, 1975).

Just as seeding rate and row width influence within-crop microclimate, other manipulations of crop architecture may produce similar microenvironmental effects. Preliminary results have shown that compass-point orientation of soybean rows (north–south vs. east–west) produces variations in both within-canopy temperature and humidity of as much as 2°C and 5%, respectively (D. C. Herzog, Univ. of Florida, North Florida Research and Education Center, unpublished data). Such depression of temperature and elevation of humidity in north-south-oriented rows is apparently due to increased duration of shading of the between-row soil surface, and should be sufficient to promote epizootics of *N. rileyi* and *Entomophthora* spp. in populations of the velvetbean caterpillar, soybean looper, and corn earworm (Kish and Allen, 1978).

### 9.9.3. Fertilization

Use of fertilizers can influence the injury to a crop from arthropod pests, primarily through alterations in crop growth or nutritional level. Some pest populations are enhanced by poor crop growth (see Anonymous, 1969), while others are enhanced by succulent crop growth. Enhancement of succulent cotton growth through fertilization renders the crop more attractive to populations of the cotton aphid *Aphis gossypii* Glover (McGarr, 1942, 1943), cotton fleahopper *Pseudatamoscelis seriatus* (Reuter) (Adkisson, 1957), and cotton bollworm *H. zea* (Adkisson, 1958). In short-season production areas, increasing levels of soil fertility delay the fruiting period of the crop, thereby reducing the potential for escape from boll weevil injury (Walker et al., 1976, 1977). Increases in soil fertility of both winter and spring wheat result in increased wheat stem sawfly injury, due (as previously discussed) to the preference of ovipositing females for large, succulent wheat plants.

The stem borers *Chilo zonellus* Swinhoe and *Sesamia inferens* Walker on corn in India are influenced by fertilization practices. Populations have been shown to increase significantly with rate of nitrogen application. No similar effect has been observed with variable rates of phosphorus (Singh and Shekhawat, 1964). Although plants grown at higher fertility supported larger stem borer populations, injury to corn at lower fertility levels resulted in more substantial yield losses. Obviously, reduction of soil fertility is not a practical management strategy for stem borer.

Some pests are influenced by fertilization through changes in plant nutritional levels. Fecundity and reproductive rate of cabbage aphids and green peach aphids [*Myzus persicae* (Sulzer)] are depressed in Brussels sprouts by reductions in soluble nitrogen levels in the crop's phloem (van Emden and Wearing, 1965). Low nitrogen or high potassium levels in the soil induce this modification in physiology of Brussels sprouts. Benefits derived from biological control of these pests are enhanced by reductions in developmental rate and fecundity of the aphids.

Biological control of other pests is affected by fertilization. Populations of some predators of bollworms in cotton are increased by nitrogen, potassium, or phosphorus fertilization (Adkisson, 1958). Application of manure to corn fields increases predatory efficiency of a mesostigmatid mite on corn rootworm larvae (Chiang, 1970).

## 9.9.4. Tillage

Tillage operations used to produce a crop include soil turning and residue-burying practices, seedbed preparation, and cultivation. The primary purpose of tillage is to establish the crop and to establish a suitable habitat for crop growth. Some forms of tillage can reduce pest populations indirectly by destroying wild vegetation and volunteer crop plants in and around crop-production habitats. Soil tillage to destroy unwanted vegetation is a cultural control practice used in many of the previously discussed examples, including the clover seed chalcid on alfalfa (Sorenson, 1930; Bacon et al., 1959), the sorghum midge on sorghum (Randolph and Montoya, 1965; Bowden, 1965; Bottrell, 1971; Huddleston et al., 1972), the wheat curl mite on wheat (Somsen and Sill, 1970), and the red-backed cutworm on asparagus (Tamaki et al., 1975).

Tillage operations also modify soil habitats where many pest and natural enemies reside during at least part of their life cycle. Frequently, these modifications may be detrimental to pests by altering survival and/or development. Some tillage operations reduce pest survival directly. Overwintering populations of *H. zea* may be greatly reduced by either fall or spring plowing operations (Barber and Dicke, 1937). Overwintering survival of the soybean stem borer *Dectes texanus* LeConte is inversely related to depth of burial of soybean crop residue following harvest (Campbell and Van Duyn, 1977). Plowing during summer months when the soil is very dry reduces wireworm populations (Landis and Onsager, 1966). Shallow tillage of wheat stubble reduces survival of wheat stem sawfly (Holmes, 1982). Injury to grapes from the grape berry moth *Endopiza viteana* can be substantially reduced by using tillage to bury overwintering cocoons under a layer of soil (Still, 1962).

Reductions in tillage usually result in increased soil surface residues. The presence of such residues has been shown to reduce certain pests. For example, populations of the greenbug *Schizaphis graminum* (Rondani) on wheat have been shown to decrease with increasing quantities of surface residue (Burton and Krenzer, 1985). It is hypothesized that surface residues serve as a reflective mulch that may either repel colonizing aphids or mask the chemical cues by which this species find its host.

Biological control agents are often affected by tillage practices. Tillage of wheat stubble not only affects wheat stem sawfly, but also its parasites (Holmes, 1982). Mortality of potato tuberworm *Phthorimaea opercullela* (Zeller) by imported and indigenous parasitoid species is increased by ridging of soil against potato vines (Watmough et al., 1973). Adults of the cereal leaf beetle *Oulema melanopus* (L.) disperse from small grain fields after harvest, but populations of its larval and pupal parasitoids remain in the field and are reduced by plowing (Carl, 1979). Disking or harrowing has fewer negative impacts on the parasite populations than does plowing.

The efficiency of natural enemies of pest mites (*Tetranychus pacificus* McGregor and *Eotetranychus willamettei* Ewing) of California grapes is reduced under dusty conditions (Flaherty et al., 1971). Consequently, outbreaks of the pests frequently occur when weeds are controlled by repeated cultivations. An alternative program of only two spring cultivations and three or four summer mowings of cover crops or weeds prevents these outbreaks.

Recently, considerable research has been conducted on arthropods in conservation tillage crop-production systems (i.e., systems where soil tillage is reduced). This research (although not directed at developing cultural control strategies) provides unique information about the effects of tillage on the arthropod fauna in crop-production habitats under reduced-tillage management. Studies of arthropod populations in various soybean tillage systems are numerous and are excellent examples. Hammond (1986) and Hammond and Funderburk (1985) review the published information. Most soybean pests and their natural enemies whose populations have been quantified in tillage studies (both ground and aboveground inhabitants) are affected to at least some extent by soil tillage practices (Table 9.2).

### 9.9.5. Water Management

In agricultural areas with abundant water resources and available water control expertise, appropriately timed flooding of fields may be useful in reducing soil-pest populations. Wireworm populations in the Pacific Northwest are destroyed by flooding the field during the hot summer months (Lane, 1941). Flooding is an important cultural management tool for key soil pests in the muck soils of the Everglades agricultural area of Florida. Flooding is frequently used to reduce populations of wireworms in vegetable and sugarcane crops (Genung, 1976). Likewise, flooding can be used to control white grubs in sugarcane, especially under conditions of high temperature (Cherry, 1984).

Irrigation timing and frequency effectively alter microclimate within and under crop canopies. Certain changes in microclimate elicited by irrigation practices can enhance dissemination and infectivity of some entomopathogenic organisms, especially fungal pathogens. Overhead sprinkler irrigation, however, may be detrimental to epizootics of *N. rileyi* on the velvetbean caterpillar in soybean by washing conidia into the soil (Kish and Allen, 1978); however, judicious timing of irrigation may also help to promote epizootics, partly through a lowering of temperature and an increase in humidity within the crop canopy. *Entomophthora* spp. infections of pests can also be enhanced under some conditions by irrigation practices (Frezzi, 1972). Direct application of entomopathogens, including microsporidia, fungal, and viral pathogens, can be applied successfully by irrigation practices to enhance epizootics of crop pests (Hamm and Hare, 1982). Irrigation may sometimes be necessary to provide suitable conditions for imported natural enemies (Stary and Erdelon, 1982).

Irrigation practices may also directly destroy some pest populations. The washing and drowning action of sprinkler irrigation activities suppress populations of the banks grass mite *Oligonychus pratensis* (Banks) on corn in the Texas High Plains (Chandler et al., 1977) and the Pacific mite *T. pacificus* on grapes in the San Joaquin Valley (Flaherty et al., 1971). Predator populations in both situ-

Table 9.2  Soybean Pests and Natural Enemies Whose Populations Are Reported to be Affected by Soil Tillage Practices

| Organism | Reference(s) |
| --- | --- |
| Pests | |
| *Anticarsia gemmatalis* | Lema (1980) |
| *Cerotoma trifurcata* | Troxclair and Boethel (1984) |
| *Delia platura* | Funderburk et al. (1983); Hammond and Jeffers (1983); Hammond (1984) |
| *Diabrotica* spp. | Troxclair and Boethel (1984) |
| *Elasmopalpus lignosellus* | All and Gallaher (1977); All et al. (1979); Cheshire and All (1979) |
| *Epilachna varivestis* | Sloderbeck and Edwards (1979) |
| *Heliothis* spp. | Roach (1981); Stinner et al. (1982) |
| *Melanoplus* spp. | Sloderbeck and Edwards (1979) |
| *Nezara viridula* | Lema (1980) |
| *Plathypena scabra* | Sloderbeck and Yeargan (1983); Troxclair and Boethel (1984); Thorvilson et al. (1985) |
| *Popillia japonica* | Hammond and Stinner (personal communication) |
| *Spissistilus festinus* | Lema (1980) |
| Slugs | Hammond and Stinner (personal communication) |
| Natural Enemies | |
| Carabids | Lema (1980); House and Stinner (1983) |
| *Geocoris* sp. | Troxclair and Boethel (1984) |
| *Lebia* spp. | Troxclair and Boethel (1984) |
| *Nabis* spp. | Troxclair and Boethel (1984) |
| *Podisus maculiventris* | Troxclair and Boethel (1984) |
| Spiders | House and Stinner (1983) |
| Staphylinids | House and Stinner (1983) |

ations are unaffected by irrigation, and thus biological control is enhanced. Applications of water to Brussels sprouts reduces soluble nitrogen levels in the phloem, thereby reducing fecundity and reproductive rate of the cabbage aphid and the green peach aphid and enhancing the effectiveness of biological control (van Emden and Wearing, 1965).

## 9.10.  SUMMARY AND CONCLUSIONS

These examples illustrate the diversity of ways that crop production and management practices affect the ecology of agricultural production systems. Yet, the impacts of cultural practices on the biology and ecology of arthropod species, whether pestiferous, innocuous, or beneficial, are rarely considered. This trend must be reversed if the inertia is not to become permanent, and if full benefits of nonchemical controls are to be realized in crop production systems. Manipulating the habitat and growing conditions of agricultural crops for insect pest management is not a new idea (Smith, 1978), and such manipulations are still useful components of several modern, successful integrated pest management programs (Stern et al., 1976; Knipling, 1979). Such manipulations have caused specific physiological changes that reduce the suitability of host plants for phytophagous insects (Hare, 1983; Rhoades, 1983).

The activities employed in producing agricultural commodities modify the ecology of an agroecosystem, with many either increasing or decreasing crop losses from arthropod pests. These activities include those modifying plant species in crop and noncrop habitats and those operations used to plant, maintain, and harvest specific crops. Frequently, these practices can be deliberately modified in ways that minimize crop loss or the need for additional control tactics. Such cultural control tactics are commonly used in integrated pest control and are usually highly desirable methods of managing arthropod pest populations. Most are compatible with other pest control tactics and are low cost. Additionally, few have undesirable ecological consequences. Consequently, cultural control should be an emphasized aspect of future pest management programs. As the examples included in this chapter should indicate, knowledge of pest and natural enemy biology aids development of cultural control strategies.

An in-depth understanding of the processes responsible for nonrandom herbivore distributions among host plants, whether in the "natural" or "monoculture" scenario, is crucial to the design of cropping systems that are subject to reduced vulnerability to herbivore outbreaks (Risch, 1980; Stanton, 1983). Spatial and temporal deployment patterns of crop and varieties (Hare, 1983; Kareiva, 1983), seasonal variation in sensitivity of plants to injury (Krischik and Denno, 1983), and the degree of synchrony between plant and insect developmental processes (Hare, 1983) will provide needed insight for sophisticated pest management. Furthermore, how agricultural practices alter host suitability and promote or depress pest outbreaks and how cultural controls, plant resistance, and biological control can be used compatibly to reduce pest populations warrants careful consideration (Denno and McClure, 1983b).

Stability of interactions among plants and herbivores is most usefully regarded as resulting from a balance of offensive and defensive power among the participants. Conversely, that instability results from perturbations of the stabilizing, opposing forces is well known (Rhoades, 1983). The application of basic research to the constant battle against agricultural and silvicultural pests should be a common objective of all scientists studying plant–herbivore interactions. Unfortunately, there remains a gap between theoretical and applied studies that impedes progress toward this goal (Denno and McClure, 1983b).

As Herzog et al. (1984) have suggested, the ultimate measure of the compatibility of any production or management tactic or strategy will be made based on biological and economic evaluations and an assessment of environmental impact. Crop production and management strategies should be tailored to be environmentally compatible, especially with the natural enemy complex that effects biological control. Of such importance is the potential effect of cultural practices on levels of biological control inflicted by natural enemies of crop pests (see, for example, Mayse et al., 1983; Allen, 1985; Shepard and Herzog, 1985) that Herzog and Funderburk (1985) were prompted to make the following recommendations:

**1.** Basic studies are urgently needed to determine ecological requirements of native and introduced biological control agents. Research is needed to determine microhabitat-microenvironmental effects of crop production practices. Then, a more precise definition of potential beneficial-detrimental impacts of cultural practices on biological control agents can be achieved.

**2.** Cultural practices designed to enhance the benefits of biological control should be economically, ecologically, and agronomically feasible. Therefore, such research efforts must be undertaken by an interdisciplinary team.

**3.** When explorations for exotic natural enemies are undertaken, the characteristics of natural enemy species needed for specific situations should receive early consideration. It is at this point that the cropping system and cultural practices employed at the sites of origin and the eventual destination are of great importance.

**4.** It is a practical impossibility to study all combinations of crops, pests, natural enemies, environments, and cultural practices to document their interrelationships. Therefore, continued development and refinement of system level crop-pest models should receive high priority to optimize coordination and implementation of research efforts.

## REFERENCES

Adkisson, P. L. (1957). Influence of irrigation and fertilizer on populations of three species of mirids attacking cotton, *FAO Plant Prot. Bull.*, **6**, 33–36.

Adkisson, P. L. (1958). The influence of fertilizer applications on populations of *Heliothis zea* (Boddie), and certain insect predators, *J. Econ. Entomol.*, **51**, 757–759.

Adkisson, P. L., G. A. Niles, J. K. Walker, L. S. Bird, and H. B. Scott. (1982). Controlling cotton's insect pests: A new system, *Science*, **216**, 19–22.

All, J. N., and R. N. Gallaher. (1977). Detrimental impact of notillage corn cropping systems involving insecticides, hybrids, and irrigation on lesser cornstalk borer infestations, *J. Econ. Entomol.*, **70**, 361–365.

All, J. N., R. N. Gallaher, and M. D. Jellum. (1979). Influence of planting date, preplanting weed control, irrigation, and conservation tillage practices on efficacy of planting time insecticide applications for control of lesser cornstalk borer in field corn, *J. Econ. Entomol.*, **72**, 265–268.

Allen, W. A. (1985). "Discussion: Status and current limits to biological control in citrus, grapes, alfalfa, cotton and soybean systems," in M. A. Hoy and D. C. Herzog, Eds., *Biological Control in Agricultural IPM Systems*, Academic, New York, pp. 558–560.

Altieri, M. A. (1981). Weeds may augment biological control of insects, *Calif. Agric.*, **35**, 22–24.

Altieri, M. A., and W. H. Whitcomb. (1979). The potential use of weeds in the manipulation of beneficial insects, *HortScience*, **14**, 12–18.

Altieri, M. A., A. van Schoonhoven, and J. Doll. (1977). The ecological role of weeds in insect pest management systems: A review illustrated with bean (*Phaseolus vulgaris* L.) cropping systems, *PANS* **24**, 185–206.

Andrewartha, H. G., and L. C. Birch. (1954). *The Distribution and Abundance of Animals*, University of Chicago Press, Chicago, 782 pp.

Anonymous (1969). *Insect Pest Management and Control*, Natl. Acad. Sci. Publ. 1956, 508 pp.

Anonymous (1978). "Alfalfa seed research pays dividends," in H. Cline, Ed., *Western Hay and Grain Grower,* Munford, Fresno, CA, 14 pp.

Apple, J. L., and R. F. Smith. (1976). "Progress, problems, and prospects for integrated pest management," in J. L. Apple and R. F. Smith, Eds. *Integrated Pest Management*, Plenum, New York, pp. 179–196.

Arkin, G. F., and H. M. Taylor, Eds., (1981). *Modifying the Root Environment to Reduce Crop Stress, Am. Soc. Agric. Eng.*, St. Joseph, Michigan, 407 pp.

Ashdown, D. (1977). *Life of the Insects,* Texas Tech Univ., Lubbock, 135 pp.

Bacon, O. G., W. D., Riley, V. E. Burton, and A. V. Sarquis. (1959). Clover seed chalcid in alfalfa, *Calif. Agric.*, **13**, 7, 11.

Barber, G. W., and F. F. Dicke. (1937). The effectiveness of cultivation as a control for the corn earworm, USDA Tech. Bull. 561, 16 pp.

Barfield, B. J., and J. F. Gerber, Eds., (1979). *Modification of the Aerial Environment of Crops*, Am. Soc. Agric. Eng., St. Joseph, MI, 538 pp.

Barnes, O. L. (1959). Effect of cultural practices on grasshopper populations in alfalfa and cotton, *J. Econ. Entomol.*, **52**, 336–337.

Bottrell, D. G. (1971). Entomological advances in sorghum production, Texas Agric. Exp. Sta. Prog. Rep. PR-2940, 28–40.

Bottrell, D. G., and P. L. Adkisson. (1977). Cotton insect pest management, *Annu. Rev. Entomol.*, **22**, 451–481.

Bowden, J. (1956). Sorghum midge, *Contarinia sorghicola* (Coq.), and other causes of grain-sorghum loss in Ghana, *Bull. Entomol. Res.*, **65**, 169–189.

Bradley, J. R., and J. W. Van Duyn. (1980). "Insect pest management in North Carolina soybeans," in F. T. Corbin, Ed., *World Soybean Research Conference II: Proceedings*. Westview Press, Boulder, CO, 897 pp.

Burleigh, J. G. (1972). Population dynamics and biotic controls of the soybean looper in Louisiana, *Environ. Entomol.*, **1**, 290–294.

Burleigh, J. G. (1975). Comparison of *Heliothis* spp. larval parasitism and *Spicaria* infection in closed and open canopy cotton varieties, *Environ. Entomol.*, **4**, 574–576.

Burton, R. L., and E. G. Krenzer. (1985). Reduction of greenbug (Homoptera: Aphididae) populations by surface residues in wheat tillage studies, *J. Econ. Entomol.*, **78**, 390–394.

Bushman, L. L., H. N. Pitre, and H. F. Hodges. (1984). Soybean cultural practices: Effects on populations of geocorids, nabids, and other soybean arthropods, *Environ. Entomol.*, **13**, 305–317.

Campbell, W. V., and J. W. Van Duyn. (1977). Cultural and chemical control of *Dectes texanus texanus* on soybeans, *J. Econ. Entomol.*, **70**, 256–258.

Capinera, J. L., T. J. Weissling, and E. E. Schweizer. (1985), Compatibility of intercropping with mechanized agriculture: Effect of strip intercropping of pinto beans and sweet corn on insect abundance in Colorado, *J. Econ. Entomol.*, **78**, 354–357.

Carl, K. P. (1979). The importance of cultural measures for the biological control of the cereal leaf beetle *Oulema melanopus* (Col. Chrysomelidae), *Mitt. Schweiz. Entomol. Ges.*, **52**, 443.

Casagrande, R. A., and F. W. Stehr. (1973). Evaluating the effects of harvesting alfalfa on alfalfa weevil (Coleoptera: Curculionidae) and parasite populations in Michigan, *Can. Entomol.*, **105**, 1119-1128.

Chandler, L. D., C. R. Ward, W. M. Lyle, and E. D. Bynum. (1977). Effects of various irrigation schedules on population densities of Banks grass mite (*Oligonychus pratensis*) and its predatory arthropods in corn, the Texas High Plains, Texas Agric. Exp. Sta. Prog. Rep. 3443c, 17 pp.

Cherry, R. H. (1984). Flooding to control the grub *Ligyrus subtropicus* (Coleoptera: Scarabaeidae) in Florida sugarcane, *J. Econ. Entomol.*, **77**, 254-257.

Cheshire, J. M., and J. N. All. (1979). Feeding behavior of lesser cornstalk borer larvae in simulations of no-tillage, mulched conventional tillage, and conventional tillage corn cropping systems, *Environ. Entomol.*, **8**, 261-264.

Chiang, H. C. (1970). Effects of manure applications and mite predation on corn rootworm populations in Minnesota, *J. Econ. Entomol.*, **63**, 934-936.

Chiang, H. C. (1973). Bionomics of the northern and western corn rootworms, *Annu. Rev. Entomol.*, **18**, 47-72.

Chiang, H. C., and R. G. Flaskerd. (1969). Northern and western corn rootworms in Minnesota, *J. Minn. Acad. Sci.*, **30**, 48-51.

Clements, F. E., and V. E. Shelford. (1939). *Bio-Ecology*, Wiley, New York, 425 pp.

Coaker, T. H. (1980). Insect pest management in *Brassica* crops by inter-cropping, *I.O.B.C. W.P.R.S. Bull.*, **3**, 117-125.

Corbet, P. S. (1976). "Pest management in ecological perspective," in J. L. Apple and R. F. Smith, Eds., *Integrated Pest Management*, Plenum, New York, pp. 51-57.

Cory, E. N., and G. S. Langford. (1955). The Japanese beetle retardation program in Maryland, Univ. Maryland Ext. Bull. 156, 12 pp.

Cromartie, W. J. (1981). "The environmental control of insects using crop diversity," in D. Pimentel, Ed., *Handbook of Pest Management in Agriculture*, Vol. II, CRC Press, Boca Raton, FL, pp. 223-251.

Dempster, J. P., and T. H. Coaker. (1974). "Diversification of crop ecosystems as a means of controlling pests," in D. P. Jones and M. E. Solomon, Eds., *Biology in Pest and Disease Control*, Wiley, New York, pp. 106-114.

Denno, R. F., and M. S. McClure. (1983a). "Variability: A key to understanding plant–herbivore interactions," in R. F. Denno and M. S. McClure, Eds., *Variable Plants and Herbivores in Natural and Managed Systems*, Academic, New York, pp. 1-12.

Denno, R. F., and M. S. McClure, Eds., (1983b). *Variable Plants and Herbivores in Natural and Managed Systems*, Academic, New York, 717 pp.

Doutt, R. L., and J. Nakata. (1965). Overwintering refuge of *Anagrus epos* (Hymenoptera: Mymaridae), *J. Econ. Entomol.*, **58**, 586.

Doutt, R. L., and J. Nakata. (1973). The *Rubus* leafhopper and its egg parasitoid: An endemic biotic system useful in grape pest management, *Environ. Entomol.*, **2**, 381-386.

Doutt, R. L., J. Nakata, and F. E. Skinner. (1966). Dispersal of grape leafhopper parasites from a blackberry refuge, *Calif. Agric.*, **20**, 14-15.

Edmunds, G. F., and D. N. Alstad. (1978). Coevolution in insect herbivores and conifers, *Science*, **199**, 941-945.

Ehrlich, P. R., A. H. Ehrlich, and J. P. Holdren. (1977). *Ecoscience: Population, Resources, Environment*, Freeman, New York, 1051 pp.

Feeny, P. P. (1975). "Biochemical coevolution between plants and their insect herbivores," in L. E. Gilbert and P. H. Raven, Eds., *Coevolution of Animals and Plants*, Univ. Texas Press, Austin, pp. 3-19.

Feeny, P. P. (1976). Plant apparancy and chemical defense, *Rec. Adv. Phytochem.*, **10**, 1-40.

Finch, S. (1980). Chemical attraction of plant-feeding insects to plants, *Appl. Biol.*, **5**, 67–143.

Flaherty, D., C. Lynn, F. Jensen, and M. Hoy. (1971). Influence of environment and cultural practices on spider mite and abundance in southern San Joaquin Thompson seedless vineyards, *Calif. Agric.*, **25**, 6–8.

Flaherty, D. L., H. Kido, V. M. Stern, and L. T. Wilson. (1985). "Biological control in San Joaquin Valley vineyards," in M. A. Hoy and D. C. Herzog, Eds., *Biological Control in Agricultural IPM Systems*, Academic, New York, pp. 501–520.

Flint, M. L., and R. van den Bosch. (1981). *Introduction to Integrated Pest Management*, Plenum, New York, 240 pp.

Fox, L. R. (1981). Defense and dynamics in plant–herbivore systems, *Am. Zool.*, **21**, 853–864.

Frezzi, M. J. (1972). Two fungi pathogenic to insects and three entomophagous insects, useful aids in Argentina in the biological control of the lucerne aphid (*Acyrthosiphon pisum* Harris), *Idia*, **291**, 21–30.

Frisbie, R. E., and J. K. Walker. (1981). "Pest management systems for cotton insects," in D. Pimentel, Ed., *Handbook of Pest Management in Agriculture*, Vol. III, CRC Press, Boca Raton, FL, pp. 187–202.

Funderburk, J. E., L. P. Pedigo, and E. C. Berry. (1983). Seedcorn maggot (Diptera: Arthomyiidae) emergence in conventional and reduced-tillage soybean systems in Iowa, *J. Econ. Entomol.*, **76**, 131–134.

Genung, W. G. (1976). Flooding in Everglades soil pest management, *Proc. Tall Timbers Conf. Ecol. Anim. Control Habitat Mgt.*, **6**, 165–172.

Gibson, K. E., and J. T. Fallini. (1963). Beet leafhopper control in southern Idaho by seeding breeding areas to range grass, USDA Agric. Res. Serv. 33–83, 5 pp.

Glass, E. H. (1975). Integrated pest management: Rationale, potential, needs and implementation, Entomol. Soc. Amer. Spec. Publ. 75-2, 141 pp.

Goodman, D. (1975). The theory of diversity-stability relationships in ecology. *Q. Rev. Biol.*, **50**, 237–266.

Grinnell, J. (1917). The niche-relationships of the California thrasher, *Auk*, **34**, 427–433.

Hagstrum, D. W. (1985). Preharvest infestation of cowpeas by the cowpea weevil (Coleoptera: Burchidae) and population trends during storage in Florida, *J. Econ. Entomol*, **78**, 358–361.

Hall, R. W. (1977). "The population biology of *Aphis nerii* B. d. F. on oleandar," Unpublished Ph.D. Dissertation, University of California, Davis, 56 pp.

Hamlin, J. C., F. V. Liebermann, R. W. Bunn, W. C. McDuffie, R. C. Newton, and L. J. Jones. (1949). Field studies on the alfalfa weevil and its environment, USDA Tech. Bull. 975, 84 pp.

Hamm, J. J., and W. W. Hare. (1982). Application of entomopathogens in irrigation water for control of fall armyworms and corn earworms (Lepidoptera: Noctuidae) on corn, *J. Econ. Entomol.*, **75**, 174–1079.

Hammond, R. B. (1984). Effects of rye cover crop management on seedcorn maggot (Diptera: Anthomyiidae) populations in soybeans, *Environ. Entomol.*, **13**, 1302–1305.

Hammond, R. B. (1986). "Pest management in reduced tillage soybean cropping systems," in G. J. House and B. R. Stinner, Eds., *Arthropods and Conservation Tillage Systems*, Entomol. Soc. Amer., College Park, MD (in press).

Hammond, R. B., and J. E. Funderburk. (1985). "Influence of tillage practices on soil-insect population dynamics in soybean," in R. Shibles, Ed., *World Soybean Research Conference III: Proceedings*, Westview Press, Boulder, CO, pp. 659–666.

Hammond R. B., and D. L. Jeffers. (1983). Adult seedcorn maggots in soybeans relay intercropped into winter wheat, *Environ. Entomol.*, **12**, 1487–1489.

Hanson, A. A. (1975). Toward ecological stability in the management of plant pests, *Iowa State J. Res.*, **49**, 447–456.

Hare, J. D. (1983). "Manipulation of host suitability for herbivore pest management," in R. F. Denno and M. S. McClure, Eds., *Variable Plants and Herbivores in Natural and Managed Systems*, Academic, New York, pp. 655-680.

Hassell, M. P. (1978). *The Dynamics of Arthropod Predator-Prey Systems*, Princeton Univ. Press, Princeton, NJ, 237 pp.

Hassell, M. P., and T. R. E. Southwood. (1978). Foraging strategies of insects, *Annu. Rev. Ecol. Syst.*, **9**, 75-98.

Hatfield, J. L., and I. J. Thomason, Eds. (1982). *Biometeorology in Integrated Pest Management*, Academic, New York, 491 pp.

Herzog, D. C. (1979). "Variety selection and planting date manipulation as cultural controls for the velvetbean caterpillar, *Anticarsia gemmatalis* (Hubner)," in F. T. Corbin, Ed., *World Soybean Research Conference II: Abstracts*, North Carolina State Univ., Raleigh, p. 23.

Herzog, D. C., and J. E. Funderburk. (1985). "Plant resistance and cultural control interactions with biological control," in M. A. Hoy and D. C. Herzog, Eds., *Biological Control in Agricultural Integrated Pest Management Systems*, Academic, New York, pp. 61-82.

Herzog, D. C., J. L. Stimac, D. G. Boucias, and V. H. Waddill. (1984). "Compatibility of biological control in soybean insect management," in P. L. Adkisson and Shi-Jun Ma, Eds., *Proceedings of the Chinese Academy of Sciences - United States National Academy of Sciences Joint Symposium on Biological Control of Insects*, September 25-28, 1982, Beijing, China, Science Press, Beijing, Peoples' Republic of China, pp. 37-60.

Herzog, D. C., J. W. Todd, and W. H. Whitcomb. (1987). "Velvetbean caterpillar," in L. D. Newsom, Ed., *Integrated Management of Insect Pests of Soybean*, Wiley-Interscience, New York (in press).

Holmes, N. D. (1982). Population dynamics of the wheat stem sawfly, *Cephus cinctus* (Hymenoptera: Cephidae), in wheat, *Can. Entomol.*, **114**, 775-788.

House, G. J., and B. R. Stinner. (1983). Arthropods in no-tillage soybean agroecosystems: Community composition and ecosystem interactions, *Environ. Mgt.*, **7**, 23-28.

Huddleston, E. W., D. Ashdown, B. Maunder, C. R. Ward, G. Wilde, and C. E. Forehand. (1972). Biology and control of the sorghum midge. 1. Chemical and cultural control studies in West Texas, *J. Econ. Entomol.*, **65**, 851-855.

Huffaker, C. B. (1982). "Overall approach to insect problems in agriculture," in J. L. Hatfield and I. J. Thomason, Eds., *Biometeorology in Integrated Pest Management*, Academic, New York, pp. 171-192.

Huffaker, C. B., A. A. Berryman, and J. E. Liang. (1984a). "Natural control of insect populations," in C. B. Huffaker and R. L. Rabb, Eds., *Ecological Entomology*, Wiley, New York, pp. 339-398.

Huffaker, C. B., H. T. Gordon, and R. L. Rabb. (1984b). "Meaning of ecological entomology—the ecosystem," in C. B. Huffaker and R. L. Rabb, Eds., *Ecological Entomology*, Wiley, New York, pp. 3-17.

Hunter, W. D. (1909). What can be done in destroying the cotton boll weevil during the winter, USDA Bur. Entomol. Circ. 107, 4 pp.

Hutchinson, G. E. (1957). Concluding remarks, *Cold Spring Harbor Symp. Quant. Biol.*, **22**, 415-422.

Isely, D. (1930). Control of the boll weevil in winter, *Ark. Agric. Exp. Sta. Bull*, **257**, 54-56.

Jensen, R. L., L. D. Newsom, and J. Gibbens. (1974). The soybean looper: Effects of adult nutrition on oviposition, mating frequency, and longevity, *J. Econ. Entomol.*, **67**, 467-470.

Johnson, M. W., R. E. Stinner, and R. L. Rabb. (1975). Ovipositional response of *Heliothis zea* to its major hosts in North Carolina, *Environ. Entomol.*, **4**, 291-297.

Jones, F. G. W. (1976). Pests, resistance and fertilizers, *Proc. Intn. Potash Inst.*, **12**, 233-358.

Kareiva, P. (1983). "Influence of vegetation texture on herbivore populations: Resource concentration and herbivore movement," in R. F. Denno and M. S. McClure, Eds., *Variable Plants and Herbivores in Natural and Managed Systems*, Academic, New York, pp. 259–290.

Kendeigh, S. C. (1961). *Animal Ecology*, Prentice-Hall, New York, 468 pp.

Kennedy, D., P. L. Adkisson, S. R., Aldrich, D. L. Dahlsten, J. E. Davies, B. E. Day, C. Gotsch, J. E. Krier, M. C. Latham, M. S. Meselson, W. W. Murdoch, K. Nair, C. E. Palm, V. R. Ruttan, and R. A. Young. (1975). *Pest Control: An Assessment of Present and Alternative Technologies*, Vol. I. Contemporary Pest Control Practices and Prospects: The Report of the Executive Committee. National Academy of Sciences, Washington, D.C., 506 pp.

King, D. R., and C. R. Weaver. (1953). The effect of meadow management on the abundance of meadow spittlebugs, *J. Econ. Entomol.*, **46**, 884–888.

Kish, L. P., and G. E. Allen. (1978). The biology and ecology of *Nomuraea rileyi* and a program for predicting its incidence on *Anticarsia gemmatalis* in soybean, Fla. Agr. Exp. Sta. Tech. Bull. 795, 48 pp.

Knipling, E. F. (1979). *The Basic Principles of Insect Population Suppression and Management*, USDA Agric. Handbook 512, 659 pp.

Krischik, V. A., and R. F. Denno. (1983). "Individual, population, and geographic patterns in plant defense," in R. F. Denno and M. S. McClure, Eds., *Variable Plants and Herbivores in Natural and Managed Systems*, Academic, New York, pp. 463–512.

Landis, B. J., and J. A. Onsager. (1966). Wireworms on irrigated lands in the West: How to control them, USDA Farmers' Bull. 2220, 14 pp.

Lane, M. C. (1941). Wireworms and their control on irrigated lands, USDA Farmers' Bull. 1866, 21 pp.

Laster, M. L. (1974). "Increasing natural enemy resources through crop rotation and strip cropping," in F. G. Maxwell and F. A. Harris, Eds., *Proceedings of the Summer Institute on Biological Control of Plant Insects and Diseases*, Univ. of Mississippi Press, Jackson, pp. 137–149.

Laster, M. L., and R. E. Furr. (1972). *Heliothis* populations in cotton-sesame interplantings, *J. Econ. Entomol.*, **65**, 1524–1525.

Lema, K. M. (1980). "Influence of no-till and conventional tillage on insect pests and soil inhabiting predator populations in Florida soybean and corn cropping systems," Ph.D. Dissertation, University of Florida, Gainesville, 230 pp.

Liebig, J. (1840). *Chemistry in Its Application to Agriculture and Physiology*, Taylor and Walton, London, 135 pp.

Litsinger, J. A., and K. Moody. (1976). "Integrated pest management in multiple cropping systems," in R. I. Papendick, P. A. Sanchez, and G. B. Triplett, Eds., *Multiple Cropping*, Amer. Soc. Agron. Spec. Publ. 27, Madison, WI, pp. 293–317.

Lowrance, R., B. R. Stinner, and G. J. House. (1984a). "Introduction," in R. Lowrance, B. R. Stinner, and G. J. House, Eds., *Agricultural Ecosystems: Unifying Concepts*, Wiley, New York, pp. 1–4.

Lowrance, R., B. R. Stinner, and G. J. House, Eds. (1984b). *Agricultural Ecosystems: Unifying Concepts*, Wiley, New York, 233 pp.

Luginbill, P., and F. H. McNeal. (1958). Influence of seeding density and row spacings on the resistance of spring wheats to the wheat stem sawfly, *J. Econ. Entomol.*, **51**, 808–809.

MacFadyen, A. (1957). *Animal Ecology: Aims and Methods*, Pitman Press, London, 264 pp.

Mattson, W. J. (1980). Herbivory in relation to plant nitrogen content, *Annu. Rev. Ecol. Syst.*, **11**, 119–161.

Mattson, W. J., and N. D. Addy. (1975). Phytophagous insects as regulators of forest primary production, *Science*, **190**, 515–522.

May, R. M. (1973). *Stability and Complexity in Model Ecosystems*, Princeton University Press, NJ, 235 pp.

Mayse, M., H. N. Pitre, and W. Whitcomb. (1983). "Effects of cultural practices on natural enemies," in H. N. Pitre, Ed., *Natural Enemies of Arthropod Pests in Soybean*, South. Coop. Ser. Bull. 285, pp. 49-55.

McClure, M. S. (1980). Foliar nitrogen: A basis for host suitability for elongate hemlock scale, *Fiorinia externa* (Homoptera: Diaspididae), *Ecology*, **61**, 72-79.

McGarr, R. L. (1942). Relation of fertilizers to the development of the cotton aphid, *J. Econ. Entomol.*, **35**, 482-483.

McGarr, R. L. (1943). Relation of fertilizers to the development of the cotton aphid in 1941 and 1942, *J. Econ. Entomol.*, **36**, 640.

McNeill, S., and T. R. E. Southwood. (1978). "The role of nitrogen in the development of insect/plant relationships," in J. B. Harborne, Ed., *Biochemical Aspects of Plant and Animal Coevolution*, Academic, New York, pp. 77-98.

Menke, W. W. (1973). A computer simulation model:: The velvetbean caterpillar in the soybean agroecosystem. *Fla. Entomol.*, **56**, 92-105.

Metcalf, C. L., W. P. Flint, and R. L. Metcalf. (1962). *Destructive and Useful Insects*, McGraw-Hill, New York, 1087 pp.

Murdoch, W. W. (1975). Diversity, complexity, stability and pest control, *J. Appl. Ecol.*, **12**, 795-807.

Newsom, L. D. (1975). "Pest management: Concept to practice," in D. Pimentel, Ed., *Insects, Science and Society*, Academic, New York, pp. 257-278.

Newsom, L. D., and D. C. Herzog. (1977). Trap crops for control of soybean pests, *La. Agric.*, **20**, 14-15.

Newsom, L. D., M. Kogan, F. D. Miner, R. L. Rabb, S. G. Turnipseed, and W. H. Whitcomb. (1980). "General accomplishments toward better pest control in soybean," in C. B. Huffaker, Ed., *New Technology of Pest Control*, New York, pp. 51-98.

Odum, E. P. (1984). "Properties of agroecosystems," in R. Lowrance, B. R. Stinner, and G. J. House, Eds., *Agricultural Ecosystems: Unifying Concepts*, Wiley, New York, 233 pp.

O'Neil, R. J. (1984). "Measurement and analysis of arthropod predation on the velvetbean caterpillar, *Anticarsia gemmatalis* Hubner," Ph.D. Dissertation, University of Florida, Gainesville, 192 pp.

Ortman, E. E., and P. J. Fitzgerald. (1964). Developments in corn rootworm research, *Proc. Annu. Hybrid Corn Ind. Res. Conf. 19th.*, **19**, 38-45.

Ottens, R. J., and J. W. Todd (1979). Effects of host plant on fecundity, longevity and oviposition rate of a whitefringed beetle, *Ann. Entomol. Soc. Amer.*, **72**, 837-839.

Ottens, R. J., and J. W. Todd. (1980). Leaf area consumption of cotton, peanuts, and soybeans by adult *Graphognathus peregrinus* and *G. leucoloma*, *J. Econ. Entomol.*, **73**, 55-57.

Owen, D. F. (1971). Species diversity in butterflies in a tropical garden, *Biol. Conserv.*, **3**, 191-198.

Pedigo, L. P., R. A. Higgins, R. B. Hammond, and E. J. Bechinski. (1981). "Soybean pest management," in D. Pimentel, Ed., *Handbook of Pest Management in Agriculture*, Vol. III, CRC Press, Boca Raton, FL, pp. 417-537.

Perrin, R. M. (1975). The role of perennial stinging nettle, *Urtica dioica*, as a reservoir of beneficial natural enemies, *Ann. Appl. Biol.*, **81**, 289-297.

Perrin, R. M. (1977). Pest management in multiple cropping systems, *Agro-Ecosystems*, **3**, 93-118.

Perrin, R. M., and R. M. Phillips. (1978). Some effects of mixed cropping on the population dynamics of insect pests, *Entomol. Exp. Appl.*, **24**, 385-393.

Poston, F. L., and L. P. Pedigo. (1975). Migration of plant bugs and the potato leafhopper in a soybean–alfalfa complex, *Environ. Entomol.*, **4**, 8-10.

Price, P. W. (1984). "The concept of the ecosystems" in C. B. Huffaker and R. L. Rabb, Eds., *Ecological Entomology*, Wiley, New York, pp. 19-50.

Rabb, R. L. (1969). Environmental manipulation as influencing populations of tobacco hornworms, *Proc. Tall Timbers Conf. Anim. Cont. Habitat Mgt.*, **1**, 175–191.

Rabb, R. L. (1978). A sharp focus on insect populations and pest management from a wide-area view, *Bull. Entomol. Soc. Amer.*, **24**, 55–61.

Rabb, R. L., G. K. DeFoliart, and G. G. Kennedy. (1984). "An ecological approach to managing insect populations," in C. B. Huffaker and R. L. Rabb, Eds., *Ecological Entomology*, Wiley, New York, pp. 697–728.

Randolph, N. M., and E. L. Montoya. (1965). Ecology, biology, and control of sorghum midge on the Texas South Plains, Texas Agric. Exp. Sta. Prog. Rep. PR-2304, 1–10.

Reagan, T. E., and J. L. Flynn. (1985). "Insect pest management of sweet sorghum in sugarcane production systems of Louisiana: Problems and integration," in W. H. Smith, Ed., *Proc. Third Southern Biomass Energy Res. Conf.* (in press).

Rhoades, D. F. (1979). "Evolution of plant chemical defense against herbivores," in G. A. Rosenthal and D. J. Janzen, Eds., *Herbivores: Their Interaction with Secondary Plant Metabolites*, Academic, New York, pp. 3–54.

Rhoades, D. F. (1983). "Herbivore population dynamics and plant chemistry," in R. F. Denno and M. S. McClure, Eds., *Variable Plants and Herbivores in Natural and Managed Systems*, Academic, New York, pp. 155–220.

Risch, S. J. (1980). The population dynamics of several herbivorous beetles in a tropical agroecosystem: The effect of interplanting corn, beans, and squash in Cost Rica, *J. Appl. Ecol.*, **17**, 593–612.

Roach, S. H. (1981). Emergence of overwintered *Heliothis* spp. moths from three different tillage systems, *Environ. Entomol.*, **10**, 817–818.

Robinson, R. R., J. H. Young, and R. O. Morrison. (1972). Strip-cropping effects on abundance of *Heliothis*-damaged cotton squares, boll placement, total bolls, and yields in Oklahoma, *Environ. Entomol.*, **1**, 140–145.

Rodriquez, J. G. (1951). Mineral nutrition of the two-spotted spider mite, *Tetranychus bimaculatus* Harvey, *Ann. Entomol. Soc. Amer.*, **44**, 511–526.

Rodriguez, J. G., Ed. (1972). *Insect and Mite Nutrition*, Elsevier/North Holland, Amsterdam, 702 pp.

Rolston, L. H., and C. E. McCoy. (1966). *Introduction to Applied Entomology*, Ronald Press, New York, 208 pp.

Root, R. B. (1973). Organization of a plant–arthropod association in simple and diverse habitat: The fauna of collards (*Brassica oleraceae*), *Ecol. Monogr.*, **43**, 95–124.

Rudd, R. L. (1971). "Pesticides," in W. W. Murdoch, Ed., *Environment, Resources, Pollution and Society*, Sinauer Assoc., Stamford, CT, pp. 279–301.

Rust, R. W. (1977). Evaluation of trap crop procedures for control of Mexican bean beetle in soybeans and lima beans, *J. Econ. Entomol.*, **70**, 630–632.

Scriber, J. M. (1982). "Nitrogen nutrition of plants and insect invasion," in R. D. Hauck, Ed., *Nitrogen in Crop Production*, Amer. Soc. Agron., Madison, WI, pp. 441–460.

Shelford, V. E. (1913). *Animal Communities in Temperate America*, University of Chicago Press, Chicago, 368 pp.

Shelton, M. D., and C. R. Edwards. (1983). Effects of weeds on the diversity and abundance of insects in soybeans, *Environ. Entomol.*, **12**, 296–298.

Shepard, M., and D. C. Herzog. (1985). "Soybean: Status and current limits to biological control in the southeastern U.S.," in M. A. Hoy and D. C. Herzog, Eds., *Biological Control in Agricultural IPM systems*, Academic, New York, pp. 503–558.

Singh, U. B., and G. S. Shekhawat. (1964). Incidence of stem borers in maize under different fertility levels, *Indian J. Agron.*, **9**, 48–50.

Slansky, F., and P. P. Feeny. (1977). Stabilization of the rate of nitrogen accumulation by larvae of the cabbage butterfly on wild and cultivated food plants, *Ecol. Monogr.*, **47**, 209-228.

Sloderbeck, P. E., and C. R. Edwards. (1979). Effects of soybean cropping practices on Mexican bean beetle and redlegged grasshopper populations, *J. Econ. Entomol.*, **72**, 850-853.

Sloderbeck, P. E., and K. V. Yeargan. (1983). Green cloverworm (Lepidoptera: Noctuidae) populations in conventional and doubled-cropped, no-till soybeans, *J. Econ. Entomol.*, **76**, 785-791.

Slosser, J. E., and E. P. Boring. (1980). Shelterbelts and boll weevils: A control strategy based on management of overwintering habitat, *Environ. Entomol.*, **9**, 1-6.

Slosser, J. E., R. J. Fewin, F. R. Price, L. J. Meinke, and J. R. Bryson. (1984). Potential of shelterbelt management for boll weevil (Coleoptera: Curculionidae) control in the Texas rolling plains, *J. Econ. Entomol.*, **77**, 377-385.

Slosser, J. E., P. W. Jacoby, and J. R. Price. (1985). Management of sand shinnery oak for control of the boll weevil (Coleoptera: Curculionidae) in the Texas rolling plains, *J. Econ. Entomol.*, **78**, 383-389.

Smith, J. G. (1969). Some effects of crop background on populations of aphids on their natural enemies on Brussels sprouts, *Ann. Appl. Biol.*, **63**, 326-329.

Smith, J. G. (1976a). Influence of crop background on aphids and other phytophagous insects on Brussels sprouts, *Ann. Appl. Biol.*, **83**, 1-13.

Smith, J. G. (1976b). Influence of crop background on natural enemies of aphids on Brussels sprouts, *Ann. Appl. Biol.*, **83**, 15-29.

Smith, R. F. (1978). "History and complexity of integrated pest management," in E. H. Smith and D. Pimentel, Eds., *Pest Control Strategies*, Academic, New York, pp. 41-53.

Somsen, H. W., and W. H. Sill. (1970). The wheat curl mite, *Aceria tulipae* Keifer, in relation to epidemiology and control of wheat streak mosiac, Kansas Agric. Exp. Sta. Res. Pub. 162, 24 pp.

Sorenson, C. J. (1930). The alfalfa-seed chalcis-fly in Utah, Utah Agric. Exp. Sta. Bull. 218, 33 pp.

Southwood, T. R. E., and M. J. Way. (1970). "Ecological background to pest management," in R. L. Rabb and F. E. Guthrie, Eds., *Concepts of Pest Management*, North Carolina State University, Raleigh, pp. 6-28.

Sprenkel, R. K., W. M. Brooks, J. W. Van Duyn, and L. L. Deitz. (1979). The effects of three cultural variables on the incidence of *Nomuraea rileyi*, phytophagous Lepidoptera, and their predators in soybeans, *Environ. Entomol.*, **5**, 205-209.

Spurr, S. H. (1969). "The natural resource ecosystem," in G. M. Van Dyne, Ed., *The Ecosystem Concept in Natural Resource Management*, Academic, New York, pp. 3-7.

Stanton, M. L. (1983). "Spatial patterns in the plant community and their effects upon insect search," in S. Ahmad, Ed., *Herbivorous Insects: Host-Seeking Behavior and Mechanisms*, Academic, New York, pp. 125-157.

Stary, P., and C. Erdelen. (1982). Aphid parasitoids (Hym.: Aphidiidae, Aphelinidae) from the Yemen Arab Republic, *Entomophaga*, **27**, 105-108.

Stern, V. M. (1969). Interplanting alfalfa in cotton to control lygus bugs and other insect pests, *Proc. Tall Timbers Conf. Ecol. Anim. Control Habitat Manag.*, **1**, 55-69.

Stern, V. M. (1981). "Environmental control of insects using trap crops, sanitation, prevention, and harvesting," in D. Pimentel, Ed., *Handbook of Pest Management in Agriculture*, Vol. I, CRC Press, Boca Raton, FL, pp. 199-209.

Stern, V. M., R. van den Bosch, and T. F. Leigh. (1964). Strip cutting alfalfa for lygus bug control, *Calif. Agric.*, **18**, 4-6.

Stern, V. M., A. Mueller, V. Sevacherian, and M. Way. (1969). Lygus bug control in cotton through alfalfa interplanting, *Calif. Agric.*, **23**, 8.

Stern, V. M., P. L. Adkisson, O. C. Beingola, and G. A., Victorov. (1976). "Cultural controls," in C. B. Huffaker and P. S. Messenger, Eds., *Theory and Practice of Biological Control*, Academic, New York, pp. 593–613.

Stevens, L. M., A. L. Steinhauer, and T. C. Elden. (1975). Laboratory rearing of the Mexican bean beetle and the parasite, *Pediobius foveolatus*, with emphasis on parasite longevity and host-parasite ratios, *Environ. Entomol.*, **4**, 953–957.

Still, G. W. (1962). Cultural control of the grape berry moth, USDA Agric. Inform. Bull. 256, 8 pp.

Stinner, R. E., J. Regniere, and K. Wilson. (1982). Differential aspects of agroecosystem structure on dynamics of three soybean herbivores, *Environ. Entomol.*, **11**, 538–543.

Summers, C. G., R. L. Coviello, W. E. Pendery, and R. W. Bushing. (1976). Effect of sorghum midge on grain sorghum production in the San Joaquin Valley relative to date of planting and plant spacing, *Hilgardia*, **44**, 127–140.

Sylkhuis, J. T. (1953). Wheat streak mosaic in Alberta and factors related to its spread, *Can. J. Agric. Sci.*, **33**, 195–197.

Tamaki, G., H. R. Moffitt, and J. E. Turner. (1975). The influence of perennial weeds on the abundance of the redbacked cutworm on asparagus, *Environ. Entomol.*, **4**, 274–276.

Tansley, A. G. (1935). The use and abuse of vegetational concepts and terms, *Ecology*, **16**, 284–307.

Tappan, W. B. (1965). The decline of tobacco hornworm populations on cigar-wrapper tobacco, *J. Econ. Entomol.*, **58**, 771–772.

Thomas, G. D., C. M. Ignoffo, C. E. Morgan, and W. A. Dickerson. (1974). Southern green stinkbug: Influence on yield and quality of soybeans, *J. Econ. Entomol.*, **67**, 501–503.

Thomas, J. G. (1969). The sorghum midge and its control, Tex. Agric. Exp. Sta. L-842, 1–4.

Thompson, H. E. (1975). Association between plant pests and cultural practices in field crop management, *Iowa State J. Res.*, **49**, 473–476.

Thorvilson, H. G., L. P. Pedigo, and L. C. Lewis. (1985). *Plathypena scabra* (F.) (Lepidoptera: Noctuidae) populations and the incidence of natural enemies in four soybean tillage systems, *J. Econ. Entomol.*, **78**, 213–217.

Tingey, W. M., and S. R. Singh. (1980). "Environmental factors influencing the magnitude and expression of resistance," in F. G. Maxwell and P. R. Jennings, Eds., *Breeding Plants Resistant to Insects*, Wiley, New York, pp. 89–113.

Troxclair, N. N., and D. J. Boethel. (1984). The influence of tillage practices and row spacing on soybean insect populations in Louisiana, *J. Econ. Entomol.*, **77**, 1571–1579.

van den Bosch, R., and V. M. Stern. (1969). The effects of harvesting practices on insect populations in alfalfa, *Proc. Tall Timbers Conf. Ecol. Anim. Control Habitat Mgt.*, **1**, 47–54.

van Emden, H. F., and C. H. Wearing. (1965). The role of the aphid host plant in delaying economic damage levels in crops, *Ann. Appl. Biol.*, **56**, 323–324.

van Emden, H. F., and G. C. Williams. (1974). Insect stability and diversity in agro-ecosystems, *Annu. Rev. Entomol.*, **19**, 455–475.

Walker, J. K., G. A. Niles, J. R. Gannaway, R. D. Bradshaw, and R. E. Glodt. (1976). Narrow row planting of cotton genotypes and boll weevil damage, *J. Econ. Entomol.*, **69**, 249–253.

Walker, J. K., J. R. Gannaway, and G. A. Niles. (1977). Age distribution of cotton bolls and damage from boll weevil, *J. Econ. Entomol.*, **70**, 5–8.

Walton, W. R., and C. M. Packard. (1936). The Hessian fly and how losses from it can be avoided, USDA Farm. Bull. 1627.

Watmough, R. H., S. W. Broodryk, and D. P. Annecke. (1973). The establishment of two imported parasitoids of potato tuber moth *(Phthorimaea operculella)* in South Africa, *Entomophaga*, **18**, 237–249.

Watson, J. R. (1930). Entomology, *Fla. Agric. Exp. Sta. Rep.*, **1928-29**, 53–58.

Wilhelm, S. (1976). "The agroecosystem: A symplified plant community," in J. L. Apple and R. F. Smith, Eds., *Integrated Pest Management*, Plenum, New York, pp. 59–70.

Wiseman, B. R., and W. W. McMillian. (1967). Relationship between planting date and damage to grain sorghum by the sorghum midge, *Contarinia sorghicola* (Diptera: Cecidomyiidae), in 1968, *J. Ga. Entomol. Soc.*, **4**, 55–58.

Woodside, A. M. (1954). Japanese beetle damage to corn as influenced by silking date, *J. Econ. Entomol.*, **47**, 349–352.

Wuensche, A. L. (1976). "Relative abundance of seven pest species and three predacious genera in three soybean ecosystems in Louisiana," M. S. Thesis, Louisiana State University, Baton Rouge, 384 pp.

Zepp, D. B., and A. J. Keaster. (1977). Effects of corn plant densities on the girdling behavior of the southwestern corn borer, *J. Econ. Entomol.*, **70**, 678–680.

# 10

## THE ECOLOGY OF INSECTICIDES AND THE CHEMICAL CONTROL OF INSECTS

*Robert L. Metcalf*

### 10.1. INTRODUCTION

Insecticides are defined by Webster as "agents for destroying insects." They may have valid roles in any program for insect control. Indeed, few if any entomologists, agronomists, or agriculturists envision modern crop production without their use. At the same time it must be emphasized that the broadest definition of integrated pest management (IPM) is "the ecological approach to insect control." The interface between these two concepts provides the subject matter for this chapter. The age of pesticides really began with the commercial introduction of DDT into agriculture in 1946, and the ensuing four decades can be divided into three distinct phases: The Era of Optimism, 1946–1962; the Era of Doubt, 1962–1976; and the Era of Integrated Pest Management, 1976– (Metcalf, 1980). These successive epochs have witnessed vast scientific and legal controversy about the use of insecticides in agriculture, most of it arising because of the enthusiastic overuse and misuse of insecticides. This has led on numerous occasions to "ecocatastrophy" and to the present-day inevitability of the development of an ecologically oriented, economically practical, and socially acceptable technology for the rational and judicious use of insecticides, that is, IPM (Metcalf, 1980; Metcalf and Luckmann, 1982).

The ecological effects of insecticide application are summarized in Table 10.1, where progression downward displays an increasing order of consequences away from the intended first-order impact on the target pest. The unwanted and ecologically disadvantageous consequences of insecticide applications (Brown, 1978) are well demonstrated in the United States where 45% of world pesticide production is applied (Furtick, 1976). The properties of most of the broad-spectrum, persistent organochlorine insecticides developed during the era of opti-

251

**Table 10.1 Ecological Effects of Insecticide Applications**

A.   Suppression of target pest population
B.   Selection of resistant pests
C.   Destruction of natural enemies
   1.   Direct suppression
   2.   Reduction of host or prey population
   3.   Contamination of food supply
D.   Promotion of pest resurgences, secondary pests
E.   Destruction of pollinators
F.   Contamination of food webs
G.   General ecotoxicity

mism insured that when widely or intensively applied they would accumulate as persisting residues in soil and groundwater and biomagnify through food chains to accumulate in residues in human adipose tissues and to be secreted in human milk and sperm (Health, Education, and Welfare, 1969). Today no human on earth is free of detectable residues of these organochlorines. The long-term consequences of this proliferation of the fabric of life by persistent organochlorine xenobiotics are, for the most part, unpredictable but in many instances this exposure has had traumatic consequences on the health and longevity of invertebrates, fish, birds, and mammals that are highly valued components of the total ecosystem.

From an ecological viewpoint the environmental experiences of the past four decades of organochlorine insecticide use have had their positive side. These insecticides have served as probes to delineate the ecological intricacies of food chain and food pyramid effects, of parasite and predator interactions with the insect pest host, and of pest insect interactions with plant hosts (Southwood, 1977). The range of properties found in the more than 200 insecticides in common use including stability, reactivity, water solubility, partition in lipid–water matrices, vapor pressure, and biochemical site of action shape the nature of the ecological probe from narrow to broad spectrum, and from short to long time duration. These probes have produced much relatively new ecological information about such phenomena as acquired pesticide resistance in host and parasite predator complexes, in pest resurgences, and in the development of secondary pests (Southwood, 1977; Metcalf, 1980).

## 10.2.  PRESENT-DAY USE OF INSECTICIDES

Use of pesticides for crop plant protection represents an almost infinitesimal portion of the 300 million years of plant–insect coevolution and a relatively insignificant segment of the emergence of the species *Homo sapiens* L. over the past 6000 years as a cultivator and domesticator of food plants. Apparently, the earliest recorded use in the western world of pesticides in agriculture was in 1790 when teas of tobacco (*Nicotiana tabaccum* L.) containing the alkaloid nicotine were recommended to control aphids without damaging the host plant. Other milestones of insecticide usage include the first use of a synthetic pesticide, Paris green or copper acetoarsenite, on potatoes to kill the Colorado potato beetle (*Leptinotarsa decemlineata* Say) about 1865, and the marketing of potassium 4,6-dinitro-*o*-cresylate as a dormant spray in Germany in 1892. The first airplane application of lead arsenate dust for the control of the catalpa sphinx [*Ceratomia catalpae* (Boisd.)] was made in Ohio in 1921 (Metcalf et al., 1962).

From this small beginning less than 200 years ago, the practice of chemical crop protection by pesticides has grown exponentially, especially since 1945, and the number of chemical compounds marketed as insecticides has increased in little over a century from 2 to about 300 today (Kenaga and End, 1974). Since 1935 the volume of insecticides produced in the United States has increased more than 20-fold from about 13.6 million kg in 1928 to more than 300 million kg today as shown in Fig. 10.1 (Metcalf, 1980). Individual insecticides with the largest volume of use in the United States in 1976 and the areas treated are shown in Table 10.2 (Eichers et al., 1978). This chemical barrage directed at crop protection from insect pests has produced ecological consequences vastly out of proportion to the time span of human preoccupation with the use of pesticides. In this Chapter I shall attempt an ecological analysis of the effects of pesticide use in crop production and then outline the rational and judicious practice of pesticide management as an important component of IPM.

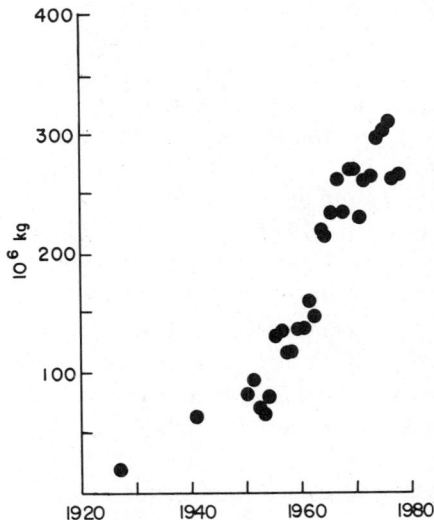

Fig. 10.1 Production of insecticides in the United States. (Data from USDA and Pesticide Reviews.)

**Table 10.2 Insecticides (AI) Most Commonly Applied to U.S. Farms—1976[a]**

| Insecticide | Type[b] | Kilograms Applied (in thousands) | Hectares Treated (in thousands) |
|---|---|---|---|
| Toxaphene | (O-Cl) | 13,935 | 1,965 |
| Methyl parathion | (OP) | 10,336 | 4,890 |
| Carbofuran | (C) | 5,268 | 4,624 |
| Carbaryl | (C) | 4,233 | 3,055 |
| Parathion | (OP) | 2,975 | 4,873 |
| Phorate | (OP) | 2,867 | 2,734 |
| EPN | (OP) | 2,834 | 830 |
| Disulfoton | (OP) | 2,493 | 2,956 |
| Fonofos | (OP) | 2,272 | 2,225 |
| Chlordimeform | | 2,036 | 1,199 |
| Terbufos | (OP) | 1,130 | 907 |
| Methomyl | (C) | 1,129 | 921 |
| Monocrotophos | (OP) | 870 | 750 |
| Malathion | (OP) | 768 | 622 |
| Diazinon | (OP) | 745 | 834 |

[a]Data from Eichers et al. (1978). [b]O-Cl (organochlorine), OP (organophosphate), C (carbamate).

## 10.2.1. Suppression of Target Pest Population

It is almost axiomatic that pesticides are highly toxic to target organisms and indeed the *modus operandi* of pesticide development over the past century has been directed toward discovery of chemical compounds of ever higher intrinsic toxicity. This evolution of the rate of application of insecticides is shown in Fig. 10.2 using data from rates of application required to produce acceptable control of insect pests attacking cotton in the United States (Braunholtz, 1976; Geiss-buhler, 1981). The rates of application have steadily declined since 1945 and have reached the 20–50 g/ha range with the synthetic pyrethroids permethrin and deltamethrin. This should be compared with the application of 10,000 g/ha of calcium arsenate during the 1930s and 1940s and a range of 675 to 1000 g/ha with the original organochlorine insecticides DDT and toxaphene. Thus, the world chemical industry continues to market more effective pesticidal products, and the use of these has decreased the net nontarget ecotoxicity.

The increasingly efficient synthetic organic insecticides developed since 1945 have been used for crop protection in huge quantities: The cumulative amount used in U.S. agriculture since 1950 is estimated at more than 2.7 billion kg of active ingredients (estimated as 50% of the accumulated production shown in Fig. 10.1). This represents an average contamination of about 16.8 kg/ha for the 162 million ha of U.S. cropland. A large variety of individual insecticides are available for use on key crops—in 1967 the registrations were cotton, 50; corn, 66; apple, 72; tobacco, 27; potato, 41; cabbage, 51; and alfalfa, 39. In U.S.

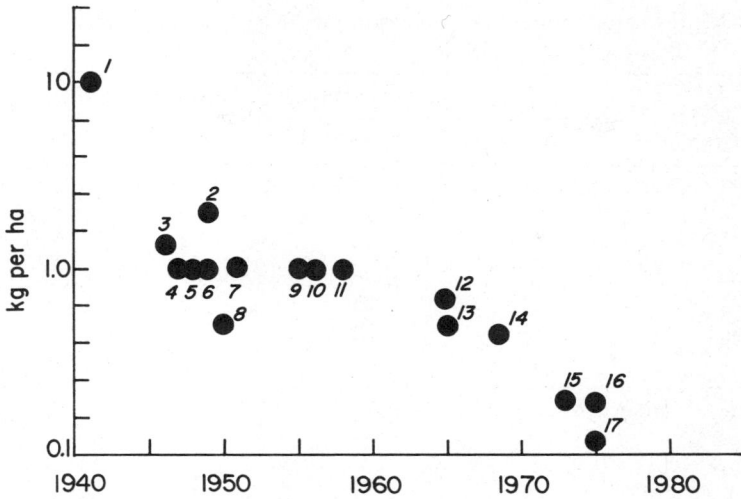

Fig. 10.2 Increased efficiency of insecticides used for cotton insect control, based on USDA recommendation rates in kg per ha. Key: (1) calcium arsenate, (2) toxaphene, (3) DDT, (4) parathion, (5) dieldrin, (6) EPN, (7) diazinon, (8) endrin, (9) azinphos methyl, (10) endosulfan, (11) carbaryl, (12) chlorpyrifos, (13) monocrotophos, (14) methomyl, (15) permethrin, (16) fenvalerate, (17) deltamethrin. [After Braunholtz (1976) and Geissbuhler (1981).]

agriculture, insecticides have had the heaviest use on a few major crops: cotton, 47%; corn, 17%; fruits, 9%; vegetables, 7%; soybeans, 4%; and tobacco, 3%;. Three crops—cotton, corn, and apples—consume 67% of the total farm usage (Pimentel, 1973; Pimentel et al., 1978). It would seem almost axiomatic from this enormous usage that insect losses would be curtailed and that insect pest problems would be minimized (Pickett, 1949; Smallman, 1964; Luck et al., 1977). However, the record indicates that neither prospect has occurred.

National surveys have been made by the U.S. Department of Agriculture (USDA) over the past 75 years, to establish crop losses from insect attacks (Marlatt, 1904; Hyslop, 1938; USDA, 1954, 1965). The data for the average crop damage on the key crops with the heaviest insecticidal use are shown in Table 10.3. They demonstrate no overall indication that crop damage from insect attack has decreased with increasing insecticide use. In fact, it has been estimated that crop losses from insect attacks in the United States have increased from 7% in the 1940s to 13% at present, despite a 10-fold increase in the annual use of insecticides (see Fig. 10.1) (Pimentel et al., 1978).

During the Age of Pesticides it was commonly supposed that the touted efficacy of the new insecticides and their enthusiastic use would suppress populations of most pest species so far below the economic threshold that they would no longer merit pest status. It is difficult to find much evidence that this has occurred. A few examples from the heavily treated crops rice, cotton, and apples are shown in Table 10.4.

**Table 10.3  Estimated Annual Losses from Insect Attacks on Crops Heavily Treated with Insecticides**

| Crop | Loss as Percentage of Total Production[a] | | | |
|---|---|---|---|---|
| | 1900–1904 | 1910–1935 | 1942–1951 | 1951–1960 |
| Cotton | 10 | 14.9 | 15 | 19 |
| Corn | 8 | 11.8 | 3.5 | 12 |
| Apple | 20 | 10.4 | 14 | 13 |
| Tobacco | — | — | 11 | 11 |
| Potato | 10 | 22 | 15.6 | 14 |
| Cabbage | 10 | 20 | 8 | 17 |
| Alfalfa | 10 | 5 | 9.3 | 15 |

[a]USDA data (Marlatt, 1904; Hyslop, 1983; USDA, 1954, 1965).

**Table 10.4  Insect Pests Virtually Eliminated by Regular Insecticide Use[a]**

| | |
|---|---|
| Rice | *Tryporyza incertulas*, paddy borer |
| | *Scotinophora lurida*, black rice bug |
| | *Oxya chinensis*, rice grasshopper |
| | *Dicladispa armigera*, rice hispa |
| | *Oulema oryzae*, rice leaf beetle |
| Cotton | *Sacadodes pyralis*, false bollworm |
| | *Alabama argillacea*, cotton leaf worm |
| Tree fruits | *Archips semiferanus*, oak leafroller |
| | *Archips argyrospilus*, fruit tree leafroller |
| | *Choristoneura fractivittana* |
| | *Choristoneura rosaceana*, obliquebanded leafroller |
| | *Spilonota ocellana*, eye-spotted bud moth |
| | *Grapholitha prunivora*, lesser appleworm |
| | *Argyrotaenia quadrifasciana* |
| | *Pandemis limitata*, threelined leafroller |

[a]Data from Kiritani (1979); Vaughan and Gladys (1976); Hikichi (1964).

## 10.2.2.  Efficiency of Application

As seen from the data in Table 10.2, the average rate of treatment with insecticides in the United States is about 1 kg/ha. At first thought, this application rate, which represents a surface contamination of about 100 mg/m², may seem to represent an efficient process. However, when contrasted with the actual amount required to kill insect pests ($LD_{99}$ in $\mu g/g$ body weight) present at populations at or above the economic threshold, it becomes apparent that present-day application rates are grossly inefficient, as demonstrated by the following examples.

*Control of the American bollworm [Heliothis armigera (Hubner)] on cotton.* Joyce (1982) has pointed out that the economic threshold for this pest on cotton in the Sudan is 10 first instar larvae per 100 plants (body weight aggregates 10 g/ha). The $LD_{99}$ for DDT is 10 $\mu$g/ha, yet 1 $\times$ $10^9$ $\mu$g DDT is applied (1 kg/ha). The efficiency or ratio of amount required to kill the target pest/amount applied = 1 $\times$ $10^{-8}$ or 0.00000001. In Joyce's (1982) terms, "it is difficult to conceive a strategy for chemical control less efficient than the use of pesticides, and a greater environmental hazard."

*Control of the adult western corn rootworm (Diabrotica virgifera LeConte) on corn.* Adult western corn rootworms often become numerous enough to damage corn silks seriously and produce underfertilized ears or "scatter corn." Carbaryl is frequently applied at 1 kg/ha to control these beetles, at an economic threshold variously estimated at one to five beetles per corn plant. At the higher figure, there are about 370,000 adult beetles per ha, weighing 3,700 g. The $LD_{99}$ for carbaryl is 0.05 $\mu$g/beetle or 18.5 mg/ha and the efficiency of application is 0.000017.

Similar calculations can be made for most insect pests and it is obvious that only a minute fraction of the insecticide applied is required for suppression of the target pest. The remainder >99.9% is essentially wasted and enters the environment in a variety of ways as suggested in Fig. 10.3. It is this general environmental contamination by insecticides that is the cause of most of the nontarget effects that result in the ecological disturbances summarized in Table 10.1.

Corrective measures are possible. By incorporating insecticides in pheromone or kairomone baits that bring the pest to the pesticide, notable gains in efficiency can be obtained. For example, the male oriental fruitfly, *Dacus dorsalis* Hendel is specifically attracted by picogram quantities of the plant-produced kairomone methyl eugenol. This insect pest was eradicated from the Island of Rota by dropping fiberboard squares impregnated with 28 g of a mixture of 97% methyl eugenol and 3% naled insecticide at the rate of 125/$mi^2$, a dosage of 3.4 g of naled insecticide per acre (Steiner et al., 1965). The plant kairomone cucurbitacins are arrestants and feeding stimulants for the western corn rootworm. Incorporation of these, from dried bitter squash, in broadcast granular baits with 0.1% methomyl insecticide has given 95–99% control of the adult corn rootworms in corn at doses of insecticide as low as 11.1 g/ha (Metcalf et al., 1986). Seed treatments with lindane at 17.3 g/ha has given 70–95% mortality of wireworms attacking a variety of field and vegetable crops (Reynolds, 1958). The red imported fire ant (*Solenopsis richteri* Forel) has been controlled by corn grits–soybean oil granular baits containing avermectin insecticide at dosages as low as 0.05–0.30 g(AI)/ha (Lofgren and Williams, 1982). Such bait technology can result in decreases of at least 100-fold in the application rates of conventional insecticides and in great improvements in specificity to the target pest. This sort of technology represents the way of the future in the chemical control of agricultural pests.

## 10.3. INSECT RESISTANCE TO INSECTICIDES

The acquired resistance or tolerance of insect pests to insecticide action has been known for over 70 years, since it was first demonstrated in 1914 in the San Jose

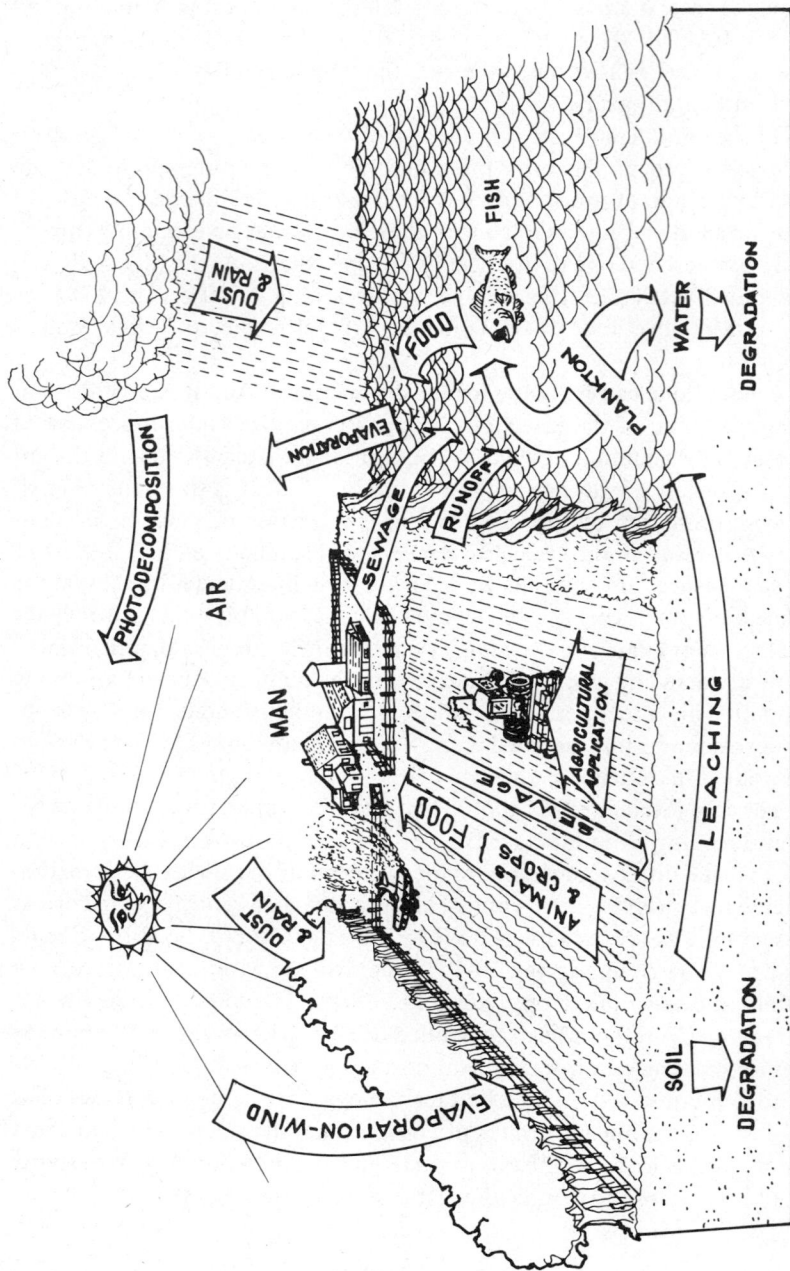

Fig. 10.3 Ways in which pesticides move away from the target site to contaminate the total environment, entering into a variety of cycles in soil, air, water and food. [Reprinted with permission from Metcalf and Sanborn (1975) *Bull. Ill. Nat. Hist. Survey.*)]

scale [*Quadraspidiotus perniciosus* (Comstock)] selected by lime sulfur spray and in 1916 in California red scale [*Aonidiella aurantii* (Maskell)] and black scale [*Saissetia oleae* (Olivier)] selected by hydrogen cyanide (Metcalf, 1955). By 1946, insecticide resistance was present in a total of 11 species: the codling moth [*Cydia pomonella* (L.)] and the peach twig borer (*Anarsia lineatella* Zeller) to lead arsenate, the citricola scale [*Coccus pseudomagnoliarum* (Kuwana)] to hydrogen cyanide, the cattle tick *Boophilus microplus* and the blue tick *B. decoloratus* to sodium arsenite dip, the citrus thrips [*Scirtothrips citri* (Moulten)] and the gladiolus thrips [*Thrips simplex* (Morison)] to potassium antimonyl tartrate, and the walnut husk fly (*Rhagoletis completa* Cresson) to cryolite (Brown and Pal, 1971). The prognostication for the future of chemical insect control was poor. However, surprisingly little scientific attention was given to insecticide resistance, which is nothing more than accelerated microevolution. Insecticide resistance began to receive the attention it deserved only after the introduction of DDT when resistant strains of the house fly (*Musca domestica* L.) appeared in Sweden and Denmark in 1946, of the mosquitoes *Culex pipiens* L. in Italy and *Aedes sollicitans* (Walker) in Florida in 1947, the bedbug (*Cimex lectularius* L.) in Hawaii in 1947, and the human body louse (*Pediculus humanus humanus* L.) in Korea and Japan in 1951 (Brown and Pal, 1971).

With the steadily widening proliferation of new insecticides and the increasing scope of their use in insect control programs, the number of scientifically documented cases of insect resistance to insecticides has increased at an exponential rate, encompassing 224 species in 1970, 364 species in 1975, 428 species in 1980, and 447 species in 1984 (Table 10.5, Fig. 10.4). Although the majority of early examples of insecticide resistance were found in insect vectors of human diseases, because of the very widespread use of DDT, lindane, and dieldrin in vector control programs, by 1980 resistance was established in 260 pests of crops, forests, and stored products as compared to 168 pests of human and animal health (Georghiou, 1981). There is every reason to believe that these figures understate the severity of the resistance problems world-wide because the sus-

**Table 10.5  Rates of Development of Insect Pest Resistance to Insecticides**

| Insecticide Type | Number of Resistant Species[a] | | | | | |
|---|---|---|---|---|---|---|
| | 1948 | 1954 | 1970 | 1975 | 1980 | 1984 |
| DDT/methoxychlor | 3 | 13 | 98 | 203 | 229 | 233 |
| Lindane/cyclodiene | 1 | 5 | 140 | 225 | 269 | 276 |
| Organophosphate | 0 | 3 | 54 | 147 | 200 | 212 |
| Carbamate | 0 | 0 | 3 | 36 | 51 | 64 |
| Pyrethroid | 0 | 0 | 3 | 6 | 22 | 32 |
| Fumigant | 3 | 3 | 3 | 9 | 7 | — |
| Other | 3 | 4 | 12 | 20 | 41 | — |

[a] Data calculated from Metcalf (1955); Brown (1971); Georghiou and Taylor (1977); Georghiou (1981, 1984).

RESISTANT
SPECIES

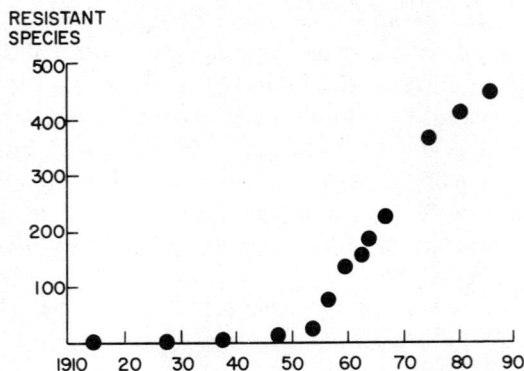

Fig. 10.4 Rate of development of insect pests resistant to insecticides (see Table 10.6).

ceptibility of many insect pest species has not been studied, is incompletely characterized, or is not reported adequately in the scientific literature.

Cases of insecticide resistance have been demonstrated in 16 orders of Arthopoda and the distribution has been recorded by Georghiou (1981) as: Acarina, 53 (12.4%); Anoplura, 6 (1.4%); Coleoptera, 64 (14.9%); Dermaptera, 1 (0.2%); Diptera, 153 (36.7%); Ephemeroptera, 2 (0.5%); Hemiptera, 20 (4.7%); Homoptera, 42 (9.8%); Hymenoptera, 3 (0.7%); Lepidoptera, 64 (14.9%); Mallophaga, 2 (0.5%); Orthoptera, 3 (0.7%); Siphonaptera, 8 (1.9%); Thysanoptera, 7 (1.6%). These data clearly reflect the relative numbers of pest species in the individual orders and the amount of insecticide pressure placed upon them.

As each new class of insecticides has been introduced and then widely deployed against insect pests, the rate of development of resistant species has followed an almost identical pattern as illustrated in Table 10.5. The curves for rates of increase in resistant species are essentially exponential until high level resistance or legal restrictions have slowed further usage. The exponential curves of resistance development are best characterized by the doubling times over the middle ranges. These can be approximated from the data in Table 10.5 as: DDT/methoxychlor, 6.3 years; lindane/cyclodienes, 5.0 years; organophosphorus, 4.0 years; carbamates, 2.5 years; and pyrethroids, 2 years (Metcalf, 1984). These data show clearly that the doubling time for development of resistant species has steadily decreased with the introduction of each new type of insecticide and demonstrate the disastrous effects of multiple resistance.

### 10.3.1. Cross-Resistance and Multiple Resistance

Listing the exponential growth in numbers of insecticides-resistant pest species does not describe adequately the impact of resistance upon applied entomology. *Cross-resistance* enables these resistant species to survive exposure to chemically related insecticides, for example, DDT and methoxychlor, lindane and dieldrin, parathion and malathion, carbaryl and carbofuran, or permethrin and fenva-lerate. Cross-resistance is generally brought about by a common detoxication pathway or by a change in susceptibility to a common biochemical lesion. *Multiple resistance* is far more serious and as shown in Table 10.6 is now found in more than 200 important pest species that display resistance to a variety of classes of insecticides with differing modes of action and different detoxication pathways (Sawicki, 1975). Cross-resistance limits the choice of available insecticides, whereas multiple resistance reflects the past history of insecticide selection and precludes a return to those used previously. Both processes seriously deplete insecticide resources (Metcalf, 1980, 1984).

Multiple resistance is produced by several types of evolutionary mechanisms arising through intense "natural selection." Altered acetylcholinesterase involves a biochemical alteration in the target-site enzyme of the organophosphorus and carbamate insecticides so that entire categories of insecticides are no longer effective, as with the cattle tick *B. microplus* in Australia (Schuntner and Thompson, 1978) and in the green rice leafhopper (*Nephotettix cincticeps* Uhler) in Japan (Hama and Iwata, 1978). This type of multiple resistance is also found in the house fly and the red spider mite (*Tetranychus urticae* Koch) (Oppennoorth and Welling, 1976).

The kdr mechanism, an insensitivity of the nerve axon, that promotes multiple resistance between DDT and the pyrethroids is well known in the house fly and is also found in *Anopheles stephensi* Liston, *A. gambiae* Giles, and *A. quadrimaculatus* Say, in *Aedes aegypti* (L.), and in *Culex tarsalis* Coquillett (Oppennoorth and Welling, 1976; Omer et al., 1980). The kdr type of multiple resistance is also found in the bedbugs *Cimex* spp. and the cattle tick *Boophilus microplus* (Canestrini), and in species of lepidopterous pests in the genera *Spodoptera*, *Plutella*, and *Heliothis* (Sawicki, 1980). Because of the very widespread nature of DDT resistance, documented in 229 pest species by 1980 (Georghiou, 1981), this type of multiple resistance is particularly ominous.

A third mechanism of multiple resistance is that produced by hydrolytic esterases whose specificity includes organophosphates, carbamates, and pyrethroids, as found in *Myzus persicae* (Sulzer) and possibly in the green rice leafhopper (Oppennoorth and Welling, 1976; Sawicki, 1980).

A fourth mechanism of multiple resistance is that produced by a generalized increase in the mixed-function oxidase (MFO) enzymes that protect insects from xenobiotic compounds and therefore prevent a variety of insecticide molecules from reaching the target site of action (Oppennoorth and Welling, 1976).

Multiple resistance is now very widely distributed in at least 44 families of 10 orders (Georghiou and Taylor, 1977; Georghiou, 1981). Pests that by 1984 had developed almost complete multiple resistance to the principal classes of insecticides (1) DDT/methoxychlor, (2) lindane/cyclodienes, (3) organophosphates, (4) carbamates, and (5) pyrethroids, include the mosquitoes *Anopheles albimanus* Wied, *A. sacharovi* Farr, and *A. stephensi*, the house fly, the German cockroach *Blattella germanica* (L.), the cattle ticks *Boophilus decoloratus* and *B. microplus*, the Colorado potato beetle, the green peach aphid *Myzus persi-*

**Table 10.6 Development of Insect Pests with Multiple Resistance to Various Groups of Insecticides**

| Year | Resistant | DDT Cyclodienes | DDT Cyclodienes Organophosphorus | DDT Cyclodienes Organophosphorus Carbamates | DDT Cyclodienes Organophosphorus Carbamates Pyrethroids | Reference |
|---|---|---|---|---|---|---|
| 1938 | 7 | 0 | 0 | 0 | 0 | Brown and Pal (1971) |
| 1948 | 14 | 1 | 0 | 0 | 0 | Brown and Pal (1971) |
| 1954 | 25 | 18 | 3 | 0 | 0 | Metcalf (1955) |
| 1970 | 224 | 42 | 23 | 4 | 0 | Brown (1971) |
| 1975 | 364 | 70 | 44 | 22 | 7 | Georghiou and Taylor (1977) |
| 1980 | 428 | 105 | 53 | 25 | 14 | Georghiou (1981) |
| 1984 | 447 | 119 | 54 | 25 | 17 | Georghiou (1984) |

*cae*, the pear psylla (*Psylla pyricola* Foerster), the diamondback moth [*Plutella xylostella* (L.)], the cotton bollworms *Heliothis armigera* and *H. virescens*, the armyworms *Spodoptera frugiperda* (Smith) and *S. littoralis*, the red flour beetle [*Tribolium castaneum* (Herbst)], and the granary weevil [*Sitophilus granarius* (L.)] (Georghiou, 1981, 1984). The rapid rates of development of multiple resistance are shown in Table 10.6. The devastating effects of cross-resistance and multiple resistance to control programs for the Colorado potato beetle attacking potatoes and the diamondback moth attacking crucifers is shown by the sequences of resistance development portrayed in Tables 10.7 and 10.8.

Since 1945 on Long Island at least 12 insecticides have been introduced to control the Colorado potato beetle and all of them have failed because of the rapid onset of resistance (Table 10.7) (Forgash, 1984). Although DDT remained effective initially for about 7 years, the onset of multiple resistance has progressively decreased the interval of effectiveness so that since 1973 no insecticide remained effective for more than 2 years. Increasing dosage rates made to counter the development of resistance have seriously contaminated groundwater resources of Long Island. Resistance development by the diamondback moth has occurred in many areas of southeast Asia as shown in Table 10.8 and over the past 30 years at least 37 insecticides, representing all classes, have failed to control this pest.

## 10.3.2. Persistence of Resistant Genes

Once selected for, resistant genes have virtually limitless persistence in wild insect populations. Although the gene frequency of a specific resistant allele may decrease upon removal of insecticide pressure, there persists a changed back-

Table 10.7 Insecticide Resistance in the Colorado Potato Beetle *Leptinotarsa decemlineata*, Long Island[a]

| Insecticide | Year Introduced | Year First Failure |
|---|---|---|
| Arsenicals | 1880 | 1940 |
| DDT | 1945 | 1952 |
| Dieldrin | 1954 | 1957 |
| Endrin | 1957 | 1960 |
| Carbaryl | 1959 | 1963 |
| Azinphos methyl | 1959 | 1964 |
| Monocrotophos | 1973 | 1973 |
| Phosmet | 1973 | 1973 |
| Phorate | 1973 | 1974 |
| Carbofuran | 1974 | 1976 |
| Oxamyl | 1978 | 1978 |
| Fenvalerate | 1979 | 1981 |
| Permethrin | 1979 | 1981 |

[a]Data from Forgash (1984).

**TABLE 10.8 Insecticide Resistance of the Diamondback Moth,** *Plutella xylostella*[a]

| Insecticide | Year Failure Reported |
|---|---|
| DDT | 1953 |
| Lindane | 1957 |
| Endrin | 1965 |
| Aldrin, dieldrin | 1967 |
| Endosulfan | 1969 |
| Isobenzan | 1974 |
| Mevinphos | 1957 |
| Malathion, parathion | 1965 |
| Diazinon | 1967 |
| Acephate, dimethoate, dichlorvos | 1974 |
| Methyl parathion, fenitrothion, leptophos, EPN | 1974 |
| Monocrotophos, phenthoate, phosphamidon, trichlorfon, naled | 1974 |
| Chlorpyrifos methyl, methamidophos | 1978 |
| Cyanophos, prothiofos | 1981 |
| Carbaryl | 1967 |
| Methomyl | 1974 |
| Cartap | 1978 |
| Propoxur, isoprocarb | 1981 |
| Resmethrin | 1978 |
| Cypermethrin, decamethrin, permethrin, fenvalerate | 1981 |

[a]Data from Georghiou (1981); Liu et al. (1982).

ground of residual inheritance in the genome that causes the strain to regain its resistance as soon as the insecticide is reapplied (Brown, 1977). Genes for DDT and cyclodiene resistance in Danish houseflies have persisted for more than 30 years, and these insecticides again became ineffective within 2 months after reapplication. Resistant genes to diazinon and dimethoate also have been shown to have very lengthy persistence (Keiding, 1977). Multiple resistance in the cotton leafworm in Egypt has shown no signs of regression over a 20-year period (El-Sebae, 1977). For the examples of resistance development by the Colorado potato beetle and the diamondback moth as shown in Tables 10.7 and 10.8, there has been no successful return to any of the insecticides that failed. Thus, there has been no successful long-term reuse of any insecticide in insect populations with resistant alleles, even though the initial resistance has apparently reverted to full susceptibility (Brown, 1977; Georghiou and Taylor, 1977; Keiding, 1977).

Insecticide resistance is believed to develop as the result of the natural selection of preadaptive mutants that possess genetically controlled mechanisms for detoxication, target site insensitivity, or other means of survival in the presence of the insecticide. For example, the major cause of DDT resistance as in the house fly, and the yellow fever mosquito is the selection of individuals that produce high levels of a detoxication enzyme DDTase that attacks the $\alpha$-H of the molecule of DDT, causing elimination of Cl and the formation of the insecticidally inactive product DDE [2,2-bis-($p$-chlorphenyl)-1,1-dichloroethylene] (Oppenoorth and Welling, 1976).

DDTase

DDT $\longrightarrow$ DDE

DDTase is readily inhibited by a variety of DDT analogs such as chlorfenethol [1,1-bis-($p$-chlorophenyl)-ethanol] and these compounds can serve as effective synergists for DDT to restore activity against the DDT-resistant house fly. This synergistic action of such DDT analogs as chlorfenethol, 1,1-bis-($p$-chlorophenyl)-2,2,2-trifluoroethanol, 1,1-bis-($p$-chlorophenyl)-ethane, bis-($p$-chlorophenyl)-chloromethane, and $p,p'$-dichlorobenzene sulfonanilide demonstrates conclusively that DDTase activity is the specific biochemical cause of a major type of DDT resistance. Genetic control of DDTase activity in the house fly results from a single gene on chromosome 2 (Brown and Pal, 1971).

Another informative example of the biochemical basis for insecticide resistance is the selective detoxication of malathion by carboxylesterase in resistant races of the mosquito *Culex tarsalis*, the house fly, the green rice leafhopper, and other species.

carboxyl esterase

The carboxylesterase selectively attacks one of the carboxyethyl groups of malathion, converting it to the nontoxic malathion monocarboxylic acid. The carboxylesterase can be inhibited by the $P = 0$ oxidation product of the insecticide EPN (*O*-ethyl *O*-*p*-nitrophenyl phenylphosphonothionate) and EPN has been used as a synergist for malathion-resistant *C. tarsalis*, thus confirming the mechanism of resistance (Oppennoorth and Welling, 1976).

It should be emphasized that there are many other biochemical processes responsible for specific insect resistance to various insecticides. It appears that almost any biochemical mechanism affecting xenobiotic molecules will be utilized as an insecticide resistance mechanism if enough selection pressure is placed upon the pest species. Furthermore, as a variety of insecticides are utilized in insect control, successive levels of biochemical resistance can be superimposed in the individual species to give very complicated resistance patterns and mechanisms (Oppennoorth and Welling, 1976).

The genetics of insecticide resistance is too complex to discuss here but has been extensively reviewed by Oppennoorth (1965) and Brown and Pal (1971).

The gravity of multiple resistance to current insect control practices relying exclusively on insecticides cannot be overstated. For example, both altered acetylcholinesterase and kdr are found in the cattle tick *B. microplus*. The crucial question today in insect pest control by chemicals is not "to what insecticides are these pests resistant" but rather "to what insecticides are they susceptible" (Keiding, 1977; Sawicki, 1975). The true dimensions of the resistance problem are not known because of lack of precise data on the primitive susceptibility levels for many important pests (Chio et al., 1976). No group of insecticides is "resistant proof" and multiple resistance can be present before new insecticides are introduced commercially (Attia and Hamilton, 1978; El-Sebae, 1977; Georghiou et al., 1975; Keiding, 1977).

### Case History    Western Corn Rootworm

The ecological implications of insecticide resistance in agricultural pests are illustrated by the history of the western corn rootworm in the U.S. corn belt. This insect was first described by LeConte from specimens collected on wild gourd, *Cucurbita foetidissima* HBK, near Fort Wallace, Kansas, in 1867 (Smith and Lawrence, 1967). The insect was sparsely distributed over the Colorado–New Mexico–Arizona area and is thought to have originally coevolved with the Cucurbitaceae and gradually transfered to the Graminaceae, especially *Trypsacum* (Smith, 1966; Metcalf et al., 1980). With the introduction of corn *Zea mays* L. by settlers migrating westward after the Civil War, a tempting new host plant became available in increasing area. *Diabrotica virgifera* was first recorded as attacking sweet corn near Loveland, Colorado, in 1909 and slowly spread eastward reaching the Missouri River by 1948, a distance of 756 km traveled in 38 years, or about 20 km/year. About this time, preplanting applications of soil insecticides were introduced to control the corn rootworm larvae. Large scale applications of BHC were made in Nebraska in 1949 and treatments with aldrin and chlordane were begun in 1952, and with heptachlor in 1954. Approximately 705,000 ha of corn soil in Nebraska were treated with these insecticides in 1954. Ineffective corn rootworm control from these soil insecticides was first noticed in 1959 in an area of south-central Nebraska where the western corn rootworm adults showed a resistance of approximately 100-fold to aldrin and heptachlor. The area infested with the resistant strain began to expand rapidly in an

128- to 160-km band along the Platte River that reached into the western edges of South Dakota, Iowa, and Missouri by 1962. Surveys made by the USDA in 1956, 1965, 1969, 1975, and 1977 have provided perhaps the best record of the migration of a species in which resistance was induced in a single locality. From 1961 to 1964 the resistant strain traveled from near Grand Island, Nebraska, to near Eau Clair, Wisconsin, about 580 km or an average of 193 km a year. The cyclodiene-resistant strain reached northwest Indiana by 1968, a distance of about 805 km in 7 years or about 115 km a year, and by 1980 had spread throughout the U.S. corn belt (Fig. 10.5) (Metcalf, 1983). The astonishing increase in the rate of migration of this species from the 19–48 km/year recorded before the onset of cyclodiene resistance to the 113–193 km/year after the resistant strain evolved appears to be the result of the increased fitness of the resistant strain and of a behavioral change associated with the R gene. The resistant beetles have become superior competitors and through competitive displacement have become the dominant rootworm pest in an area where the northern corn rootworm (*Diabrotica barberi* Smith and Lawrence) that inhabits an almost identical ecological niche was formerly the dominant pest. In Illinois, for example, in 1967 the western corn rootworm comprised 9% of the corn rootworm population, in 1975 35% and in 1977 65%, reaching 89–93% in some localities (Wedberg and Black, 1970).

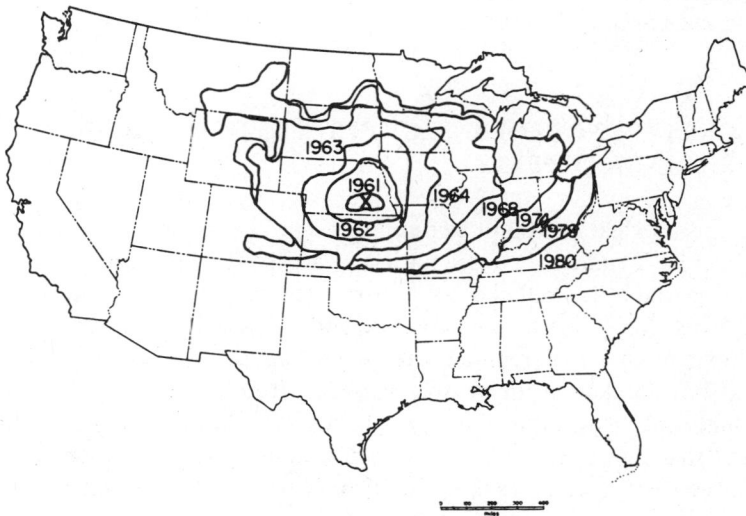

Fig. 10.5 Spread of the cyclodiene-resistant western corn rootworm *Diabrotica virgifera* from the original mutant appearing in south central Nebraska (X) in 1959, to encompass the entire corn belt [Reprinted with permission from Metcalf and Luckmann (1982).]

### 10.3.3. Migrations and Introductions of Multiple-Resistant Insect Pests

The migration of multiple-resistant races of insect pests from one crop to another, from one part of the country to another, or from one country to another adds a significant new dimension to the resistance problem. Resistant species may originate in areas of intensive crop production and of insecticide application and subsequently migrate to other areas where cropping practices are very different, with disastrous effects on IPM programs. Newsom (1980) has described the migration of the soybean looper [*Pseudoplusia includens* (Walker)], originating on heavily sprayed ornamental crops in South Florida, across the Gulf of Mexico to Louisiana soybeans bearing a complement of genes for resistance to methomyl, acephate, and pyrethroids. The southern corn rootworm (*Diabrotica undecimpunctata howardi* Barber) migrates northward into the upper Mississippi Valley each year bearing an assortment of genes for resistance to DDT, methoxychlor, cyclodienes, malathion, and carbaryl selected by insecticide applications in the overwintering areas of the southern states (Chio et al., 1976). Other multiple-resistant migrants up the Mississippi Valley as far as southern Canada include the corn earworm and the fall armyworm, both with R genes to all classes of insecticides.

The brown planthopper (*Nilaparvata lugens* Stål) migrates annually from China to Japan, Taiwan, and the Philippines where its multiple R genes add immensely to the difficulties for control (Kiritani, 1979). Malathion R races of the stored grain pests *Tribolium castaneum* Jac. du Val and *T. confusum, Sitophilus oryzae* (L.), and *S. granarius* (L.), have appeared near major seaports throughout the world, and pyrethroid resistance is now spreading in an analogous way (Georgiou, 1983).

## 10.4. RESURGENCES OF INSECT PESTS AND DEVELOPMENT OF SECONDARY PESTS

From an ecological viewpoint a resurgence of an insect pest attacking a cultivar is defined as a sudden and dramatic upward shift in the general equilibrium position so that it lies well above the economic injury level. Resurgences typically result from alterations in the action of both density-independent and density-dependent factors that regulate the general equilibrium position. Thus, they may arise from changes in weather patterns, from large-scale adoption of new agronomic practices, for example, reduced tillage, or from wholesale changes in the gene structure of cultivars that eliminate antibiotic factors, for example, the "green revolution" changes of rice cultivars from *indica* progenitors to *japonica* progenitors. Resurgences may also result from mutations of insect pest biotypes to produce relative immunity to suppressive factors present in the cultivar through antibiosis or antixenosis, or in the agroecosystem as pesticides.

In this chapter we are concerned about the role of pesticide use in promoting resurgences of target insect pest population and in the associated phenomenon

where the use of pesticides results in the resurgence of nontarget insect herbivores so that they become important secondary pests. Such resurgences are
almost always the result of repeated applications of broad-spectrum insecticides
so that resistant races of the resurgent pests are selected coincidentally with depletion or virtual elmination of natural enemies. The overall ecological phenomena of resistance, resurgence, and development of secondary pests occur most
frequently on crops heavily treated with insecticides, for example, cotton, rice,
deciduous and citrus fruits, vegetables, and ornamentals.

Pickett and Patterson (1953) demonstrated in Nova Scotia apple orchards
that the application of flotation sulfur fungicides to control apple scab *Venturia
inaequalis* (Cke), resulted in marked destruction of the chalcid parasite *Aphytis
mytilaspidis* (Le Baron) and the predatory mite *Hemisarcoptes malus* (Shimer)
regulating populations of oystershell scale [*Lepidosaphes ulmi* (L.)]. Thus, the
latter became an increasingly severe pest over a 15-year period in orchards
sprayed with sulfur. A suitable biological balance was restored by substituting
copper fungicides that did not adversely affect the natural enemies.

The proliferation of the use of DDT and other new broad-spectrum insecticides after 1946 and the steadily increasing scale of their utilization greatly increased the occurrences of this phenomenon; Ripper (1956) tabulated more than
50 species of phytophagous arthropods whose populations had undergone resurgences after insecticidal treatments with diverse insecticides such as calcium arsenate, cryolite, DDT, BHC, aldrin, toxaphene, and parathion. Two types of
resurgences were characterized: (1) pests whose populations were initially suppressed by the insecticide application but which rebounded to excessive levels
within a relatively short time, and (2) potential pests, that is, economically unimportant nontarget species that developed into serious pests after insecticide applications to control other target species. For both types of resurgence there is
overwhelming evidence of a negative correlation between the population density
of the resurging species and their natural enemies (Reynolds, 1971). The cabbage aphid [*Brevicoryne brassicae* (L.)], when first sprayed commercially with
para-oxon, provided a good example of a resurgent pest. Although the initial kill
was very high, the destruction of natural enemies resulted within 2 weeks in the
most enormous cabbage aphid outbreak ever seen in England. The first use of
DDT sprays in California citrus orchards in 1946–1947 to control citricola scale
and other pests resulted in an instructive example of the resurgence of a secondary pest, the rarely seen cottony cushion scale (*Icerya purchasi* Maskell). DDT
almost eliminated its specific predator *Rodolia cardinalis* (Mulsaut) and within
a few months the cottony cushion scale became so abundant that the citrus trees
dipped with honeydew and total defoliation and loss of crop often resulted. Ultimately thousands of acres were damaged and DDT became an anathema to citrus growers (Ripper, 1956).

The use of the modern, broad-spectrum insecticides DDT, parathion, carbaryl, and their relatives in deciduous and citrus fruit orchards and on cotton
and other crops resulted in drastic changes in the populations of phytophagous
mites. These were rarely abundant enough in the period before 1946 to require

pesticide applications and were kept in biological balance by effective natural enemies, including the predaceous mites *Typhlodromus* spp. and *Amblyseius* spp., the Coccinellidae predator *Stethorus*, and the predaceous thrips *Scolothrips*. The evidence is now overwhelming that the use of the broad-spectrum pesticides is responsible for the enormous outbreaks of red spider mites that have occurred worldwide since the introduction of DDT (Ripper, 1956; McMurtry et al., 1970).

The impact of these resurgent epidemics of Tetranychidae on pest control programs is demonstrated by the sudden appearance in the entomological literature of the terms miticide and acaricide in 1947 (see indices of the Review of Applied Entomology). Insect control recommendations of state agricultural experiment stations are revealing. Circular 212 "Directions for spraying fruit in Illinois," 1918, recommended eight spray treatments for apples using lime sulfur, Bordeaux mixture, oil lead arsenate, and nicotine. Circular 568 in 1944 (a revision) recommended 14 spray treatments with the same pesticides plus ferbam fungicide. Circular 610 in 1947 first included DDT for codling moth control and recommended 15 spray treatments. Red spider mites were mentioned as specific pests for the first time and DNOC was recommended as an acaricide. Circular 623 in 1948 recommended 17 spray treatments and included chlordane and BHC plus two special "mite sprays." Circular 821 in 1960 recommended 29 pesticides including the organochlorines DDT, DDD, methoxychlor, BHC, dieldrin, and endrin; the organophosphates parathion, EPN, azinphos methyl, carbophenothion, and ethion; the carbamate carbaryl; and the acaricides chlorobenzilate and chlorbenside. For the first time, a subheading "insect and mite control by rotation," suggested a need for pesticide management. Circular 1151 in 1968 recommended 43 pesticides, including the selective acaricides chloropropylate, dicofol, ovex, propargite, tetradifon, orthothioquinox, and organotins; for the suppression of red spider mites with minimal effects on their predators.

Today, the usage of broad-spectrum pesticides has almost inevitably been followed by pest resistance, pest resurgence, and by outbreaks of secondary pests (Georghiou and Taylor, 1977; Reynolds, 1971). An inventory of the 25 most serious insect pest of agriculture in California, each judged to cause more than $1 million in damage in 1970, disclosed that 17 were resistant to one or more classes of insecticides and 24 were either secondary pest outbreaks, or pest resurgences aggravated by the use of insecticides (Luck et al., 1977). The widespread development of this "man-made" or entomologenic pest outbreak is one of the most serious indictments of our present-day pest control technology (Luck et al., 1977; Van den Bosch, 1978). Resurgences and outbreaks of secondary pests on the crops most heavily treated with insecticides, including cotton, rice, deciduous fruits, and citrus fruits, are summarized in Table 10.9. Invariably, as indicated in this table, the ecological imbalance has been related to the suppression of key natural enemies by the injudicious application of insecticides. It is also clear from Table 10.9, that these resurgences and outbreaks of secondary pests

**Table 10.9 Development of Resurgent and Secondary Pests after Insecticide Applications**

| Pest | Inducing Insecticide | Key Natural Enemy |
|---|---|---|
| **Cotton** | | |
| *Heliothis virescens*[a] | DDT, toxaphene, endrin, monocrotophos, | *Chrysopa* |
| *H. zea*[a] | dicrotophos, methyl parathion, azinphos | *Orius* |
| *H. armigera*[b] | methyl, dimethoate, carbaryl, aldicarb, | *Geocoris* |
| *Spodoptera littoralis*[c] | methomyl | |
| *S. exigua*[d] | | *Nabis* |
| *S. sunia*[e] | | |
| *Trichoplusia ni*[f] | | |
| *Bucculatrix thuberiella*[g] | DDT, OP, carbaryl | *Apanteles* |
| | | *Cirrospilus* |
| | | *Closterocerus* |
| *Bemisia tabaci*[h] | DDT, endrin, monocrotophos, fenvalerate, permethrin | *Encarsia* |
| | | *Eretmocerus* |
| **Rice** | | |
| *Chilo suppressalis*[i] | BHC, parathion | *Trichogramma* spp. |
| *Nilaparvata lugens*[j] | BHC, methyl parathion, diazinon, carbofuran | *Cyrtorhinus lividipennis* |
| *Sogatella furcifera* | pyrethroids | *Spiders* |
| *Nephotettix cincticeps*[k] | DDT, BHC, methyl parathion | *Lycosa pseudoannulata* |
| | | *Oedothorax insecticeps* |
| *Laodelphax striatellus* | | *Paracontrolia andoi* |
| **Deciduous Fruits** | | |
| *Panonychus ulmi*[l] | DDT, OP, pyrethroids | *Amblyseius fallacis, Zetzellia mali* |
| *Tetranychus urticae*[l] | DDT, OP, pyrethroids | *Typhlodromus occidentalis Stethorus punctum* |
| *Psylla pyricola*[m] | DDT | *Chrysopa, Orius* |
| *Eriosoma lanigerum*[n] | DDT, fenvalerate | *Aphelinus mali* |
| *Argyrotaenia velutinana*[o] | DDT | *Epirhysallis atriceps, Glypta vulgaris* |
| **Citrus Fruits** | | |
| *Icerya purchasi*[p] | DDT, malathion | *Rodolia cardinalis* |
| *Coccus hesperidum*[q] | Parathion | *Metaphycus luteolus* |
| *Aonidiella aurantii*[r] | Temephos, dioxathion | *Aphytis africanus* |

[a]Adkisson (1971); Reynolds et al. (1982); Van Steenwyk et al. (1976).
[b]Eveleens (1983).
[c]El Sebae (1977).
[d]Eveleens et al. (1973).
[e]Vaughan & Gladys (1976).
[f]Falcon et al. (1968).
[g]Reynolds (1971); Smith (1970).
[h]Eveleens (1983); Johnson et al. (1982).
[i]Kiritani (1979).
[j]Chelliah & Heinrichs (1980).
[k]Hirano & Kiritani (1976).
[l]Ripper (1956); McMurtry and Huffaker (1970); Hull and Starner (1981).
[m]Westigard et al. (1968).
[n]Penman and Chapman (1980).
[o]Hough (1927); Glass and Chapman (1948).
[p]DeBach and Bartlett (1951).
[q]Bartlett and Ewart (1951).
[r]Bedford (1971).

can be caused by applications of a wide variety of insecticides of all the common types.

The catastrophic ecological and economic consequences of the overdependence on broad spectrum insecticides can be judged from the following case histories.

### Case History   Cotton Bollworms

World-wide, cotton is the crop most heavily treated with insecticides and the ecological consequences have been dramatic. The primary pests for which insecticidal control was applied were the cotton boll weevil (*Anthonomus grandis* Boheman) and the pink bollworm [*Pectinophora gossypiella* (Saunders)]. The modern era of intensive insecticide treatment began with the introduction of aircraft dusting with calcium arsenate in 1923 (Reynolds et al., 1982). After 1946 the advent of DDT, BHC, and toxaphene and the subsequent introduction of the other organochlorine insecticides aldrin, dieldrin, endrin, and heptachlor (Fig. 10.2) provided cotton growers with highly effective controls for all of the arthropod pests of the cotton crop. Production levels were forced by high fertilizer applications, irrigation, and long-season varieties.

In the United States the cotton boll weevil developed resistance in Louisiana and Mississippi to most of the organochlorines in the mid-1950s, and the resistant strains spread rapidly to encompass all of the infested area by 1960. The organophosphorus insecticides methyl and ethyl parathion, azinphos methyl, and EPN were substituted for the organochlorines but these did not give adequate control of the bollworms *Heliothis zea* and *H. virescens*, which had become secondary pests because of destruction of their predators (Table 10.9). To control the bollworms, mixtures of the organochlorines and organophosphates were applied: DDT-toxaphene or DDT-endrin together with the parathions, azinphos methyl, or EPN. These mixtures sterilized the cotton fields of both pests and beneficial insects. By the early 1960s the bollworms developed high levels of resistance to organochlorine and carbamate insecticides and became more important pests than the boll weevil. Organophosphorus insecticides such as methyl parathion and monocrotophos were introduced at high dosages to control the bollworms. By the late 1960s the bollworms had developed resistance to the organophosphorus insecticides, and growers in the lower Rio Grande Valley were spraying their cotton fields 15–18 times a season with methyl parathion and still suffering high losses. Adkisson (1971) has recorded the destruction of a flourishing cotton industry in northern Mexico by an outbreak of the tobacco budworm. The excessive costs of insecticide applications and the economic damage to the crop caused a decline in the cotton acreage from 288,000 ha in 1960 to ~490 nonproductive ha in 1970.

Similar experiences with *Heliothis* bollworms occurred in most cotton-growing areas of the world—Peru, Australia, and Central America. In Nicaragua, cotton production flourished during the 1950s and reached a peak by 1965, but during the next 5 years decreased at an annual rate of 15.9% due to the development of explosive populations of *H. zea* and *Spodoptera sunia*. The conventional remedy of higher dosages and more frequent applications occurred and treatments with insecticides averaged 25–30 per season and in single fields reached 48–50, sometimes with five different insecticides in a single application (Vaughan and Gladys, 1976). The costs of insect control reached 26–40% of total production costs and other secondary pests such as the beet armyworm [*Spodoptera exigua* (Hubner)] and the cabbage looper [*Trichoplusia ni* (Hubner)] caused severe damage. Among the unfortunate environmental consequences were thousands of cases of human poisoning from the heavy and frequent applications of organophosphorus insecticides, the highest recorded residues of DDT in human milk, and a resurgence in malaria transmission because of resistance developed by drift of cotton insecticides in the vector *Anopheles albimanus* (Vaughan and Gladys, 1976).

In Egypt, the cotton leafworm [*Spodoptera littoralis* (F.)] attained epidemic population levels in 1961, after 6 years of toxaphane applications to control the pink bollworm and the spiny bollworm (*Erias insulana* Boisd.). Between 1961 and 1975 more than 368,000,000 kg AI of toxaphene, carbaryl, trichlorfon, endrin, DDT, lindane, methyl parathion, fenitrothion, monocrotophos, dicrotophos, phospholan, mephospholan, leptophos, chlorpyrifos, stirifos, methamidophos, azinphos methyl, methomyl, and diflubenzuron were applied to the Egyptian cotton crop in attempts to control the cotton leafworm. This pest now exhibits multiple resistance to virtually every insecticide, including the synthetic pyrethroids (El-Guindy et al., 1982), and no new insecticide has remained effective for more than 2–4 years nor have resistance levels shown any signs of reversion over more than 20 years (El-Sebae, 1977).

### Case History    Cotton Whitefly

The cotton whitefly [*Bemisia tabaci* (Gennadius)] has been recorded as an occasional secondary pest of cotton in the Gezira area of Sudan since the development of large-scale irrigation schemes after 1925. Typically before insecticide spraying, it was an early or midseason pest, appearing in large numbers in October, declining precipitously in November, and remaining at low levels until March (Fig. 10.6). This pattern remained constant for the ensuing 25–30 years. It was the emergence of the cotton jassid *Empoasca lybica* (de Berg) as an important cotton pest that began the bizarre ecological perturbations that ultimately resulted in what Eveleens (1983) has described as a crisis in cotton insect control. During the exploitative phase of cotton culture in the Gezira during the 1950s, heavy use of fertilizer and frequent spraying with DDT and other broad-spec-

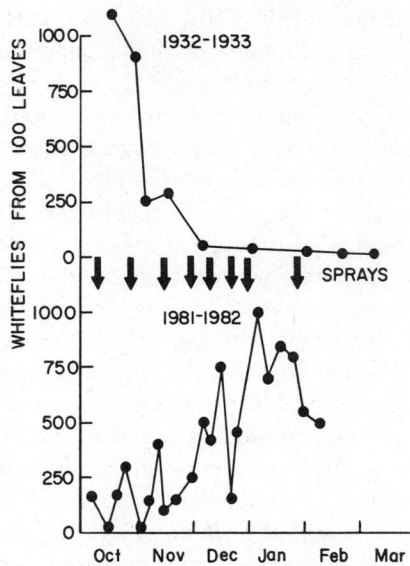

Fig. 10.6 Change in the seasonal life history of the cotton whitefly *Bemisia tabaci* in the Sudan, from pattern before spraying 1932–1933 to that caused by repeated applications of DDT and monocrotophos, 1981–1982 (after Eveleens, 1983).

trum persistent insecticides transformed the cotton bollworm *Heliothis armigera* into a major pest during the 1960s. In attempts to control the bollworm, the initial single-season spray of DDT was increased to as many as eight seasonal sprays with DDT, dimethoate, quinalphos, and endosulfan. The cotton whitefly is very difficult to kill with conventional insecticides because the nymphs congregate on the lower sides of the leaves and are protected by a dense waxy integument. However, the intensive spray application almost obliterated the density-dependent regulating Aphelinidae parasites *Encarsia formosa* Gahan and *Eretmocerus* spp. Parasitism of the white fly by these efficient natural enemies may approach 100% in the late cotton season in the Sudan (Joyce, 1955). As a result of the destruction of these efficient regulating factors by excessive applications of insecticides, the population of *B. tabaci* has assumed a completely different seasonal pattern in the Gezira (Fig. 10.6) and the resurgent population has seemed almost literally to explode. In the Sudan, the use of DDT was prohibited in 1981 and area-wide ULV spraying with monocrotophos was instituted to control *H. armigera* (Eveleens, 1983). By the 1980–1981 season, much of the irrigated cotton suffered from almost uncontrollable whitefly outbreaks, cotton production declined from about an average of 1653 kg seed cotton per ha during 1970–

1975 to 1020 kg/ha for 1975–1981: this despite an increase in spraying costs of over 600% from 1972–1973 to 1980–1981 (Eveleens, 1983). This astounding resurgence of the cotton white fly has brought about expenditures for insect control that were greater than the average net profit from cotton culture during recent years and has contributed to a crisis in the future development of the Gezira scheme.

The experience with the cotton whitefly in the irrigated Imperial Valley of California is very similar (Johnson et al., 1982). In cotton culture in this area, repeated applications of organochlorine, carbamate, and organophosphorus insecticides directed at the control of the pink bollworm, which invaded from Arizona in 1966, produced resistant and resurgent populations of the tobacco budworm. The overenthusiastic introduction of the synthetic pyrethroids permethrin and fenvalerate in 1975 to control the bollworm and the pink bollworm had little effect on the immature whiteflies, although their important Aphelinidae parasites *E. formosa* and *Eretmocerus haldemani* (How.) were almost destroyed. As a result the resurgence of *B. tabaci* reached catastrophic levels, as shown in Fig. 10.7. Insect pest control costs are reported to have been as much as $865/ha during 1981.

The whitefly is a vector of the virus diseases cotton leaf crumple, infectious yellows, and cucurbit leaf curl. These are readily spread to lettuce and melon crops, reducing production by as much as 50%. Total damages to growers and consumers from the cotton whitefly resurgence in the Imperial Valley in 1981 are stated to have approached $100 million (Duffus and Flock, 1982).

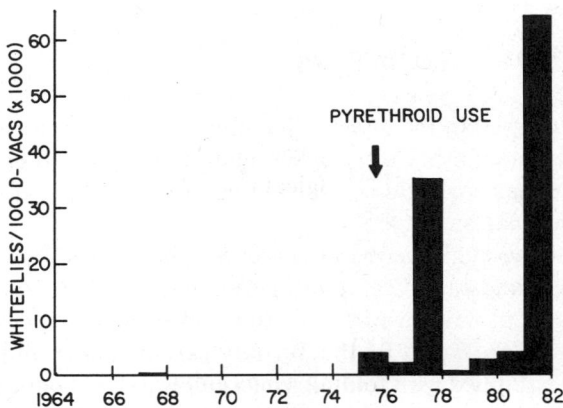

Fig. 10.7 Resurgence of the cotton whitefly *Bemisia tabaci* on cotton in the Imperial Valley, California, produced by applications of fenvalerate and permethrin. [After Johnson et al. (1982).]

## 10.5.  ENHANCED MICROBIAL DEGRADATION OF INSECTICIDES IN SOILS

The deliberate application of insecticides to soils at planting time is widely used to control subterranean insect pests such as corn rootworm larvae *Diabrotica* spp., wireworms, white grubs, and cutworms. Such applications of carbamate and organophosphorus insecticides were made to approximately 14 million hectares of U.S. cornland in 1976 (Eichers et al., 1978). Recently, after repeated applications of these carbamates and organophosphates, the residual action against soil insects has been inadequate and treated corn fields have experienced damaged roots, extensive lodging, and yield reductions (Felsot et al., 1982). Laboratory studies with [$^{14}$C]carbofuran, for example, have shown that "problem soils" evolved $^{14}CO_2$ as much as 10 times more rapidly than nonproblem soils (Kaufman and Edwards 1983). Enhanced degradation induced by carbofuran treatment extended to a variety of other *N*-methylcarbamate insecticides, including aldicarb, carbaryl, and propoxur (Harris et al., 1984). This enhanced degradation of soil insecticides by "problem soils" is drastically reduced by heat treatment, freezing, gamma-irradiation, or addition of antibiotics to the soil indicating an active microbial population as the cause (Kaufman and Edwards, 1983; Harris et al., 1984). All these factors indicate that even a single soil application of an insecticide may induce changes in soil microorganisms so that they utilize the insecticide as a carbon and nitrogen source, thus destroying its residual effectiveness. The phenomenon is obviously one of "natural selection" and has been shown to extend to organophosphorus insecticides such as isofenphos and to a variety of soil applied herbicides (Kaufman and Edwards, 1983). This phenomenon has adverse implications for the future use of soil insecticides in crop protection.

## 10.6.  EFFECTS ON POLLINATORS

The mutalism involved in the insect pollination of many species of plants is one of the most fascinating aspects of the coevolution of these two groups of organisms. Pollination has a central ecological role in maintaining the diversity and biomass of many ecosystems.

The origin of the angiosperms or flowering plants must have occurred between the Jurassic and early Cretaceous periods, 130 million year B.P. and these plants sprang into prominence in the latter half of the Cretaceous and dominated the floras of the world by the Tertiary period (Baker and Hurd, 1968). Insects suitably equipped with folding wings and suitable mouthparts to exploit the pollen and nectar of the angiosperms began to appear about the end of the Permian 225 million year B.P. It is thought that anthophily originated with the Coleoptera, progressed through the Lepidoptera and Diptera, and culminated in the Hymenoptera during the Tertiary. Today there are some 250,000 species of angiosperms and more than 20,000 species of bees alone are involved in the synomomy.

Cultivated crop plants that are dependent upon or benefit from insect pollination comprise about 15% of the U.S. dietary and have a value, estimated in 1971, of about $11,500,000,000 (USDA, 1976). The major cultivars that are almost exclusively insect pollinated include alfalfa, cotton, peanut, sugarbeet, citrus and deciduous fruits, and virtually all vegetables.

## 10.6.1. Effects of Insecticides

Pollinating insects, especially wild and domestic bees, are very sensitive to the toxic actions of insecticides. Foraging honey bees from a single colony may visit flowers over several square kilometers and may travel 5.6 to 6.4 km from the hive (Levin et al., 1968). Thus, they have an extensive opportunity to contact the ubiquitous insecticides heavily applied to crops such as cotton, alfalfa, fruits, and vegetables. Damage to bee colonies can also result from insecticides applied to rangelands and forests, and even to their use in suburbia. The damage to a colony that results is influenced not only by the inherent toxicity of the insecticide to the bee but also by its rapidity of action, its formulation, its residual properties, the number of applications, the time of day, and the weather. Toxic amounts of almost any insecticide can be carried by worker bees back to the hive to cause serious damage to the queen and her offspring. Such insecticide residues are stored in pollen and comb wax where they persist for 7 to 14 months (Erickson and Erickson, 1983).

Bee poisoning by insecticides originated with the use of calcium and lead arsenates applied to tree fruits during the early 1900s. As a result, legislation was enacted to prohibit spraying of fruit trees while in bloom. The arsenical problem was exacerbated by the introduction of aircraft spraying of calcium arsenate for cotton insect control in 1923 and the increased use of this insecticide from production of 1,096,000 kg in 1921 to 38,114,000 kg in 1942. The replacement of the arsenicals by DDT after 1946 partially alleviated the bee poisoning problem as DDT with an $LD_{50}$ of 114 $\mu$g/g was substantially less toxic. However, the succeeding cyclodiene, organophosphorus, and carbamate insecticides were much more toxic to bees, and bee poisoning reached disastrous levels when the use of the organochlorine insecticides were restricted during the 1970s and they were replaced in most agricultural uses by the organophosphates and carbamates. The most damaging insecticides to bees today are parathion, methyl parathion, carbaryl, and carbofuran. The annual total loss to bees and bee products in the United States is estimated at $135 million annually and averaged 6.2% of the total value of bees and bee products from 1967 to 1978 (Erickson and Erickson, 1983).

Eckert (1944) estimated that 1500 colonies of honey bees were poisoned in the Imperial Valley, California, by the drift of arsenical dusts from tomatoes and corn. Carbaryl widely applied in the same area to combat the pink bollworm in 1966, is alleged to have killed as many as 30,000 colonies of bees. Levin (1970) estimated that 76,000 colonies were killed by insecticides in California in 1967 and 83,000 colonies were killed in 1968. More than 300,000 colonies in the United States were damaged by insecticides in 1977–1978, and the problem of bee damage from insecticides is no nearer solution now than it was 40 years ago (Erickson and Erickson, 1983).

The acute toxicities of a variety of insecticides to *Apis mellifera* L. workers have been classified by Anderson and Atkins (1968) as:

**1.** Highly toxic, $LD_{50}$ 0.001–1.99 $\mu$g per bee: aldicarb, aldrin, arsenicals, azinphos methyl, carbaryl, chloropyrifos, diazinon, dicrotophos, dieldrin, dimethoate, EPN, heptachlor, lindane, malathion, methyl parathion, mevinphos, monocrotophos, naled, parathion, phosphamidon, and tetraethyl pyrophosphate. The synthetic pyrethoids, cypermethrin, deltamethrin, fenvalerate, flucythrinate, and permethrin should be included in this category.

**2.** Moderately toxic, $LD_{50}$ 2.0–10.99 $\mu$g per bee: carbophenothion, chlordane, DDT, demeton, demeton methyl, disulfoton, endosulfan, endrin, phorate, and temephos.

**3.** Relatively nontoxic, $LD_{50}$ > 11.00 $\mu$g per bee: binapacryl, BT, chlorobenzilate, cryolite, dicofol, dioxathion, ethion, menazon, methoxychlor, ovex, propargite, rotenone, ryania, toxaphene, and trichlorfon.

## 10.7. CONTAMINATION OF FOOD WEBS

The imperative need to base the use of insecticides upon sound ecological principles is most clearly illustrated by our present knowledge of the effects of massive use of insecticides in United States agriculture upon the total environment. The general inefficiency of insecticide application is discussed in Section 10.2.2. — much less than 0.1% of the treatment directly affects the target pest and more than 99.9% enters the environment in a variety of ways as suggested in Figure 10.3. In 1976 about 73,500,000 kg AI of insecticides were applied to 30,335,000 ha of U.S. cropland for an average contamination of about 2.4 kg/ha (Eichers et al., 1978). For cotton, the most heavily treated crop, 29,500,000 kg AI were applied to 2,835,000 ha, an average of 10.3 kg/ha.

The insecticides that have produced the greatest impact are the "uncontrollable" organochlorines and their effects in causing perturbations throughout the various trophic levels leading to humans, their domestic animals, and their cherished wildlife have been major factors in stringent Federal regulation of the use of these insecticides in the United States: DDT (1973), aldrin and dieldrin (1974), heptachlor and chlordane (1976), chlordecone (1980), endrin (1981), mirex (1982), and toxaphene (1983).

The extent to which the nontarget portions of insecticide applications permeate the various ecosystems of North America is a function of the method of application, that is, air or ground spraying, soil treatment, and so forth, and the

physico-chemical properties of the individual insecticides, that is, water solubility, lipid–water partitioning, and stability to hydrolysis and oxidation in both the
living and nonliving components of the environment. Increasing knowledge
about these ecosystem effects has substantially enhanced the understanding of
food chains and food web interactions in the total biosphere and the insecticides
themselves have become useful probes for ecological investigations (Southwood,
1977). As a result, new terms and concepts have entered ecological and environmental lexicons, including micropollutant, xenobiotic, biomagnification, and
biodegradability.

### 10.7.1. Quantitative Aspects of Bioconcentration of Insecticides

The important physico-chemical properties of insecticides that are responsible
for major ecological impacts upon food webs are lack of reactive centers or de-
gradophores that promote biodegradability, and relatively low water insolubility
and high solubility in lipids that promote absorption and storage in the lipid
tissues of living organisms. Both of these properties are maximized in the organochlorine insecticides that typically have a multiplicity of C–Cl bonds that
are environmentally and biologically stable, and are of very low water solubility
and high lipid solubility. The propensity of insecticides for bioconcentration and
magnification through food chains can be estimated by their water insolubility
(which is inversely proportionate to their lipid solubility) and by their octanol/
$H_2O$ partition coefficient (Metcalf, 1977b). The effects of these properties upon
bioconcentration of various insecticides in the fish *Gambusia affinis* (Baird and
Girard) in standardized laboratory model ecosystems (Metcalf, 1977b) are
shown in Fig. 10.8.

#### 10.7.1.1. Bioconcentration Directly from Water.

Absorption of lipid-soluble xenobiotics by aquatic organisms can occur directly through integument or gills and is generally a first-order process, that is,
the amount penetrating per unit time is a function of the amount absorbed at the
lipid interface. The absorption process can be expressed as a form of Fick's law
(Buerger, 1966)

$$\frac{dS}{dt} = -DA \frac{dc}{dx} \tag{1}$$

where $D$ is the diffusion coefficient, $A$ the cross-sectional area of surface, $x$ the
surface thickness (constant), $c$ the concentration of insecticide, and $S$ the total
amount of insecticide present at the surface. Buerger (1966) has extended this
process to the absorption of insecticides into aquatic organisms through a series
of biological barriers and concluded that the penetration of the integument will
follow first-order kinetics.

The absorption and metabolism of xenobiotic insecticides by aquatic organisms living in a large pool of constant contamination represents an example of

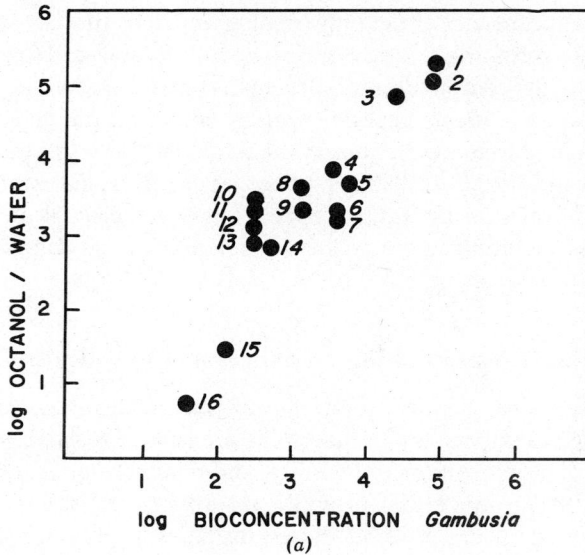

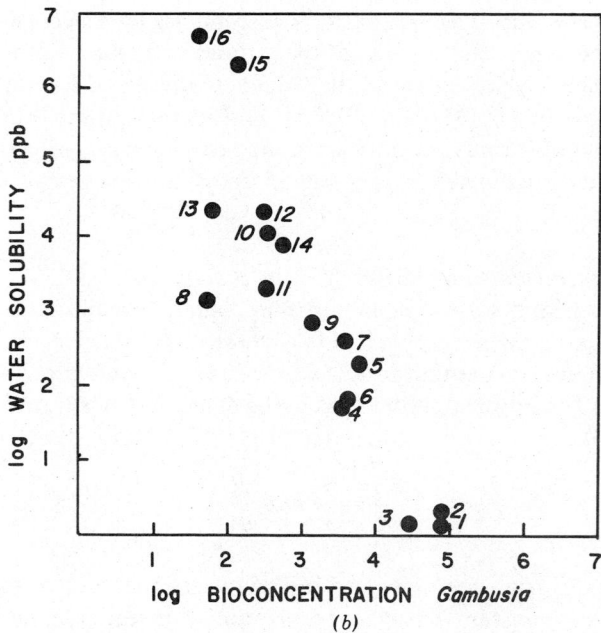

Fig. 10.8 (a) Relationship of octanol–water partition and (b) water solubility to bioaccumulation of insecticides in the mosquito fish *Gambusia affinis*. Data from laboratory model ecosystem studies carried out with radiolabeled insecticides under standardized conditions (Metcalf, 1977). Key: (1) DDT, (2) DDD, (3) DDE, (4) heptachlor, (5) dieldrin, (6) chlordane, (7) toxaphene, (8) leptophos, (9) methoxychlor, (10) EPN, (11) chlorpyrifos, (12) parathion, (13) isofenphos, (14) lindane, (15) propoxur, (16) aldicarb.

280

compartmental analysis of the pharmacokinetics of drug absorption and metabolism (Goldstein et al., 1969; Kerr and Vass, 1973). The aquatic organism is assumed to be represented by a single homogenous compartment of constant size $V$ (organism mass or volume), and the insecticide to be absorbed at a constant zero-order rate $k_a$, determined by the lipid–water partition coefficient, from an aquatic environment contaminated at a constant level $C_e$. The concentration in the organisms $C_o$ can be represented as

$$\frac{dC_o}{dt} = \frac{C_e k_a}{V} - k_c C_o \tag{2}$$

Here $k_c$ is the rate constant for clearance of the insecticide by the combination of the processes of degradation, elimination, and "growth dilution." This occurs at a rate proportional to the internal concentration $C_0$ and follows first-order kinetics.

Integration of Eq. (2) gives the concentration in the organism at any time $t$ as

$$C_o = \frac{C_e k_a}{k_c V} (1 - e^{-k}c^t) \tag{3}$$

If the rate of absorption of the insecticide into the living organism is constant and the rate of output is exponential [Eq. (3)], then the concentration in the organisms $C_o$ will increase until a steady state is attained where clearance is equal to absorption. Thus, as $t$ becomes infinite $1 - e^{-k}c^t$ becomes unity, and a "steady state" is reached where

$$C_o = \frac{C_e k_a}{k_c V} \tag{4}$$

This is called the plateau principle (Goldstein et al., 1969).

However, in poikilothermic aquatic organisms such as fish in abyssal lakes, $k_c$ determined by enzymatic processes influenced by environmental temperatures becomes very small in comparison with the rate of absorption $k_a$, which is largely independent of temperature except as a function of respiratory rate. Therefore the length of time required to reach a plateau or stable body residue level will often exceed the longevity of the organism. Thus, with the stable organochlorine insecticides where lipid–water partitioning is very high and degradation and elimination are very slow as in the lake trout [*Salvelinis namaycush* (Walbaum)] in Lake Michigan, the bioconcentration of DDT and dieldrin is a linear function with the age of the organism (Table 10.10) and mature fish can bioconcentrate these insecticides from $10^5$- to $10^6$-fold from parts per trillion (ppt) residues in water (Table 10.11).

**Table 10.10 Bioconcentration of DDT and Dieldrin
in Lake Trout *Salvelinis namaycush* in
Lake Michigan (EPA Data 1972)[a]**

| Lake Trout Size (cm) | Whole Body Residues (ppm) | |
| --- | --- | --- |
| | DDT-T | Dieldrin |
| 5–15 | 0.89 | 0.03 |
| 15.2–25.1 | 2.24 | 0.12 |
| 25.4–40.4 | 6.00 | 0.14 |
| 40.6–55.6 | 8.00 | 0.21 |
| 55.9–68.33 | 14.62 | 0.23 |
| 68.6–83.6 | 19.23 | 0.26 |

[a]After Metcalf (1977).

**Table 10.11 Bioconcentration of DDT and Dieldrin in
Lake Trout *Salvelinus namaycush* of
Lake Michigan (EPA Data 1972)[a]**

| Insecticide | Average Concentration (ppm) | | Bioconcentration Factor |
| --- | --- | --- | --- |
| | Open Lake | Lake Trout | |
| DDT-(T) | 0.0000060 | 18.80 | $3.13 \times 10^6$ |
| Dieldrin | 0.0000020 | 0.26 | $1.30 \times 10^6$ |

[a]After Metcalf (1977).

### 10.7.1.2. *Bioconcentration from Food.*

The rate of xenobiotic accumulation from ingestion of contaminated food or prey can be treated as a modification of Eq. (1), where the factor for accumulation of the xenobiotic is the daily intake or concentration in prey $C_p$ times the weight of prey, $W_p$ times a suitable factor $c$ to measure efficiency of extraction of the insecticide from the food. Thus,

$$\frac{d\,C_o}{dt} = \frac{c(C_p\,W_p)}{V} - k_c C_o \qquad (5)$$

Reinbold et al. (1971) contaminated *Daphnia magna* (Strauss) with DDT by placing them in water containing 0.003 ppm for 48 hr until they contained 25–26 ppm DDT (dry weight). These daphnia were fed to the guppy *Lebistes reticulatus* (Peters) in clean water. As shown in Fig. 10.9, the rate of storage of DDT in the guppy was nearly linear with time over a 20-day interval, but methoxychlor,

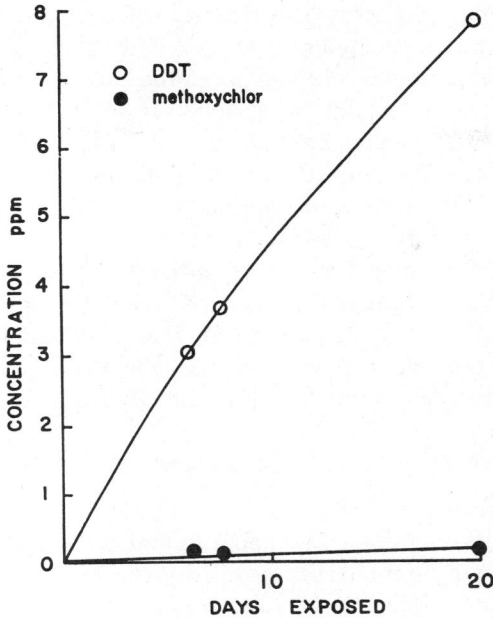

Fig. 10.9 Bioconcentration of [$^{14}$C] DDT and [$^{3}$H] methoxychlor by the guppy *Lebistes reticulatus* fed on *Daphnia magna* containing 25 ppm DDT and 22 ppm methoxychlor [Reprinted with permission from Reinbold et al. (1971) *Bull. Ill. Nat. Hist. Survey.*]

with $CH_3O$-degradophores, was readily cleared by the guppy and did not bioconcentrate appreciably when fed at approximately the same rate, 21–22 ppm, in daphnia.

The following examples from natural food webs demonstrate the hazardous nature of lipid soluble micropollutant insecticides to a variety of living animals.

### Case History    DDD in Clear Lake

The insecticide DDD, a very effective larvicide, was used in 1949 in Clear Lake, California at 0.014–0.02 ppm to control the larvae of the Clear Lake gnat *Chaoborus astictopus* (Dyar and Shannon). This treatment of the entire lake achieved 99% control of the pestiferous gnat without producing an obvious effect on the fish population of the lake. The treatment was effective for about 5 years, when it was repeated at 0.020 ppm in 1954. The second application achieved 99% control for 3 years, but a third application at 0.020 ppm in 1957 was less effective, apparently due to resistance in the *Chaoborus* larvae. Clear Lake was a breeding ground for the western grebe *Aechmophorus occidentalis* (Lawr.) that visited the lake to feed on the fish. After the first application of DDD, there was no further breeding of the birds until 1962 when a single chick hatched and 1963 when three chicks hatched. After the second application of DDD in

1954, sick and dying birds were observed whose adipose tissues contained DDD(T) residues averaging 1600 ppm and as high as 2133 ppm. The insecticide had permeated the biota of the entire lake and was found in phytoplankton up to 5.3 ppm, whereas the visceral fat of several species of fish contained much higher amounts: the blue-gill [*Lepomis pallidus* (Mitchill)], 125–254 ppm, the brown bullhead [*Ameirus nebulosus* (Le Suer)], 342–2700, and the predaceous largemouth bass [*Micropterus salmoides* (Lacépeden)], 1550–1700 ppm.

The overall biomagnification from the 0.02 ppm in the water of the lake to the visceral fat of the largemouth bass and grebes was approximately 80,000-fold. It was concluded that the DDD had accumulated through the food chain of water < plankton < herbivorous fish < predaceous fish < fish-eating bird (Cottam, 1965; Hunt, 1966).

### Case History    DDT Spraying in Orchards, Door County, Wisconsin

In the years between 1950 and 1965, DDT was applied annually at about 30 tons and its relative DDD at about 15 tons to deciduous fruit orchards. Run-off from the orchards caused micropollution of the Green Bay, Lake Michigan, and residues of DDT, its principal degradation product DDE, and of DDD were accumulated by biota in the bay; they reached a maximum in the fish-eating herring gull (*Larus argentatus* Pont.) at the top of the food chain, as shown in Table 10.12 (Hickey et al., 1966). The overall biomagnification of DDT from bottom sediments of Green Bay to the fat of the herring gull was about 170,000 and of DDE was about 1,280,000-fold. As shown in Table 10.12 DDT and DDE residues increased about

**Table 10.12  Biomagnification of DDT and Its Metabolites DDD and DDE in the Food Chain of the Herring Gull, Green Bay, Wisconsin**[a]

|  | Residues in ppm | | |
|---|---|---|---|
|  | DDT | DDD | DDE |
| Bottom mud | 0.0052 | 0.0012 | 0.0015 |
| Shrimp, *Pontoporeia affinis* | 0.14 | 0.03 | 0.24 |
| Chub, *Leucichthys* | 1.83 | 0.39 | 2.3 |
| Whitefish, *Corregonus clupeaformis* | | | |
| Muscle | 1.8 | 0.76 | 3.0 |
| Long-tailed duck, *Clangula hyemalis* | | | |
| Muscle | 0.8 | 0.68 | 4.8 |
| Fat | 34 | 9.0 | 94 |
| Herring gull, *Larus argentatus* | | | |
| Muscle | 14.1 | 4.8 | 79.9 |
| Fat | 390 | 126 | 1925 |

[a]Hickey et al. (1966).

10-fold over each trophic level. Moreover, the ratio of DDE to DDT increased progressively with trophic level as DDT was progressively degraded to DDE.

## 10.8. GENERAL ECOTOXICITY

Specific insecticides that are highly persistent and bioaccumulative have been found not only to permeate virtually every niche in specific agroecosystems to which they are applied but also through their persistent presence as trace contaminants of water, air, soil, and food to become detectable throughout the total human environment. The production and extent of use of a number of the organochlorine insecticides over a 30-year period between 1946 and 1976 has been estimated as (AI): DDT, $>2.3 \times 10^9$ kg; toxaphene, $4.5 \times 10^8$ kg; cyclodienes including aldrin, dieldrin heptachlor, chlordane, and endrin, $>2.7 \times 10^8$ kg (Metcalf, 1976). These insecticides have entered the environment in a variety of ways as depicted in Fig. 10.3 and because of their persistence and high lipid–water partitioning (Fig. 10.8) now are ubiquitous micropollutants of all living systems. This is demonstrated by the extent of contamination of the living and nonliving environment of the United States during the period of maximum use of the organochlorines 1960–1970 (Table 10.13). Legal restrictions of the use of these organochlorines during the 1970s have resulted in slow diminution of the total residues, but these will be detectable in the various substrates for decades to come. The extent to which such contamination poses a severe threat to human health and to other species of the living world remains debatable (HEW, 1969). However, the ubiquitous presence of detectable amounts of these organochlorines in human tissue cannot be considered benign. All of the organo-

**Table 10.13 Average U.S. Levels of Contamination by Organochlorine Insecticides**

|  | Cropland Soil (ppm)[a] | Water (ppb)[b] | Human Fat (ppm)[c] | Human Milk (ppm)[d] | Food Daily Intake (mg)[e] |
|---|---|---|---|---|---|
| DDT | 0.17 | 0.008–0.144 | 2.3 | 0.022 | 0.037 |
| DDD |  | 0.004–0.080 |  | 0.0099 | 0.016 |
| DDE | 0.06 | 0.002–0.011 | 8.0 | 0.041 | 0.024 |
| Aldrin | 0.02 | 0.001–0.006 |  |  | 0.002 |
| Dieldrin | 0.03 | 0.008–0.122 | 0.29 | 0.0073 | 0.006 |
| Endrin | 0.01 | 0.008–0.214 | 0.03 |  | Trace |
| Heptachlor epoxide | 0.01 | 0.001–0.008 | 0.1–0.24 | 0.0027 | 0.003 |
| Toxaphene | 0.07 |  |  |  |  |
| BHC | 0.01 | 0.003–0.022 | 0.60 | 0.0024 | 0.004 |

[a]Wiersma et al. (1972).          [d]Curley and Kimbrough (1969).
[b]Breidenback et al. (1967).      [e]Duggan (1969).
[c]Durham (1969).

chlorines listed in Table 10.13 are suspect or demonstrated carcinogens in laboratory animals (Metcalf, 1976). There can be no doubt that bioaccumulation in specific food webs can pose a threat to the continued existence of species at the top of food chains as shown in Section 10.7 with the western grebe in Clear Lake or the herring gull in Green Bay (Table 10.12). The following case histories demonstrate the approach to ecocatastrophy that can be caused by particularly injudicious use of insecticides. Many others could be cited.

### Case History    Dieldrin in Illinois

Dieldrin was applied in 1957 to 7,226.8 ha of Illinois farm land at 2.2–3.4 kg/ha to "eradicate" the Japanese beetle (*Popillia japonica* Newman). "Everything was treated including farms, houses, streams, and roads" (Luckmann and Decker, 1960). Devastating effects were observed on wildlife and domestic animals throughout the treated area. Ground squirrels, muskrats, and cotton tails were almost annihilated and short-tailed shrews, fox squirrels, woodchucks, and meadow mice suffered heavy losses. Meadowlarks, robins, brown thrashers, starlings, grackles, and ringnecked pheasants were virtually eliminated from the treated area. A total of 12 species of mammals were killed, with tissue residues of dieldrin ranging up to 18 ppm, and 19 species of birds (residues up to 36 ppm) also died. In the town of Sheldon, it was estimated that 1840 birds and 440 mammals were killed by a single application of dieldrin (Scott et al., 1959).

The effects on domestic animals were equally severe. Beef animals were poisoned and 90% of the cats and some dogs on treated farms were killed. Six shorthorn heifers confined to a supposedly untreated pasture became ill with milk containing 1.14 ppm dieldrin. On another farm, ewes whose milk contained 2.2 and 8.7 ppm dieldrin died together with their two sets of twin lambs. Autopsy of the ewe with 2.2 ppm dieldrin milk residue showed the following tissue residues of dieldrin: flesh, 8.9; liver, 12; heart, 24; omental fat, 99; and renal fat, 130 ppm (Luckmann and Decker, 1960).

The uncontrollable nature of organochlorine insecticide contamination is shown by the effects of aldrin applications to corn soil for control of the western corn rootworm. Residues of the oxidation product dieldrin average 0.11 ppm in Illinois and reached a maximum of 1.42 ppm (Wiersma et al., 1972). Soybeans planted in old corn soil accumulated dieldrin at 0.01–0.02 ppm (Petty and Bruce, 1972) and when processed for oil, left dieldrin-contaminated soybean cake that became an ingredient of chicken feed. Chickens absorb and store dieldrin in their body fat up to 40–50 times the rate of intake (Gannon et al., 1960) and in 1974, as many as 10 million broiler chickens in Mississippi were destroyed because of dieldrin tissue residue that exceeded the 0.3 ppm "safe level" by 15-fold or more (Washington Post, 1974).

### Case History    Heptachlor in Hawaii

Heptachlor was widely used on pineapple tops in Hawaii since 1958 to control ants that interfered with the biological control of the pineapple mealybug [*Dysmicoccus brevipes* (Cock.)]. The amounts of heptachlor used were probably less than 4535 kg (AI) over a 10-year period. However, the leafy pineapple tops were marketed as "green chop" for dairy cattle feed. Although heptachlor and heptachlor epoxide residues were detected in the green chop as early as 1970, nothing was done to monitor the contamination systematically, nor was this use of heptachlor regulated either by the EPA or the State of Hawaii, despite a federal EPA ban on the general uses of heptachlor in 1978. By 1980 it was discovered that residues of heptachlor epoxide, a carcinogen in laboratory animals, pervaded the entire Hawaiian ecosystem. Millions of gallons of milk were destroyed in 11 successive recalls because they contained heptachlor epoxide levels greater than the 0.3 ppm EPA action level (the EPA tolerance level for heptachlor and heptachlor epoxide in both milk and pineapple is zero). This level was later lowered by EPA to 0.1 ppm. Testing of human milk samples showed that 62% contained heptachlor epoxide residues above the EPA action level. Milk producers were awarded $28,400,000 from a law suit against the pineapple industry and the manufacturer of heptachlor (Honolulu Star Bulletin, 1985).

## 10.9.  AN ECOLOGICAL APPROACH TO INSECTICIDE USE

The methodology of insecticide use in agriculture during the past 40 years has been largely one of eradication. Applications were aimed at sterilizing the field or the orchard so that only the cultivars remained. The insecticides employed in largest amounts (Table 10.2) were broad-spectrum biocides and these were generally selected for maximum persistence in the environment. The timing of applications of these insecticides was characteristically based on a prophyllactic or insurance philosophy without consideration of the nature of the insect pest complex or the population dynamics of the infestations. The results of this eradication philosophy have been ecologically disruptive, as discussed above, and have often led in the short term to economic disaster and in the long term to irreparable environmental damage.

Integrated pest management (IPM) was developed as a rational methodology to rectify the grievous results of the overuse and misuse of insecticides. IPM has been described as the ecological approach to insect control. It provides for (1) determining how the life system of the pest needs to be modified to reduce its numbers to tolerable levels, that is, *below the economic injury level*; (2) applying biological knowledge and current technology to achieve the desired modification, that is, *applied ecology*; and (3) devising procedures for pest control compatible with economic and environmental quality aspects, that, is *economic and social acceptance* (Geier, 1966; Metcalf and Luckmann, 1982).

The use of insecticides can be an important part of IPM practice, but this use must be judicious and integrated with other biotic and abiotic factors that regulate the population dynamics of the target pest as well as those of the surrounding agroecosystem. IPM is based around the population dynamics of the insect pest and its cornerstone is the economic threshold. This is defined as the population density of the target pest at which control measures should be applied to prevent an increasing pest population from reaching the economic injury level (Stern et al., 1959). This concept clearly rejects the eradication philosophy of pest control and extends management options to include reliance on density-dependent factors such as natural enemies and diseases and upon density-independent factors such as host plant resistance, agronomic practices, and weather. Under the IPM philosophy, "doing nothing" is a valid alternative.

A synoptic model of the theoretical growth rates of insect pest populations in relation to their habitat stability, their food web relationships, and their role in ecological successions has been developed by Southwood (1977) (Fig. 10.10). This model is based on the actual population dynamics of some 30 insect pests occupying a continuum ranging from *r*-selected species or early colonizers including aphids, thrips, and spider mites to *K*-selected species or strong competitors such as the codling moth, and plum curculio (MacArthur and Wilson, 1967, Pianka 1970). The natural enemy ravine is a prominent feature of the synoptic model of insect population growth (Fig. 10.10) and its dimensions are indicative of the capacity of natural enemies to regulate pest population growth. The popu-

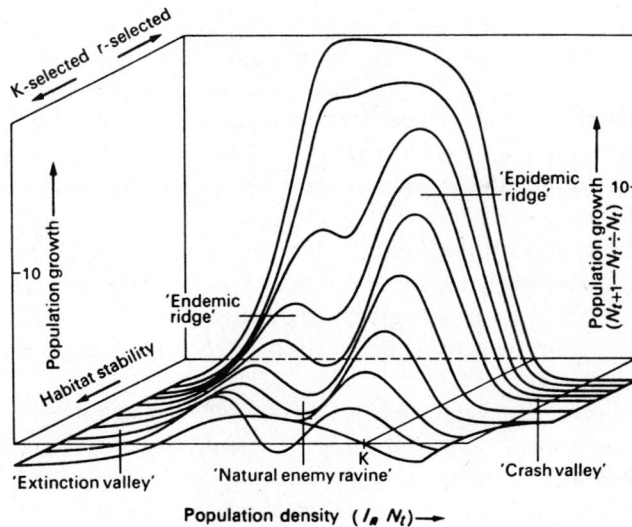

Fig. 10.10 Synoptic model of insect population growth (from Southwood, 1977). (Reprinted with permission from the *Journal of Animal Ecology*, **45**, 949–965. Copyright © by Blackwell Scientific Publications Limited.)

lation growth rates of strongly $r$-selected species in new habitats are so high that natural enemies seldom prevent them from exceeding the economic threshold. These $r$-selected species will almost always achieve pest status if they invade in numbers adequate to colonize the crop and if there is sufficient duration between invasion and harvest. At the other extreme of the continuum, strongly $K$-selected species in stable, well-established habitats invest much energy in defense rendering natural enemies less effective. The natural enemy ravine is highly developed in the intermediate portion of the continuum, where most insect pest species are found. In this region the density-dependent regulating action of natural enemies stabilizes pest population below epidemic levels. Thus, the model accounts for the destabilizing action of injudicious insecticide applications in producing resurgences of insect pests in typical agroecosystems.

The Southwood model (Fig. 10.10) provides a theoretical basis for an ecologically oriented view of the role of insecticide applications in IPM (Southwood, 1977; Conway, 1981). The enormous problems of pest resurgences and development of secondary pests caused by typical prophyllactic applications of insecticides have demonstrated that natural enemy conservation and enhancement are the key factors in planning any IPM strategy for the control of intermediate pests. Realistic economic thresholds must be established and any use of insecticides must be highly selective in regard to the fundamental properties of the chemical, the method of application, and its timing. Where a continuum of pests and potential pests coexist in a crop agroecosystem, the use of insecticides against $r$ or $K$ pests must not impact adversely on the natural enemy complexes present. The early and massive onslaughts of $r$ pests represent the most difficult IPM situations. Biological control is generally ineffectual and the application of insecticides based on sound forcasting is the most effective tool. This use should be incorporated with polygenic-horizontal host plant resistance, and by cultural controls such as crop rotation, timing of planting, and harvesting.

For $K$-selected pests, the use of insecticides must be based on realistic economic thresholds and applications should be specifically aimed at the target pest. Selective insecticide applications should incorporate refugia by spot treatments and treatments of alternate strips or rows either in space or time. This insecticide use should be incorporated with strong oligogenic-vertical host plant resistance and cultural methods such as destruction of crop refuse, destruction of alternative hosts, and changes in agronomic practices. An evaluation of the effectiveness of the five major control technologies for pests of the $r,K$ continuum is presented in Table 10.14.

## 10.9.1. Selectivity of Insecticides

The problems of insect pest resurgences and the development of secondary pests could be ameliorated if insecticides were highly selective to target pests without appreciably damaging natural enemy populations. Selectivity can be achieved (1) physiologically by use of chemicals with specific activity against small groups of target pests, (2) ecologically by exploiting significant vulnerabilities in various

Table 10.14 Impact of Control Strategies on Insect Pest of the *r, K* Continuum[a]

| Method | *r* | Intermediate | *K* |
|---|---|---|---|
| Chemical | +++ | + | ++ |
| Biological | ? | +++ | ++ |
| Cultural | ++ | + | +++ |
| Reproductive | | + | +++ |
| Host resistance | ++ | ++ | ++ |

[a]From Southwood (1977).

life stages of the insect pest, (3) by improving application techniques so that the insecticide affects as little of the nontarget environment as possible, and (4) behaviorally to bring the pest to the insecticide by exploiting its instinctual reactions (Metcalf, 1982).

Physiological selectivity is best illustrated by the microbial or viral insecticides such as *Bacillus thuringiensis* toxin (BT) that has an almost specific effect as a stomach poison for lepidopterous larvae such as cabbageworms, bagworms, cankerworms, tent caterpillars, and gypsy moths. The nuclear polyhedrosis virus (NPV) of the cotton bollworm is a viral insecticide effective orally against only species of *Heliothis*. These microbial insecticides seem to be devoid of toxicity to natural enemies and other organisms.

Many studies of the comparative toxicity to beneficial insects have shown a wide spectrum of variation among the array of conventional insecticides. These have been extensively reviewed by Croft and Brown (1975). Predators, including 15 species of Coccinellidae, were ranked in five classes of decreasing toxicity both by number and position:

1. Parathion, methyl parathion, malathion, azinphos methyl, carbaryl.
2. Mevinphos, phosphamidon, diazinon, dimethoate, ethion.
3. Demeton, demeton methyl, carbophenothion, trichlorfon, thiometon.
4. Lindane, toxaphene, endrin, DDT, endosulfan.
5. Chlorobenzilate, schradan, binapacryl, tetradifon, dicofol.

Limited data for hemipteran, chrysopid, and acarine predators followed roughly similar rankings. Parasites, including 17 species of Hymenoptera and a Tachinidae, were ranked as follows:

1. Endrin, dieldrin, aldrin, lindane, DDT.
2. Parathion, methyl parathion, malathion, carbaryl, azinphos methyl, toxaphene, phosphamidon.
3. Carbophenothion, trichlorfon, demeton methyl, endosulfan, thiometon.

Insecticides of short persistence may be expected to be less hazardous than those of long persistence, and systemic insecticides generally have a greatly reduced hazard to both predators and parasites.

The newer synthetic pyrethroids present a considerable spectrum of toxic properties and residual actions and represent a new challenge to the development of selective toxicity. The selective toxicity to predators and parasites has been reviewed by Croft and Whalon (1982). Only permethrin, cypermethrin, de-camethrin, and fenvalerate have been widely evaluated and these showed vari-able toxicity and selectivity to representative species of most families of natural enemies. In general, the pyrethroids displayed low to moderate toxicity and fa-vorable selectivity to most hemipteran predators and unfavorable selectivity to phytoseiid mites in relation to their prey. The pyrethroids were more selective to natural enemies of cotton pests over their prey or host than for apple insects. However, see Fig. 10.7 for pyrethroid-induced resurgence of the cotton whitefly.

A remarkable degree of selectivity covering an 18-fold range has been demon-strated for the Mediterranean flour moth and its ichneumonid parasite *Venturia canescens* as shown in Table 10.15 (Elliott et al., 1983).

## 10.9.2. Resistant Natural Enemies

Parasites and predators of insect pests are generally slower to develop resistance to specific insecticides than the pests themselves, and this has been definitive in the problems of pest resurgences (Section 10.4). The key pests of apple, the cod-ling moth *Cydia pomonella* (L.), the apple maggot [*Rhagoletis pomonella* (Walsh)], the plum curculio [*Conotrachelus nenuphar* (Herbst)], and the red-banded leafroller [*Argyrotaenia velutinana* (Walker)], however, have remained susceptible to azinphos methyl for 20 years although the important mite preda-tors *Amblyseius fallacis* (Garman) and *Typhlodromus occidentalis* (Nesbitt) have become highly resistant to this and other organophosphorus insecticides. Thus, elements of stability have been introduced into apple IPM programs (Croft, 1982). These are based around the use of organophosphorus insecticides for the control of the key pests together with biological control of the red spider mites supplemented with selective acaricides such as tricytin and propargite (Croft, 1982). The resistant predatory mites have been introduced successfully into a number of apple IPM programs where resistance management of both pests and natural enemies is an important feature. There is optimism that this

**Table 10.15 Selective Toxicity of Pyrethroids**[a]

|  | LD$_{50}$ ($\mu$g/g) | | |
|  | *Ephestia kuhniella* | *Venturia canescens* | Ratio P/H |
| --- | --- | --- | --- |
| Kadethrin | 0.07 | 10 | 145 |
| Bioallethrin | 0.11 | 11 | 97 |
| Flucythrinate | 0.026 | 1.8 | 69 |
| Biopermethrin | 0.16 | 2.3 | 14 |
| Deltamethrin | 0.032 | 0.39 | 12 |
| Bioresmethrin | 0.25 | 2.4 | 9.8 |
| Fenvalerate | 0.23 | 1.9 | 8.1 |

[a]From Elliott et al. (1983).

philosophy of resistance management can be incorporated into other IPM programs for agricultural crops (Croft and Whalon, 1982).

## REFERENCES

Adkisson, P. L. (1971). "Objective uses of insecticides in agriculture," in J. E. Swift, Ed., *Agricultural Chemicals: Harmony or Discord for Food, People, Environment*, Proc. Symp. Univ. Calif. Div. Agr. Sciences, Sacramento, California, pp. 43–51.

Anderson, L. D., and E. L. Atkins. (1968). Pesticide usage in relation to bee keeping, *Annu. Rev. Entomol.*, **13**, 213–238.

Attia, F. I., and J. T. Hamilton. (1978). Insecticide resistance in *Myzus persicae* in Australia, *J. Econ. Entomol.*, **71**, 851–853.

Baker, H. G., and P. D. Hurd, Jr. (1968). Intrafloral ecology, *Annu. Rev. Entomol.*, **13**, 385–414.

Bartlett, B. R., and W. H. Ewart. (1951). Effect of parathion on parasites of *Coccus hesperidum*, *J. Econ. Entomol.*, **44**, 344–347.

Bedford, E. C. G. (1971). The effect of Abate and Delnav on populations of red scale *Aonidiella aurantii* Mask in citrus orchards under integrated control, *J. Entomol. Soc. Southern Africa*, **34**, 159–178.

Braunholtz, J. T. (1976). Pesticide development and the chemical manufacture, *Proc. XV International Congress, Entomology*, Washington, D.C., pp. 747–755.

Breidenbach, A. W., C. G. Gunnerson, F. K. Kawahara, J. S. Lichtenberg, and K. S. Green. (1967). Chlorinated hydrocarbon pesticides in major river basins, 1957–65, *Publ. Health Rep.*, **82**, 139–156.

Brown, A. W. A. (1971). "Pest resistance to pesticides," in R. White-Stevens, Ed., *Pesticides in the Environment*, Vol. I, pt. 2, Dekker, New York, pp. 458–552.

Brown, A. W. A. (1977). Resistance as a factor in pesticide management, *Proc. XV Int. Congr. Entomol.*, Washington, D.C., pp. 816–824.

Brown, A. W. A. (1978). *Ecology of Pesticides*, Wiley, New York, 525 pp.

Brown, A. W. A., and R. Pal. (1971). *Insecticide Resistance in Arthropods*, WHO Monograph Series No. 38, Geneva, Switzerland, 491 pp.

Buerger, A. A. (1966). A model for the penetration of integument by nonelectrolytes. *J. Theoret. Biol.*, **11**, 131–139.

Chelliah, S., and E. A. Heinrichs. (1980). Factors affecting insecticide-induced resurgence of the brown planthopper, *Nilaparvata lugens*, *Environ. Entomol.*, **9**, 773–777.

Chio, H., C.-S. Chang, R. L. Metcalf, and J. Shaw. (1976). Susceptibility of four species of *Diabrotica* to insecticides, *J. Econ. Entomol.*, **71**, 389–393.

Conway, G. (1981). "Man vs. pests," in R. M. May, Ed., *Theoretical Ecology*, 2nd ed., Blackwell, Oxford, pp. 356–386.

Cottam, C. (1965). The ecologist's role in the problems of pesticide pollution, *Bioscience*, **15**, 457–463.

Croft, B. A. (1982). Developed resistance to insecticides in apple arthropods: A key to pest control failures and successes in North America, *Entomol. Exp. Appl.*, **31**, 88–110.

Croft, B., and A. W. A. Brown. (1975). Responses of arthropod natural enemies to insecticides, *Annu. Rev. Entomol.*, **20**, 285–535.

Croft, B. A., and M. E. Whalon. (1982). Selective toxicity of pyrethroid insecticides to arthropod natural enemies and pests of agricultural crops, *Entomophaga*, **27**, 3–21.

Curley, A., and R. Kimbrough. (1969). Chlorinated hydrocarbon insecticides in plasma and milk of pregnant and lactating women, *Arch. Environ. Health*, **18**, 156–164.

DeBach, P. O., and B. R. Bartlett. (1951). Effects of insecticides on biological control of insect pests of citrus, *J. Econ. Entomol.*, **44,** 372-383.

Duffus, J. E., and R. A. Flock. (1982). Whitefly transmitted disease complex in the desert southwest, *Calif. Agr.*, **36,** 4-6.

Duggan, R. E. (1969). Pesticide residues in foods, *Ann. N.Y. Acad. Sci.*, **160,** 173-182.

Durham, W. F. (1969). Body burden of pesticides in man, *Ann. N.Y. Acad. Sci.*, **160,** 183-195.

Eckert, J. E. (1944). The poisoning of bees with methods of prevention, *J. Econ. Entomol.*, **27,** 551-552.

Eichers, T. R., P. A. Andrilenas, and T. W. Anderson. (1978). Farmers' use of pesticides in 1976, U.S.D.A. Agr. Econ. Rept. No. 418, Washington, D.C., 57 pp.

El-Guindy, M. A., S. M. Madi, M. E. Keddis, Y. H. Issa, and M. M. Abdel Sattar. (1982). Development of resistance to pyrethroids in field populations of the Egyptian cotton leafworm *Spodoptera littoralis*, *Int. Pest Control*, **24,** 6, 8, 10-11, 16-17.

Elliott, M., N. F. James, J. H. Stevenson, and J. H. H. Waters. (1983). Selectivity of pyrethroid insecticides between *Ephestia kuhniella* and its parasite *Venturia canescens*, *Pesticide Sci.*, **14,** 423-426.

El-Sebae, A. H. (1977). "Incidents of local pesticide hazards and their toxicological interpretation," in *Proc. UC/AID Univ. Alexandria*, Seminar Pesticide Management, Alexandria, Egypt, pp. 137-152.

EPA. (1972). The pollution potential in pesticide manufacture, Office of Water Programs Study Series 5, TS-00-72-04, Washington, D.C.

Erickson, E. H. Jr., and B. J. Erickson. (1983). Honey bees and pesticides, *Am. Bee J.*, **123,** 724, 726-730, 797-800, 802-805, 814.

Eveleens, K. G. (1983). Cotton-insect control in the Sudan-Gezira: Analysis of a crisis, *FAO Crop Protection*, **2,** 273-287.

Eveleens, K. G., R. van den Bosch, and L. E. Ehler. (1973). Secondary outbreak induction of beet armyworm by experimental insecticide application in cotton in California, *Environ. Entomol.*, **2,** 497-503.

Falcon, L. A., R. van den Bosch, C. A. Ferris, L. K. Stromberg, L. K. Etzel, R. E. Stinner, and T. F. Leigh. (1968). A comparison of season long pest control programs in California during 1966, *J. Econ. Entomol.*, **61,** 633-642.

Felsot, A. S., J. C. Wilson, D. E. Kuhlman, and K. L. Steffy. (1982). Rapid dissipation of carbofuran as a limiting factor in corn rootworm control in fields with histories of continuous carbofuran use, *J. Econ. Entomol.*, **75,** 1098-1103.

Forgash, A. (1984). History, evolution, and consequences of insecticide resistance, *Pestic. Biochem. Physiol.*, **22,** 178-186.

Furtick, W. R. (1976). "Insecticides in food production," in R. L. Metcalf and J. J. McKelvey, Jr., Eds., *The Future for Insecticides: Needs and Prospects*, Wiley, New York.

Gannon, N., R. P. Link, and G. C. Decker. (1960). Storage of dieldrin in tissues of steers, hogs, lambs, and poultry fed dieldrin in their diets, *J. Agric. Food Chem.*, **7,** 826-828.

Geier, P. W. (1966). Management of insect pests, *Annu. Rev. Entomol.*, **11,** 471-490.

Geissbuhler, A. (1981). The agrochemical industry approach to integrated pest control, *Phil. Trans. R. Soc. Lond.*, **295,** 111-123.

Georghiou, G. P. (1981). *The Occurrence of Resistance to Pesticides in Arthropods*, FAO, Rome, Italy.

Georghiou, G. P. (1983). "Pesticide resistance in time and space," in G. P. Georghiou and T. Saito, Eds., *Pest Resistance to Pesticides*, Plenum, New York, pp. 1-46.

Georghiou, G. P. (1984). "The consequences of evolutionary adaptation of pests to pesticides," in *Symposium on Pesticide Resistance Management*, National Academy Sciences, Washington, D.C., Nov. 27-28, pp. 14-43.

Georghiou, G. P., V. Ariaratnam, M. E. Pasternak, and C. S. Lin. (1975). Organophosphorus multiresistance in *Culex pipiens quinquefasciatus* in California, *J. Econ. Entomol.*, **68**, 461–67.

Georghiou, G. P., and C. E. Taylor. (1977). "Pesticide resistance as an evolutionary phenomenon," in *Proc. XV Internat. Congr. Entomol.*, Washington, D.C., pp. 749–785.

Glass, E. H., and P. J. Chapman. (1948). Red-banded leaf roller problem in New York, *J. Econ. Entomol.*, **42**, 29–35.

Glass, E. H., and S. E. Lienk. (1971). Apple insect and mite populations developing after discontinuance of insecticides: 10 year record. *J. Econ. Entomol.*, **64**, 23–27.

Goldstein, A., L. Aranow, and S. M. Kalman. (1969). *Principles of Drug Action*, Harper & Row, New York.

Hama, H., and T. Iwata. (1978). Studies on the inheritance of carbamate resistance in the green rice leafhopper, *Appl. Entomol. Zool.* (Jpn.), **13**, 190–202.

Harris, C. R., R. A. Chapman, C. Harris, and C. M. Tu. (1984). Biodegradation of pesticides in soil: Rapid induction of carbamate degrading factors after carbofuran treatment, *J. Environ. Sci. Health B*, **19**, 1–11.

Health, Education, and Welfare. (1969). *Report of the Secretary's Commission on Pesticides and Their Relationship to Environmental Health*, Washington, D.C., 677 pp.

Hickey, J. J., J. A. Keith, and F. B. Coon. (1966). An exploration of pesticides in a Lake Michigan Ecosystem, *J. Appl. Ecol.*, **3** (suppl.), 141–154.

Hikichi, A. (1964). The status of apple leaf rollers in Norfolk Co., Ontario. *Proc. Entomol. Soc. Ont.*, **94**, 38–40.

Hirano, C., and K. Kiritani. (1976). "Paddy ecosystem affected by nitrogenous fertilizers and insecticides," in *Science for Better Environment*, Proc. Int. Congress Human Development Kyoto 1975. Asaki Evening News, Tokyo, pp. 197–206.

Honolulu Star-Bulletin. (1985). *Honolulu Star-Bull.*, **74**, Jan. 9, p. A1.

Hough, W. S. (1927). A study of the biology and control of the red banded leaf roller, Va. Agr. Exp. Sta. Bull. 259.

Huffaker, C. B. (1971). "The ecology of pesticide interference with insect populations," in J. E. Swift, Ed., *Agricultural Chemicals—Harmony or Discord for Food, People, Environment*, Univ. Cal. Div. Agr. Sci., 151 pp.

Hull, L. A., and V. R. Starner. (1981). Impact of four synthetic pyrethroids on major natural enemies and pests of apple in Pennsylvania, *J. Econ. Entomol.*, **76**, 122–130.

Hunt, E. G. (1966). "Biological magnification of pesticides," in *Scientific Aspects of Pest Control*, Publ. 1402, Natl. Acad. Sci Washington, D.C., pp. 251–262.

Hyslop, J. A. (1938). Losses occasioned by insects, mites, and ticks in the U.S., USDA Bur. Entomol. E-444, 57 pp.

Johnson, M. W., N. C. Toscano, H. T. Reynolds, E. S. Sylvester, K. Kido, and E. T. Natwick. (1982). Whiteflies cause problems for Southern California cotton growers, *Calif. Agr.*, **36** (9110), 24–26.

Joyce, R. J. V. (1955). Cotton spraying in the Sudan Gezira. Entomological problems arising from spraying and spraying methods, *FAO Plant Protec. Bull.*, **3**, 97–103.

Joyce, R. J. V. (1982). "Review of the role of chemical pesticides in *Heliothis* management," in *Proc. Intern. Workshop on Heliothis management*, ICRISAT Center, India, Nov. 15-20, 1981, pp. 173–188.

Kaufman, D. D., and D. F. Edwards. (1983). "Influence of environmental conditions on pesticide residues," in J. Miyamoto, Ed., *Human Welfare and the Environment*, IUPAC Pesticide Chemistry, Pergamon, New York, pp. 177–182.

Keiding, J. (1977). "Resistance in the housefly in Denmark and elsewhere," in D. L. Watson and A. W. A. Brown, Eds., *Pesticide Management and Insecticide Resistance*, Academic, New York, pp. 261–302.

Kenaga, E. E., and C. S. End. (1974). Commercial and experimental organic insecticides, Entomol. Soc. Amer. Spec. Publ. 74-1 (Oct.)., 77 pp.

Kerr, S. R., and W. P. Vass. (1973). "Pesticide residues in aquatic invertebrates," in C. A. Edwards, Ed., *Environmental Pollution by Pesticides*, Plenum, London, Chapt. 4.

Kiritani, K. (1979). Pest management in rice, *Annu. Rev. Entomol.*, **24**, 279-312.

Levin, M. D. (1970). The effects of pesticides on beekeeping in the U.S., *Am. Bee J.*, **110**, 8-9.

Levin, M. D., W. B. Forsyth, G. L. Fairbrother, and F. B. Skinner. (1968). Impact on colonies of honey bees of ultra-low volume (undiluted) malathion applied for control of grasshoppers, *J. Econ. Entomol.*, **61**, 58-62.

Liu, M.-Y., Y. J. Tzeng, and Chih-ning Sun. (1982). Insecticide resistance in the diamondback moth, *J. Econ. Entomol.*, **75**, 153-155.

Lofgren, C. S., and D. F. Williams. (1982). Avermectin $B_1a$: A highly potent inhibitor of reproduction by queens of the imported red fire ant (Hymenoptera: Formicidae), *J. Econ. Entomol.*, **75**, 798-803.

Luck, R. E., R. van den Bosch, and R. Garcia. (1977). Chemical insect control—a troubled management strategy, *BioScience*, **27**, 606-611.

Luckmann, W. H., and G. C. Decker. (1960). A five year report on observations in the Japanese beetle control area of Sheldon, Illinois, *J. Econ. Entomol.*, **53**, 821-827.

MacArthur, R. H., and E. O. Wilson. (1967). *The Theory of Island Biogeography*, Princeton University Press, Princeton, NJ, 203 pp.

Marlatt, C. L. (1904). Annual losses occasioned by destructive insects in the U.S., USDA Agric. Yearb. Agric., pp. 461-474.

McMurtry, J. A., C. B. Huffaker, and M. van de Vrie. (1970). I. Tetranychid enemies: Their biological characters and the impact of spray practices, *Hilgardia*, **40**, 331-390.

Metcalf, R. L. (1955). Physiological basis for insecticide resistance to insecticides, *Physiol. Revs.*, **35**, 192-232.

Metcalf, R. L. (1976). "Organochlorine insecticides, survey and prospects," in R. L. Metcalf and J. J. McKelvey, Jr., Eds., *The Future for Insecticides: Needs and Prospects*, Wiley, New York, pp. 223-285.

Metcalf, R. L. (1977a). "Biological fate and transformation of pollutants in water," in I. M. Suffet, Ed., *Fate of Pollutants in the Air and Water Environments*, Vol. II, Wiley, New York, pp. 195-221.

Metcalf, R. L. (1977b). Model ecosystem approach to insecticide degradation, *Annu. Rev. Entomol.*, **22**, 241-261.

Metcalf, R. L. (1980). Changing role of insecticides in crop protection, *Annu. Rev. Entomol.*, **25**, 219-256.

Metcalf, R. L. (1982). "Insecticides in pest management," in R. L. Metcalf and W. H. Luckmann, Eds., *Introduction to Insect Pest Management*, Wiley, New York, pp. 217-277.

Metcalf, R. L. (1983). "Implications and prognosis of resistance to pesticides," in G. P. Georghiou and T. Saito, Eds., *Pest Resistance to Pesticides*, Plenum, New York, pp. 703-733.

Metcalf, R. L. (1984). "Trends in the use of chemical insecticides," in *Judicious and Efficient Use of Insecticides on Rice*, International Rice Research Institute, Los Baños, Phillipines, pp. 69-89.

Metcalf, R. L., and W. H. Luckmann, Eds. (1982). *Introduction to Insect Pest Management*, 2nd ed., Wiley, 577 pp.

Metcalf, R. L., and J. R. Sanborn. (1975). Pesticides and environmental quality in Illinois, *Bull. Ill. Nat. Hist. Survey*, **31**, Art. 9, 381-436.

Metcalf, C. L., W. P. Flint, and R. L. Metcalf. (1962). *Destructive and Useful Insects: Their Habits and Control*, McGraw-Hill, New York, 1085 pp.

Metcalf, R. L., R. A. Metcalf, and A. M. Rhodes. (1980). Cucurbitacins as kairomones for diabroticite beetles, *Proc. Natl. Acad. Sci. USA*, **77**, 3769-3772.

Metcalf, R. L., J. E. Ferguson, R. Lampman, and J. F. Andersen. (1986). Dry cucurbitacin-containing baits for controlling Diabroticina beetles, *J. Econ. Entomol.*, (in press).

Newsom, L. D. (1980). The next rung up the integrated pest management ladder, *Bull. Entomol. Soc. Amer.*, **26**(3), 369–374.

Omer, S. M., G. P. Georghiou, and S. N. Irving. (1980). DDT/pyrethroid resistance interrelationships in *Anopheles stephensi*, *Mosq. News*, **40**, 200–209.

Oppenoorth, F. J. (1965). Biochemical genetics of insecticide resistance, *Annu. Rev. Entomol.*, **10**, 185–206.

Oppenoorth, F. J., and W. Welling. (1976). "Biochemistry and physiology of resistance," in C. W. Wilkinson, Ed., *Insecticide Biochemistry and Physiology*, Plenum, New York, pp. 507–551.

Penman, D. R., and R. B. Chapman. (1980). Woolly apple aphid outbreak following use of fenvalerate in apples in Canterbury, New Zealand, *J. Econ. Entomol.*, **73**, 49–51.

Petty, H. B., and W. N. Bruce. (1972). Insecticide residues in soybeans, Proc. 24th Ill. Custom Spray Operators Training School, Urbana, IL, Jan. 26, 27.

Pianka, E. M. (1970). On *r*- and *K*-selection, *Am. Nat.*, **104**, 592–297.

Pickett, A. D. (1949). A critique on insect chemical control methods, *Can. Entomol.*, **81**, 67–76.

Pickett, A. D., and N. A. Patterson. (1953). The influence of spray programs on the fauna of apple orchards in Nova Scotia IV. A review, *Can. Entomol.*, **85**, 472–98.

Pimentel, D. (1973). Extent of pesticide use, food supply, and pollution, *J. N. Y. Entomol. Soc.*, **81**, 13–33.

Pimentel, D., J. Krummel, D. Gallahan, J. Hough, A. Merrill, I. Schreiner, P. Vittum, F. Koziol, E. Boch, D. Yen, and S. Fiance. (1978). Benefits and costs of pesticide use in U.S. food production, *BioScience*, **28**, 772–784.

Reinbold, K. A., I. P. Kapoor, W. F. Childers, W. N. Bruce, and R. L. Metcalf. (1971). Comparative uptake and biodegradability of DDT and methoxychlor by aquatic organisms, *Bull. Ill. Nat. Hist. Survey*, **30**, Art. 6, 405–417.

Reynolds, H. T. (1958). Research advances in seed and soil treatment with systemic and non-systemic insecticides, *Adv. Pest. Control Res.*, **2**, 135–182.

Reynolds, H. T. (1971). "A world review of the problem of insect population upsets and resurgences caused by pesticide chemicals," in J. E. Swift, Ed., *Agricultural Chemicals—Harmony or Discord for Food, People, Environment*, Univ. Calif. Div. Agr. Sci., pp. 108–112.

Reynolds, H. T., P. L. Adkisson, R. F. Smith, and R. E. Frisbie. (1982). "Cotton insect pest management," in R. L. Metcalf and W. H. Luckmann, Eds., *Introduction to Insect Pest Management*, 2nd ed., New York, Wiley, pp. 375–441.

Ripper, W. E. (1956). Effect of pesticides on the balance of arthropod populations, *Annu. Rev. Entomol.*, **1**, 403–443.

Sawicki, R. M. (1975). "Effects of sequential resistance on pesticide management, in *Proc. 8th British Insecticide and Fungicide Conference*, pp. 799–811.

Sawicki, R. M. (1980). Is pyrethroid resistance unavoidable? *Tables Rondes Roussel Uclef.*, **37**, 51–53.

Schuntner, C. A., and P. G. Thompson. (1978). Mechanisms of resistance to bromophos-ethyl in two strains of the cattle tick *Boophilus microplus*, *Austr. J. Biol. Sci.*, **31**, 317–325.

Scott, T. G., Y. L. Willis, and J. A. Elis. (1959). Some effects of a field application of dieldrin on wildlife, *J. Wildlife Manag.*, **23**, 409–427.

Smallman, B. N. (1964). Perspectives in insect control, *Can. Entomol.*, **96**, 167–169.

Smith, R. F. (1966). The distribution of Diabroticites in western North America, *Bull. Entomol. Soc. Amer.*, **12**, 108–110.

Smith, R. F. (1970). "Pesticides: Their use and limitations in pest management," in *Concepts of Pest Management*, North Carolina State University, pp. 103–118.

Smith, R. F., and J. F. Lawrence. (1967). *Clarification of the Status of the Type Specimens of Diabroticites*, University of California Press, Berkeley and Los Angeles, 174 pp.

Southwood, T. R. E. (1977). Entomology and mankind, *Am. Sci.*, **65**, 30–39.

Steiner, L. F., W. C. Mitchell, E. J. Harris, T. T. Kozuma, and M. S. Fujimoto. (1965). Oriental fruit fly eradication by male annihilation, *J. Econ. Entomol.*, **58**, 961–964.

Stern, V. M., R. F. Smith, R. van den Bosch, and K. S. Hagen. (1959). The integrated control concept, *Hilgardia*, **29**, 81–154.

USDA. (1954). Losses in agriculture, Agric. Res. Service Rep. 20-1, Washington, D.C., 190 pp.

USDA. (1965). Losses in agriculture, Agr. Handbook 291, Agric. Res. Service, Washington, D.C., 120 pp.

USDA. (1976). Insect pollination of cultivated crop plants, Agr. Handbook No. 496, Agr. Res. Service, Washington, D.C., 411 pp.

van den Bosch, R. (1978). *The Pesticide Conspiracy*, Doubleday, Garden City, NY, 226 pp.

Van Steenwyk, R. A., N. C. Toscano, G. R. Ballmer, K. Kido, and H. T. Reynolds. (1976). Increased insecticide use in cotton may cause secondary pest outbreaks, *Calif. Agr.*, **30**, 14–15.

Vaughan, M. A., and L. Q. Gladys. (1976). Pesticide management on a major crop with severe resistance problems, in *Proc. XV Int. Congr. Entomol.*, Washington, D.C., pp. 812–815.

*Washington Post*. (1974). July 18, p. C5.

Wedberg, D. L., and K. D. Black. (1970). Insect situation and outlook and insecticide usage, 30th Illinois Custom Spray Operators Training School, Urbana, IL, pp. 119–146.

Westigard, P. H., L. G. Gentner, and D. W. Berry. (1968). Present status of biological control of the pear psylla in southern Oregon, *J. Econ. Entomol.*, **61**, 740–743.

Wiersma, G. B., H. Tai, and P. F. Sand. (1972). Pesticide residue levels in soils FY 1969, *Pestic. Monit. J.*, **6**, 194–228.

# 11

## ACROECOLOGY
## AND ECONOMICS

*David Pimentel*

## 11.1. INTRODUCTION

Each person in the nation eats three meals per day and each year consumes about 700 kg (1500 lb) of food (USDA, 1983). Thus, a total of 162 million tonnes of food is consumed annually in the United States. To purchase this much food, the people of this nation spend $321 billion for food annually (USBC, 1983), making the food system one of the largest industries in the nation, which employs about 20% of the total labor force. Although about $321 billion is spent for food, the value of farm products totals about $144 billion, with crops amounting to about $84 billion and livestock $61 billion (USDA, 1983). Thus, farm value of crops and livestock accounts for about 45% of total value. For grain products, however, farmers receive only 13% of the retail value of the crop (USBC, 1983).

The production of such large quantities of food affects our natural resources including land, water, energy, and biota (Pimentel, 1984). Agriculture uses more land, water, and energy and its influence on natural biota is more widespread than any other industry in the nation. Clearing vast land areas for crops and livestock, pumping immense quantities of water, use of agricultural chemicals, and other manipulations all contribute to the ecological impact of agriculture.

Of the agricultural chemicals, pesticides have the greatest ecological effects. Despite the use of about 450 million kg of pesticides annually in the United States, pests destroy more than 40% of the food. About 37% of food crop is lost before harvest (Pimentel, 1981), plus an additional 9% postharvest (USDA, 1965). Both the food lost and cost of control are significant economic losses to the nation. The objective of this paper is to examine the ecology of agriculture and economics of pest losses and control. Included in this study will be an assessment of the ecological causes of pest problems and the immense environmental and social costs in attempting to control the pests of agriculture.

299

## 11.2. ECOLOGY AND ECONOMICS OF NATURAL RESOURCES USED IN AGRICULTURE

### 11.2.1. Land

Over half (62%) of the total land area (911 million ha) of the United States is managed for various kinds of agricultural production. Of this, about 405 million ha are pasture and range for livestock (USDA, 1983), and about 160 million ha are in crops. In general, croplands are more ecologically altered than are either pasture or rangelands. However, land clearing by humans and overgrazing by livestock may cause environmental changes that can be more severe than alterations occurring in croplands.

In part because of changes imposed on native habitats, soil erosion is now a major environmental problem in the United States. Erosion diminishes the productivity of agricultural land, and is a serious threat to maintaining a sustainable agriculture in the future (SCS, 1977; Foster and Meyer, 1977; Larson, 1981; Timmons, 1981; OTA, 1982; GAO, 1983). Average soil loss from U.S. croplands by both water and wind is estimated to be about 17 t/ha yr (Lee, 1984), while soil is being reformed under agricultural conditions at a rate of only 0.5–2 t/ha yr (Swanson and Harshbarger, 1964; Larson, 1981). Some regions of the United States, such as Iowa, which has some of the most fertile soils, have lost half of their topsoil since 1885 (Risser, 1981).

Soil loss adversely affects crop productivity by (1) reducing organic matter and fine clays, which lessens the nutrient holding capacity of the soil; (2) reducing water holding capacity; and (3) restricting rooting depth because the soil layer thins (OTA, 1982). Diminished yields have been observed in the southern United States where reductions range from 25 to 50% for crops such as corn, soybeans, cotton, oats, and wheat (Adams, 1949; Buntley and Bell, 1976; Langdale et al., 1979). Nonetheless, crop yields in most regions have been rising because energy inputs (e.g., fertilizers, pesticides, and irrigation) have been increased and high-yielding crop varieties are now available (Pimentel et al., 1976; OTA, 1982).

### 11.2.2. Water

Water is the major limiting factor for all food and forest production. A large portion of the water used by plants is lost through transpiration as well as evaporation. A forest stand, for example, will transpire from 4 to 7 million l/ha yr (Spurr and Barnes, 1980), while a corn crop will take up and transpire about 2.4 million l/ha of water during the growing season (Penman, 1970). In areas where rainfall is undependable or scarce, crops are irrigated. Agricultural irrigation consumes about 83% of the water that is withdrawn from streams, lakes, and underground (Fig. 11.1).

About 22% of the water used for irrigation is pumped from underground aquifers (Murray and Reeves, 1977). Already serious overdrafts of these water re-

WATER
WITHDRAWN

FRESHWATER
CONSUMED

Rural
(1 percent)

Public Supplies
(7 percent)

Rural
(4 percent)

Public Supplies
(7 percent)

Industry
(6 percent)

INDUSTRY
(58 percent)

IRRIGATION
(34 percent)

IRRIGATION
(83 percent)

1,600 billion liters
per day withdrawn

360 billion liters
per day consumed

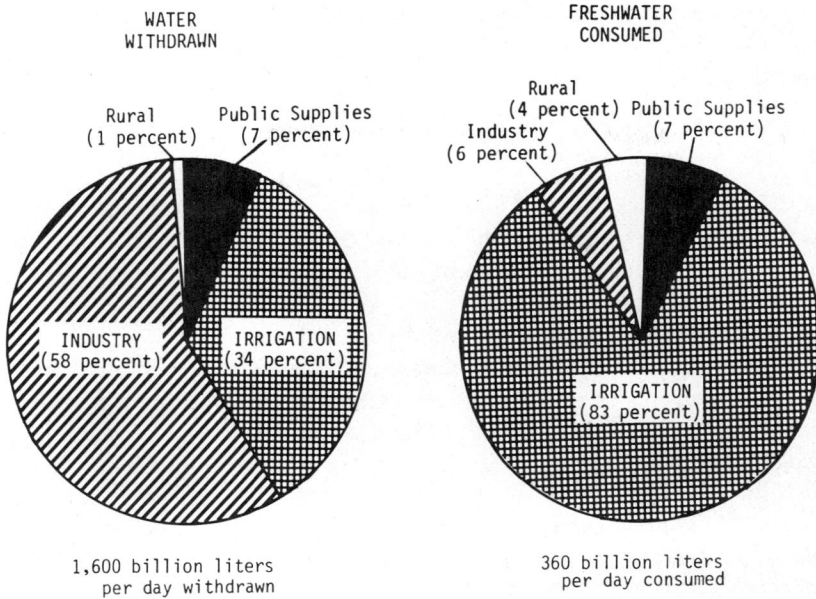

Fig. 11.1 Off-channel water withdrawals in the United States and the proportion of freshwater consumed in 1975 (Pimentel et al., 1982a).

sources are occurring, and in some areas like Texas overdraft has reached as high as 77% (USWRC, 1979). This has necessitated deeper mining of aquifers, plus the expenditure of more energy to lift the water onto the cropland. Eventually, of course, the underground aquifers will either be dry or so low that farmers cannot afford to pump the water. This has already happened in several regions of the western United States, and irrigated farming in some of these regions has already been abandoned (Pimentel et al., 1982a).

### 11.2.3. Energy

In addition to land and water, solar and fossil energy are essential resources in agricultural production. Solar energy is free but fossil energy is expensive. From the total input of $1.1 \times 10^{15}$ kcal, agriculture produces about $6.9 \times 10^{15}$ kcal in crops and forest products (Pimentel et al., 1978a). About 17% of the total fossil energy expended in the United States is used in the food system (Pimentel and Wen Dazhong, 1985). Of this, about 6% is expended for production, 6% for processing and packaging, and 5% for distribution and home preparation. This 17% amounts to 1500 l of oil equivalents per person per year. However, we consume in the form of food an equivalent of 160 l of oil. Thus, for each kilo-calorie that we consume, about 10 kcal of fossil energy are required.

Although agriculture and forestry produce more biomass energy than fossil energy consumed, the fossil energy inputs are enough to be a concern. For example, to produce a hectare of corn about 1100 l of oil equivalents are utilized (Pimentel and Wen Dazhong, 1985). About a third of this energy is expended for fertilizers, a third for mechanization, and about 15% for pesticides.

### 11.2.4. Biological Resources

Some 200,000 species of plants and animals exist in this country where they perform many essential functions (Pimentel et al., 1980a). Primarily, plants and animals are used for human food. In the United States just 10 species of crop plants and 8 species of livestock provide about 80% of the food consumed. Forest products provide the materials for shelter and fuel.

Biological resources contribute many additional vital functions. These include preserving genetic material for crop and livestock breeding; degrading the immense amounts of wastes produced by humans and animals; cleaning water and soil of pollutants; recycling vital chemical elements contained in wastes back to the soil; and retarding soil erosion and water run-off. The average biomass per hectare of some of these useful natural biota ranges from 1000 kg for earthworms and insects to as much as 2700 kg for fungi (Fig. 11.2). This volume of biomass is in stark contrast to humans, which average only 18 kg/ha.

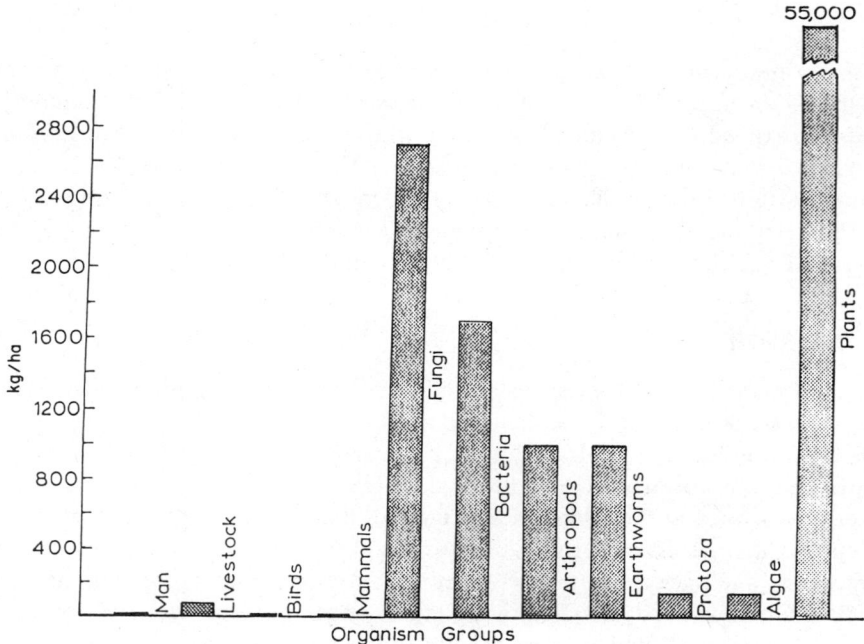

Fig. 11.2 The average biomass weight of man and his livestock per acre in the United States and the estimated biomass of natural biota species groups of birds, mammals, arthropods, earthworms, protozoa, bacteria, algae, and fungi in the environment (Pimentel et al., 1980a).

Some species of natural biota "fix" nitrogen, thus providing vital nitrogen for crop production. Estimates are that about 14 million tonnes of nitrogen are fixed biologically in the United States each year (Delwiche, 1970). Considering that 11 million tonnes of nitrogen fertilizer, worth about $4 billion, also are used each year, increased production and use of biological nitrogen would be of immense value. To do this, agronomists, microbiologists, horticulturists, and ecologists would have to join the investigation to integrate more legumes into the current cropping system and also find ways to increase nitrogen produced by the micro-organisms.

Another biological resource that performs an essential task in the human eco-system are the bees. Cross-pollination by bees makes possible the production of many fruit and vegetable crops. In this country, the yields from some 90 crops, worth about $20 billion, are dependent upon bee pollination. Improved pollina-tion of crops is possible if greater care were practiced in the use of pesticides. At present, less than 1% of the pesticides reach the target pests while 99% pollutes the environment—some of which causes bee kills (Pimentel et al., 1980b). Some bee species require some type of protected structure in which to rear their young. Providing hives and other suitable structures for wild bees will attract bees to rest close to agricultural fields.

## 11.3. ECOLOGICAL CAUSES OF PEST PROBLEMS

We can better understand the ecology and economics of pest control if we under-stand changes in the agroecosystem that cause pest outbreaks. Although this assessment of the ecological basis for pest problems focuses on individual ecosys-tem alterations, it must be emphasized that pest outbreaks are usually the result of a complex combination of changing factors in the ecological system.

### 11.3.1. Monocultures

Plant communities in natural ecosystems normally consist of a complex group of species. In crop production, however, the natural plant community is removed or destroyed, and replaced with a single crop plant species. Frequently, out-breaks of insect pests have followed this practice of crop monoculture. For ex-ample, Marchal, as early as 1908, noted that when humans planted a vast area of a country with certain crops while excluding others, they offered insect pests feeding on these plants favorable conditions for their explosive increase.

Although some experiments have confirmed that a diverse plant community may reduce some insect herbivores associated with a target plant host (Pimentel, 1961a; Tahvanainen and Root, 1972; Root, 1973, 1975; Cromartie, 1981), cer-tain combinations of crops may enhance pest problems. Planting cotton and corn together, for example, increases certain pests like *Heliothis zea* (Boddie) (Pimentel et al., 1977). Clearly, some knowledge of the ecology of pest popula-tions can help reduce pest problems.

### 11.3.2. New Crop Introductions

When a crop plant is newly introduced into a region, some native insects and plant pathogens may begin to feed on the new host plant. This happens because the introduced crop has never been exposed to the organisms in its new biotic community, and therefore lacks natural resistance to them. Thus, a new attacking species may become a serious pest on the introduced crop. For example, when the potato (*Solanum tuberosum* L.), which originated in Bolivia and Peru (Hawkes, 1944), was introduced into the southwestern United States, it was attacked by the Colorado potato beetle [*Leptinotarsa decemlineata* (Say)]. This beetle, which is native to the United States, had originally coevolved with and fed on wild sand bur (*Solanum rostratum* Dunal.) (Elton, 1958). When the potato was introduced into the Southwest, the beetle spread onto the potato plants, which lacked any natural resistance to it. Soon this insect became the most serious pest of the potato in Colombia and subsequently throughout the world.

### 11.3.3. Introduced Pests

New species of insects, pathogens, plants, mammals, or birds may be introduced into an ecosystem and become the major pest species in their new habitats (Simmonds and Greathead, 1977). This occurs because new associations are formed between exploiting species and victims (prey and hosts). If the host lacks natural resistance, then the introduced, exploiter population can and often does increase to outbreak levels (Pimentel, 1961b; Pimentel et al., 1975).

A classic example of this is the damage done by the gypsy moth [*Lymantria dispar* (L.)], which was brought from Europe into the eastern United States. There it reached outbreak levels on its new plant hosts, primarily oaks and related hardwoods (USDA, 1973). Subsequently, it has spread westward in the country.

Dutch elm disease, now a problem throughout the United States, is the result of the chance introduction of a fungal pathogen from European elms to the American elm (*Ulmus americana* L.). Unfortunately the American elm lacked resistance to this new pathogen, which has caused widespread destruction of these trees. In addition, the spread of this disease was hastened by the introduction of the European bark beetle (*Scolytus multistriatus* Marsham), which also increased quickly to outbreak levels (Matthysse, 1959). Together, these two introduced species have been responsible for decimating the beautiful elms of our cities and countryside.

### 11.3.4. Climatic Regions

Sometimes, the distribution of plants and animals in nature is the result of the differential survival of parasite, predator, and host in different climatic regions (Elton, 1927; Andrewartha and Birch, 1954). This means some plants and animals are able to escape severe attack from their usual parasites and predators if they can grow in a climatic region unsuited to their natural enemies.

In agriculture, as well as in the wild, climatic conditions influence the degree of pest attack on a particular crop or animal. For example, potatoes grown in northern Maine or in North Dakota have fewer insect pests and experience less damage than potatoes grown in warmer regions of the United States. The potato aphid [*Macrosiphum euphorbiae* (Thomas)], potato stalk borer [*Trichobaris trinotata* (Say)], and potato tuberworm [*Phthorimaea operculella* (Zeller)] are more serious pests in the warmer Southeast than in the cooler northern and mountainous regions. One indication of the differences in the severity of the pest problem is that almost 100% of the potato acreage in the Southwest is treated with insecticides, compared with 65% in the cooler mountain region (USDA, 1975).

### 11.3.5. Breeding Crops

The differences in pest resistance levels that may exist in crop plants and their effectiveness to resist pest attack is well illustrated with pea aphids [*Acyrtho-siphon pisum* (Harris)] associated with alfalfa (*Medicago sativa* L.). Five young pea aphids placed on a common crop variety of alfalfa produced a total of 290 offspring in 10 days, whereas the same number of aphids placed on a resistant variety of alfalfa produced a total of only two offspring per aphid in the same time (Dahms and Painter, 1940). Obviously, a pest population that has a 145-fold greater rate of increase on a host plant will inflict significantly greater damage than one with an extremely low rate of increase.

One of the major ecological factors that has increased pest problems is the breeding of susceptible crop genotypes (Lupton, 1977). In past years, the prime aim of the plant breeder has been to increase crop yields. When altering the genetic makeup of the crop plant to increase yield, if little or no attention is given to resistance to pest attacks, natural resistance can be greatly reduced or lost. Therefore, it is vital that plant breeding focuses not only on yields but equally on maintaining or improving natural pest resistance of the crops.

### 11.3.6. Genetic Diversity

Parasites associated with hosts in nature appear to be fairly stable in terms of genetics and population outbreaks (Pimentel, 1984). However, in agricultural ecosystems, when a parasite is stressed by a single factor, it often evolves to overcome the single factor resistance and cause serious damage to crops. For example, both parasitic stem rust and crown rust have been able to overcome genetic resistance bred into their oat host. To prevent this from happening, the cultivated oat varieties are changed approximately every 5 years to counter the changes evolving in the races of stem rust and crown rust (Stevens and Scott, 1950; Van der Plank, 1968). Perhaps in breeding resistant strains with sufficient diversity, the instability of host plant resistance in crops can be reduced.

### 11.3.7. Plant Spacings

The feeding intensity of herbivores depends upon the relative abundance of herbivores per unit of crop biomass. Thus, plant survival may depend on the spatial pattern of the plants in the field because the pattern affects herbivore density per plant.

In cultivated fields, densities of crop plants are carefully controlled to ensure optimal growth and thus a maximum economic yield. The spacings of cultivated plants, seldom similar to those in the wild, alter the natural pattern of the ecological habitat and may make it easy for herbivores to find their host plants.

Plants and other hosts, for example, existing in relatively dense stands, may limit the number of herbivores per plant. Lebedev (1942) reported that planting dense stands of turnips reduced the numbers of flea beetles (*Phyllotreta* spp.) on each turnip plant. Another field experiment examined the impact of three different planting densities (dense, sparse, and dispersed) on animals feeding on several *Brassica* crops (Pimentel, 1961c). Although the total number of herbivores was greatest in the dense planting, herbivores per plant surface area in both the sparse and dispersed planting were more than five times greater than in the dense planting. Overall damage was greatest on the dispersed plants.

### 11.3.8. Continuous Crop Culture

When crops are grown in the same location year after year, the numbers of insects and plant pathogens may increase and cause damage to crops planted. For example, when cole crops are cultured for several years in the same soil, club root (*Plasmodiophora brassicae* Wor.) organisms increase rapidly and eventually ruin production (Walker et al., 1958). One effective control of these pests is to plant non-*Brassica* crops on the land for several years to break the growth cycle of the pest organism.

The culture pattern commonly used with corn (*Zea mays* L.) illustrates the effect of continuous culture on pest density. Before the late 1940s, when most corn was planted in rotation after soybeans, other legumes, or small grains, the corn rootworm complex was not a problem. Now only about 40% of the corn is in rotation and this change has been associated with an increasing density of the rootworm complex (Pimentel et al., 1977).

### 11.3.9. Nutrients

All organisms, including crop pests, have specific nutrient requirements for their survival. Altering the nutrient level in the soil and, subsequently, the nutrients in the host plant, can influence the density of the pests feeding on it.

For example, Haseman (1946) reported that the grain aphid [*Macrosiphum granarium* (Kirby)], feeding on small-grain plants with high nitrogen content, produced significantly more progeny than those on plants with low levels of nitrogen. This was substantiated by Barker and Tauber (1951), who observed similar associations with pea aphid feeding on the garden pea.

## 11.3.10. Planting Date

Some plants can avoid pest attack if grown early in the spring before the pests have become abundant. For instance, wild radishes germinate early in the spring and make most of their growth before the cabbage maggot fly emerges and can attack them. Damage to the radishes under these conditions is usually minimal (D. Pimentel, unpublished data). Established crop growth early in the season can help minimize pest damage.

## 11.3.11. Crop Associations

Some pests are able to move from one host plant species to another when availability of one host plant declines (Klass, 1978). In the United States, mirid plant bugs feed on both alfalfa and cotton, and when alfalfa is mowed for hay, the bugs move to cotton in large numbers (Stern, 1969). Plant bugs generally prefer alfalfa and therefore if the alfalfa is managed carefully by planned cutting, the invasion of bug populations onto cotton can be minimized. The bugs remain on the attractive alfalfa and generally avoid cotton.

## 11.3.12. Pesticides and Crop Physiology

Some pesticides alter the physiology of crops and make them more susceptible to pest attack. For example, increased insect and pathogen problems have been observed on corn following the use of herbicides. When corn plants were treated with a regular dosage of 2,4-D, aphid numbers on the corn were twice those of aphids on untreated corn (Oka and Pimentel, 1976). Also, corn borer infestations average 28% on a 2,4-D-treated corn population compared with only 16% on an untreated corn population.

In laboratory investigations of the impact of 2,4-D on the relative resistance of corn plants to pathogens, exposed corn plants were significantly more susceptible to corn smut disease [*Ustilago maydis* (Beckm.) Ung.] (Oka and Pimentel, 1974) and to southern corn leaf blight (*Helminthosporium maydis* Nisik. and Miyake) (Oka and Pimentel, 1976) than unexposed plants. The relationship of pesticide use and crop physiology needs more research to avoid problems.

## 11.3.13. Ecology of Pests and Crops

In summary, it is apparent that humans have altered their natural environments and crop environments in many ways. Sometimes the result has been advantageous, but all too often the changes have enabled pests to increase the damage they inflict on valued crops.

## 11.4. ECONOMIC LOSSES DUE TO PESTS

To control pest damage better, more pesticides are used each year and these increase the cost of raising crops. Each year in the United States about 37% of all crops are lost to pests (13% to insects, 12% to pathogens, and 12% to weeds) despite the use of all pesticidal and nonchemical controls (Pimentel, 1981). From 1942–1951 to the present the estimated losses to weeds have fluctuated and are now slightly less than the first estimate of 13.8%, whereas losses to plant pathogens have increased slightly from 10.5% to 12% (Table 11.1).

Crop losses inflicted by insect pests, however, have nearly doubled (7% to about 13%) from the 1940s to the present (Table 11.1). This occurred despite a 10-fold increase in insecticide use. Fortunately, the impact of this loss has been effectively offset by increased crop yield, obtained through the use of higher yielding varieties and increased use of fertilizers and other fossil energy inputs.

The substantial increase in crop losses due to insect damage despite increased insecticide use can be accounted for by some of the major changes that have taken place in agricultural technology since the 1940s. These include the planting of some crop varieties that are more susceptible to insect pests; destruction of natural enemies of certain pests with insecticides, which created the need for additional pesticide treatments (e.g., cotton); increase in insecticide resistance; reduced crop rotations and crop diversity; reduced Food and Drug Administration tolerance for insects and insect parts in foods; increased "cosmetic standards" for fruits and vegetables; reduced field sanitation, including less attention to the destruction of infected fruit and crop residues; reduced tillage; culturing some crops in climatic regions favorable to pests; use of herbicides and fungicides that increase insect pest problems (Pimentel et al., 1978b).

Crop losses from birds and mammals overall are estimated to be less than 1%. However, with certain crops located where birds or mammals are abundant, losses may range from 10 to 50%. For example, corn and small grain losses may be significant near lakes and swamps where large numbers of blackbirds nest.

Table 11.1  **Comparison of Annual Pest Losses (dollar) in the United States for the Periods 1904, 1910, 1942–1951, 1951–1960, 1974, 1980, and an Estimate of Losses if no Pesticides Were Used and Some Nonchemical Alternatives Were Employed**[a]

| | Percentages of Pest Losses in Crops | | | |
|---|---|---|---|---|
| | Insects | Diseases | Weeds | Total |
| Without pesticides | 18 | 15 | 9 | 42 |
| 1980 | 13 | 12 | 12 | 37 |
| 1974 | 13 | 12 | 8 | 33 |
| 1951–1960 | 12.9 | 12.2 | 8.5 | 33.6 |
| 1942–1951 | 7.1 | 10.5 | 13.8 | 31.4 |
| 1910–1935 | 10.5 | NA | NA | NA |
| 1904 | 9.8 | NA | NA | NA |

[a]Pimentel, 1981; Chandler, 1981.

Harvested crops annually are valued at about $83 billion (USDA, 1983). This figure is after the 37% loss. Therefore, the potential value of all crops before loss is calculated be about $135 billion. Thus, the total monetary loss attributed to pest damage is about $50 billion per year.

Losses in livestock due to diseases in North America are estimated to be 17.5% despite all control efforts (Drummond et al., 1978). This figure often includes losses due to the reactions of livestock to infestations or arthropod and helminth pests. With livestock losses estimated at 17.5%, then the value of livestock losses to pests is calculated to be about $13 billion. This livestock loss added to the $50 billion in crop loss, suggests that total agricultural annual losses to pests are about $63 billion. This about equals the value of the annual oil imports. Clearly, this is a substantial loss even though it does not include the costs of pest control.

Another way to estimate the direct benefits of pesticides to crop production is to calculate crop losses that occur when the crops are grown without pesticides. Such an analysis has limitations (Pimentel et al., 1978b). The data for current crop loss to pests, despite pesticide use, are based primarily on USDA data (especially USDA, 1965). These estimates of crop loss without pesticide use are probably exaggerated because of the nature of experimental field research and because of an unknown "overlap" factor (Pimentel et al., 1978b). Most field experiments are run testing the effectiveness of either an insecticide, a fungicide, or an herbicide. Adding the results of these separate tests with individual pesticides suggests, for example, that apple losses without pesticides would be 126%!

Even with the limitations mentioned, some clarification of pesticide value is possible. At the current level of insecticide use the total dollar value of crop losses caused by insects is estimated, as mentioned, to be 13% of the potential crop value. When insecticides were withdrawn and a few readily available nonchemical controls were substituted when possible, dollar values of crop losses due to insects were calculated to increase an estimated 5% (Pimentel et al., 1978b). The nonchemical alternatives considered were cultural control methods and the use of pathogenic insect viruses and bacteria. The alternatives were limited to those useful only on corn, cotton, lettuce, and cole crops. While additional alternative controls might be employed, it is doubtful that they would have substantially reduced crop losses from insects (Pimentel et al., 1978b).

The crop loss due to plant pathogens if fungicides were withdrawn was calculated to increase from the current 12 to 15% (Pimentel et al., 1978b). The additional loss of 3% is not as large as the loss due to the withdrawal of insecticides (5%), mainly because less than 1% of crop acreage is treated with fungicides. No substitute nonchemical controls were considered to be readily available to control any of the plant pathogens now controlled by fungicides. Of course, there are some alternatives, such as sanitation, the increased use of resistant varieties, and the elimination of alternate hosts, but these were not considered in this analysis (Pimentel et al., 1978b).

If herbicides were withdrawn, increased crop losses were not as great as with either insects and diseases (Pimentel et al., 1978b). Weed control, especially in row crops, can be accomplished relatively effectively with mechanized and cultural methods. Indeed, mechanical cultivation has been practiced for many years and continues to be effectively used today in most crops.

Together, mechanical and cultural weed controls were calculated to cost 50% more than herbicidal control. This greater cost associated with substituting mechanical and cultural control increased the dollar losses associated with weeds by an estimated 1% over current losses (Pimentel et al., 1978b).

In summary, assuming no overlap in the increased losses from insects, diseases, and weeds, the total increase in loss was calculated to be 9% (5%, 3%, and 1%) higher than current losses. This 9% represents $12 billion in crops. With current pesticide treatment costs (material and application) estimated to be about $3 billion, then the return per dollar invested is about $4.

## 11.5. COSTS OF PEST CONTROL

### 11.5.1. Pesticides

The $4 return per dollar invested for pesticidal control is an excellent return and agrees with earlier estimates of $3 to $5 return per dollar invested for pesticide controls (PSAC, 1965; Headley, 1971; Pimentel, 1973).

### 11.5.2. Biological Controls

Likewise the dollar return when parasite and predator species have been introduced into the United States for control of insect and weed pests has been excellent. In some cases, like the cottony cushion scale of citrus and Klamath weed, control was spectacular and effective (PSAC, 1965). No further attention to these pests has been necessary as long as pesticide use did not reduce the natural enemy populations (see Metcalf, Chapter 10).

The only costs for these and other biological controls were the research costs to introduce the appropriate natural enemy. However, on the basis of available literature, it is apparent that in only one introduction in 20 is there potential to have an effective biological control success (Hokkanen and Pimentel, 1984). But even when the failures over the years are taken into account, the successes still provide about a $40 return per dollar invested in research and transporting the biocontrol agent into the United States (PSAC, 1965). In California, ratios of gains have ranged up to $30 to $100 (Huffaker et al., 1976).

Assessing the costs of research and total benefits accruing from biological control in Australia for four biocontrol projects over a 25-year period, Marsden et al. (1980) calculated that the return per dollar invested was about $30. This agrees well with returns achieved in the United States.

### 11.5.3. Host-Plant Resistance

Another way to reduce loss and increase return in dollar investment has been to select host plants for resistance to insect and plant pathogens. Notable are the control of insect pests of wheat, alfalfa, barley, sorghum, and corn that have resulted from research and development on host plant resistance (Gallun et al., 1975; Schalk and Ratcliffe, 1976). Using this technique over a 10-year period has provided a $300 return per dollar invested in research.

Research costs are the only dollar investment when host plant resistance is used as a pest deterrent, because resistance is inherent in the seeds themselves. The technique has a great many advantages. The few associated environmental risks, which include the possible increase of toxic chemicals in the crop plant or altering the nutritional content of the food, can be minimized if they receive attention during the genetic selection process (Pimentel et al., 1984).

### 11.5.4. Crop Rotations

Costs of pest control can also be decreased by rotating crops (PSAC, 1965; NAS, 1975). For example, corn rootworms in corn are controlled when corn is planted after a legume or small grain (Pimentel et al., 1977). However, several factors, especially economics, may determine whether or not it is feasible to rotate crops.

Planting a legume or small grain as a rotation crop with corn may cause a net reduction in profit per hectare. Estimates are that planting an alternative crop to corn would cost about $37 per additional hectare. This is because the profits per hectare with corn are greater than with a legume or small grain crop. Therefore, using crop rotation would cost about $37/ha (Pimentel et al., 1979). With the control of the rootworm using an insecticide costing only about $8/ha, then the nonchemical control will not be economically feasible. Of course, what is not included in this assessment are the environmental costs of using an insecticide, such as development of resistance, destruction of natural enemies, and human poisonings (Pimentel et al., 1980b).

### 11.5.5. Sanitation

Destroying crop residues that harbor insect pests and plant pathogens is an effective pest control strategy. For example, shredding cotton stalks and other cotton plant debris controls the boll weevil (Sterling, 1984), but costs about $50/ha. Also, plowing the soil and burying the pests destroys insects, plant pathogens, and weeds and is 90% effective in destroying such pests as the corn borer (Pimentel et al., 1982b). Again, the costs are higher than for pesticides based on an estimated $50 cost ($30 for manpower, $15 for fuel, and $5 for equipment) to till a hectare. In general, pesticide controls are made less costly than various sanitation practices, but as mentioned, numerous environmental risks are associated with the use of chemicals.

### 11.5.6. Planting Time

Crop planting times can be manipulated to minimize pest attack. Delaying the planting of a crop until after the pest has emerged and died can control certain pests. For example, delayed seeding of wheat has proven to be effective in controlling the fall brood of the Hessian fly pest of wheat. Although research developing this technique and its implementation cost about $2 million per year, the resulting increase in yield was valued at $114 million per year. Thus, the net return per dollar invested in control was about $55, an excellent return for dollar invested (PSAC, 1965).

### 11.5.7. Integrated Pest Management (IPM)

Limited information is available on the costs and benefits of integrated pest management programs. In Central American cotton production, IPM returned a net profit of two-fold that of conventional pest control (ICAITI, 1977). This resulted in part from reducing insecticide use by one-third while increasing cotton yields 15%.

Similarly, the Texas High Plains Boll Weevil control program is reported to provide $82 million annual benefits to the state (Frisbie and Adkisson, 1985). In the Lower Rio Grande Valley Cotton Program, an investigation employing combined short-season and manipulated irrigation, provided a net profit return of $121.65/ha greater than conventional pesticide and irrigation systems (Frisbie and Adkisson, 1984). These and other carefully managed IPM programs have the potential to provide Texas with more than $250 million in economic benefits each year.

Net profits in producing apples in northeastern orchards employing IPM instead of conventional central state programs ranged from $64 to $243/ha per year (White and Thompson, 1982; Whalon and Croft, 1983). At the same time, there were reduced external costs because of the reduction in insecticide effects on the environment.

Another successful IPM program was carried out in Wisconsin for control of the alfalfa weevil and increased profits by as much as $69/ha (McGuckin, 1983) and reduced insecticide use by 75% over conventional control (Frisbie and Adkisson, 1984). The economic, as well as the ecological benefits, of such an IPM program should be clear to both farmers and the public.

### 11.6. ENVIRONMENTAL AND SOCIAL COSTS OF PESTICIDE CONTROLS

### 11.6.1. Human Poisonings

In the United States, the best estimate is that about 45,000 human pesticide poisonings occur annually, with about 3,000 of these serious enough to hospitalize the patient (Pimentel et al., 1980b). In addition, an estimated 200 fatalities are attributed to pesticides each year. Of the fatalities, about 50 are due to accidental poisoning (EPA, 1974).

Pesticides have also been implicated in the incidence of cancer (Radomski et al., 1968; Cassarett et al., 1968; Wassermann et al., 1976, 1978; Infante et al.,

1978). Although Clark et al. (1977) reported that cotton and vegetable farming in the southeastern United States accounted for 1.6–6.7% of the cancer in their sample, Schotterfeld (personal communication, 1978) estimated that the fraction of human cancer attributable to pesticides is probably less than 1%. If it is assumed that only 0.5% of all human cancer is caused by pesticides, then the annual social cost to the nation for this cancer is about $125 million (Pimentel et al., 1980b). Adding this cost to the costs associated with hospitalization and human deaths, the total rises to a yearly cost of $184 million (Pimentel et al., 1980b). Many would contend this places a low value on a human life.

## 11.6.2. Livestock Destruction and Contamination

Many livestock are poisoned by pesticides and a wide array of livestock products, including meat, milk, and eggs, are contaminated with pesticides. An assessment of the annual economic losses due to animal poisonings as contaminated products, indicated that the total value lost was about $12 million (Pimentel et al., 1980b).

## 11.6.3. Destruction of Natural Enemies and Pesticide Resistance

Many insect and mite populations present in agroecosystems remain as minor and unimportant pests in crops as long as their natural enemies are present to keep them under control (DeBach, 1964; Huffaker, 1971). But when insecticides or other pesticides are employed for control of a major pest, some natural enemy populations may be reduced or eliminated. Over the years this has contributed to outbreaks of pests that were previously not a problem (Pimentel and Edwards, 1982).

An analysis was made of what the additional control costs would be if natural enemies were destroyed, and only the cost of additional sprays and/or more expensive insecticides used to bring the new pest outbreaks under control was included. The calculation suggested an estimated annual cost of $153 million in treatments to compensate for the loss of the natural enemy controls (Pimentel et al., 1980b). Certainly these costs could be substantially reduced if more careful pesticide application procedures were used.

In addition to destroying natural enemies, the widespread use of pesticides often has resulted in pests evolving resistance. An estimated 428 insects and mites are now resistant to pesticides (Georghiou, 1983). The development of resistance in crop-pest populations often requires additional sprays and/or the use of more effective, more expensive pesticides substituted for the previously used pesticides.

In one cost analysis, it was calculated that about $128 million is spent each year to control resistant insects and mites in field crops, vegetables, and fruit (Pimentel et al., 1980b). For the control of livestock and public health pests that have evolved resistance, estimates are that $15 million is spent for additional pesticidal controls. Thus, the total cost attributed to the development of resistance in various pest populations is about $133 million each year.

### 11.6.4. Honey Bee and Pollination Losses

Honey bees and wild bees, essential to the pollination of fruits, vegetables, and forage crops are highly susceptible to insecticides used in agriculture and forestry (McGregor, 1976). Estimates of annual agricultural losses due to poor pollination caused by pesticides range from about $80 million (Atkins, personal communication, 1977) to a high of $4 billion (S. E. McGregor, personal communication, 1977).

In addition to these decreased crop yields because of diminished pollination, honey bee colonies are destroyed and honey production is reduced following improper pesticide application. The annual cost is reported to be about $55 million. Combined, these losses total about $135 million each year (Pimentel et al., 1980b).

### 11.6.5. Crop Destruction by Pesticides

Pesticide use normally protects crops from pests and raises yields; however, at times crops are damaged after pesticide use. This can occur when pesticide drifts from a treated crop to another more sensitive and/or persists in the soil at levels toxic to a sensitive crop planted in the soil the following year. This occurred in Texas when 2,4-D being applied by aircraft to wheat drifted from the wheat acreage onto highly sensitive cotton crops nearby. As a result, an estimated $10.5 million of cotton has been destroyed in the Lubbock area this summer and a wide array of lawsuits is in process because of this drift problem (Silbert, 1984; R. E. Frisbie, personal communication, 1984). The episode in Texas during the 1984 growing season further illustrates the problem of drift and crop destruction. Note, only about 50% of the pesticide lands in the target area, the other half drifts onto nearby agroecosystems or natural ecosystems (Ware et al., 1970). The value of this loss is about $10.5 million. Estimates are that on average about $70 million or more crops are destroyed annually because of drift and soil contamination from pesticide applications (Pimentel et al., 1980b).

### 11.7. SUMMARY

An estimated 37% of all crops is lost annually to pests despite the use of all pesticidal and nonchemical controls. The dollar value of crop losses is calculated to be $50 billion annually, and when livestock losses are added the annual loss total rises to $63 billion. This only represents the great decrease in food and fiber yield and does not include external costs to the environment.

Despite the fact that insecticide use has risen 10-fold during the last 40 years, losses due to insects have nearly doubled during this same period. These increased losses occur because of numerous changes that have taken place in agricultural practices. This confirms the need for ecological analyses of agroecosystems to determine what factors in each cropping system are contributing to pest outbreaks. Such an approach will allow us to employ sound ecological principles to manage pest populations for the benefit of society and the environment.

At present, the estimated $3 billion spent for pesticidal controls annually prevents an estimated $12 billion in crop losses. However, as mentioned, most of this $12 billion in potential loss occurs because of situations created by current agricultural practices that encourage pest outbreaks. Thus, in one sense the benefits claimed for pesticide use exist because we have created agricultural ecosystems that require heavy pesticide use.

Pesticidal controls are an excellent technology when employed as a part of an ecological approach in pest management. They have the potential to give a $4 return per dollar invested in control. In addition, various nonchemical controls like biological controls and host plant resistance have provided returns in terms of reduced crop losses of $30 to $300 per dollar invested in control. This indicates the potential such controls have in the future.

Even when the environmental and social costs of pesticidal controls, estimated to be at least $1 billion annually, are considered, pesticidal controls remain economically profitable—returning about $3 per dollar invested in control.

Although future research in pest control has the potential to reduce the economic costs of pest control, it is vital that new techniques focus on reducing the high environmental and social costs currently borne by society because of the use of pesticide chemicals. More careful use of pesticides coupled with a greater knowledge of agroecosystems will enable agriculturists to reduce substantially the damage loss while preserving the integrity of the environment.

## REFERENCES

Adams, W. E. (1949). Loss of topsoil reduced crop yields, *J. Soil Water Conserv.*, **4**, 130–133.

Andrewartha, H. G., and L. C. Birch. (1954). *Distribution and Abundance of Animals*, University of Chicago Press, Chicago, 782 pp.

Barker, J. S., and O. E. Tauber. (1951). Fecundity of and plant injury by the pea aphid as influenced by nutritional changes in the garden pea, *J. Econ. Entomol.*, **44**, 1010–1012.

Buntley, G. J., and F. F. Bell. (1976). Yield estimates for the major crops grown on soils of West Tennessee, Tenn. Agr. Exp. Sta. Bull. 561, 124 pp.

Cassarett, L. J., G. C. Fryer, W. L. Yauger, and H. W. Klemmer. (1968). Organochlorine pesticide residues in human tissues—Hawaii, *Arch. Environ. Health*, **17**, 306.

Chandler, J. M. (1981). "Estimated losses of crops to weeds," in D. Pimentel, Ed., *CRC Handbook of Pest Management in Agriculture*, Vol. I, CRC Press, Boca Raton, FL, pp. 95–109.

Clark, L. C., C. M. Shy, B. M. Most, J. W. Florin, and K. M. Portier. (1977). Cancer mortality and agricultural pesticide use in the southeastern United States, 8th Int. Sci. Mtg., Internatl. Epidemiol. Assoc. San Juan, Puerto Rico, 25 pp.

Cromartie, W. J. (1981). "The environmental control of insects using crop diversity," in D. Pimentel, Ed., *Handbook of Pest Management in Agriculture*, Vol. I, CRC Press, Boca Raton, FL, pp. 223–251.

Dahms, R. G., and R. H. Painter. (1940). Rate of reproduction of the pea aphid on different alfalfa plants, *J. Econ. Entomol.*, **33**, 688–692.

Debach, P. H., Ed. (1964). *Biological Control of Insect Pests and Weeds*, Reinhold, New York, 844 pp.

Delwiche, C. C. (1970). The nitrogen cycle, *Sci. Am.*, **223**(3), 137–158.

Drummond, R. O., R. A. Bram, and N. Konnerup. (1978). "Animal pests and world food production," in D. Pimentel, Ed., *World Food, Pest Losses, and the Environment*, Westview Press, Boulder, CO, pp. 63-93.

Elton, C. S. (1927). *Animal Ecology*, Sidgwick and Jackson, London, 207 pp.

Elton, C. S. (1958). *The Ecology of Invasions by Animals and Plants*, Methuen, London, 159 pp.

EPA. (1974). *Strategy of the Environmental Protection Agency for Controlling the Adverse Effects of Pesticides*, Environmental Protection Agency, Office of Pesticide Programs, Office of Water and Hazardous Materials, Washington, D.C., 36 pp.

Foster, G. R. and L. D. Meyer. (1977). "Soil erosion and sedimentation by water—An overview," in *Proceedings of the National Symposium on Soil Erosion and Sedimentation by Water*, American Society of Agricultural Engineers, ASAE Publication 4-77, 151 pp.

Frisbie, R. E., and P. L. Adkisson, (1985). "IPM: Definitions and current status in U.S. agriculture" in M. A. Hoy and D. C. Herzog, Eds., *Biological Control in Agricultural Integrated Pest Management Systems*, Academic, New York, pp. 41-51.

Gallun, R. L., K. J. Starks, and W. D. Guthrie. (1975). Plant resistance to insects attacking cereals, *Annu. Rev. Entomol,* **20,** 337-357.

GAO. (1983). *Agriculture's Soil Conservation Programs Miss Full Potential in the Fight Against Soil Erosion*, U.S. General Accounting Office, U.S. Government Printing Office, Washington, D.C., 79 pp.

Georghiou, G. P. (1983). "Management of resistance in arthropods," in G. P. Georghiou and T. Saito, Eds., *Pest Resistance to Pesticides*, Plenum, New York, pp. 769-792.

Haseman, L. (1946). Influence of soil minerals on insects, *J. Econ. Entomol.,* **39,** 8-11.

Hawkes, J. G. (1944). *Potato Collecting Expeditions in Mexico and South America*, Imperial Bureau of Plant Breeding & Genetics, School of Agriculture, Cambridge, England, 142 pp.

Headley, J. C. (1971). "Productivity of agricultural pesticides," in *Economic Research on Pesticides for Policy Decision Making*, Proc. Symp. Econ. Res. Serv., USDA, pp. 80-88.

Hokkanen, H., and D. Pimentel. (1984). New approach for selecting biological control agents, *Can. Entomol.,* **116,** 1109-1121.

Huffaker, C. B., Ed. (1971). *Biological Control*, Plenum, New York, 511 pp.

Huffaker, C. B., J. F. Simmonds, and J. E. Laing. (1976). "The theoretical and empirical basis of biological control," in C. B. Huffaker and P. S. Messenger, Eds., *Theory and Practice of Biological Control*, Academic, New York, pp. 41-78.

ICAITI. (1977). *An Environmental and Economic Study of the Consequences of Pesticide Use in Central American Cotton Production*, Final Report, Instituto Centro-Americano de Investigacion y Tecnologia Industrial, Guatemala, 295 pp.

Infante, P. F., S. S. Epstein, and W. A. Newton. (1978). Blood dyscrasias and childhood tumors and exposure to chlordane and heptachlor, *Scand J. Work Environ. Health,* **4,** 137-150.

Klass, D. C. L. (1978). Polyculture cropping systems: Review and analysis, Cornell Internatl. Agr. Bull. 32, 69 pp.

Langdale, G. W., J. E. Box, Jr., R. A. Leonard, A. P. Barnett, and W. G. Fleming. (1979). Corn yield reduction on eroded southern piedmont soils, *J. Soil Water Conserv.,* **34,** 226-228.

Larson, W. E. (1981). Protecting the soil resource base, *J. Soil Water Conserv.,* **36,** 13-16.

Lebedev, V. A. (1942). On means for combatting garden insects of the genus *Phyllotreta* and on the effect of these on the growth and yield of plants, *Zachsh. Rast.,* **I,** 131-138 (in Russian; translated by E. Matthews).

Lee, L. K. (1984). Land use and soil loss: A 1982 update, *J. Soil Water Conserv.,* **39,** 226-228.

Lupton, F. G. H. (1977). "The plant breeders' contribution to the origin and solution of pest and disease problems," in J. M. Cherrett and G. R. Sagar, Eds., *Origins of Pest, Parasite, Disease and Weed Problems*, Blackwell, Oxford, pp. 71-81.

Marchal, P. (1908). The utilization of auxiliary entomophagous insects in the struggle against insects injurious to agriculture, *Pop. Sci. Mon.*, **72**, 352-370, 406-419.

Marsden, J. S., G. E. Martin, D. J. Parham, T. J. Ridsdill Smith, and B. G. Johnston. (1980). *Returns on Australian Agricultural Research*, The Joint Industries Assistance Commission-CSIRO benefit-cost study of the CSIRO Division of Entomology, Canberra, 107 pp.

Matthysse, J. G. (1959). An evaluation of mist plowing and sanitation in Dutch elm disease control programs, Cornell Misc. Bull. 30, N.Y.S. Coll. Agr., 16 pp.

McGregor, S. E. (1976). Insect pollination of cultivated crop plants, Agricultural Handbook No. 496, USDA, Agr. Res. Service, Washington, D.C., 412 pp.

McGuckin, T. (1983). Alfalfa management strategies for Wisconsin dairy farms—application of stochastic dominance, *North Central J. Agr. Econ.*, **5**, 43-49.

Murray, R. C., and E. B. Reeves. (1977). Estimated use of water in the United States in 1975, U.S. Geol. Surv. Circ. 765, 39 pp.

NAS. (1975). *Pest control: An Assessment of Present and Alternative Technologies*, Vols. I-V, National Academy of Sciences, Washington, D.C., 506 pp., 169 pp., 139 pp., 170 pp., 282 pp.

Oka, I. N., and D. Pimentel. (1974). Corn susceptibility to corn leaf aphids and common corn smut after herbicide treatment, *Environ. Entomol.*, **3**, 911-915.

Oka, I. N., and D. Pimentel. (1976). Herbicide (2,4-D) increases insect and pathogen pests on corn, *Science*, **193**, 239-240.

OTA. (1982). *Impact of Technology on U.S. Cropland and Rangeland Productivity*, Office of Technology Assessment, U.S. Government Printing Office, Washington, D.C., 266 pp.

Penman, H. L. (1970). The water cycle, *Sci. Am.*, **223**(3), 99-108.

Pimentel, D. (1961a). Animal population regulation by the genetic feedback mechanism, *Am. Nat.*, **95**, 65-79.

Pimentel, D. (1961b). Species diversity and insect population outbreaks, *Ann. Ent. Soc. Am.*, **54**, 76-86.

Pimentel, D. (1961c). The influence of plant spatial patterns on insect populations, *Ann. Entomol. Soc. Am.*, **54**, 61-69.

Pimentel, D. (1973). Extent of pesticide use, food supply, and pollution, *J. N.Y. Entomol. Soc.*, **81**, 13-33.

Pimentel, D., Ed. (1981). *Handbook of Pest Management in Agriculture*, Vols. I-III, CRC Press, Boca Raton, FL, 597 pp., 501 pp., and 656 pp.

Pimentel, D. (1984). "Genetic diversity and stability in parasite-host systems," in B. Shorrocks, Ed., *Evolutionary Ecology*, Blackwell, Oxford, 418 pp.

Pimentel, D., and C. A. Edwards. (1982). Pesticides and ecosystems, *BioScience*, **32**, 595-600.

Pimentel, D., and C. W. Hall, Eds. (1984). *Food and Energy Resources*, Academic, New York, 268 pp.

Pimentel, D., and W. Dazhong. (1986). "Technological changes in energy use in agricultural production," in S. R. Gliessman, Ed., *Research Approaches in Agricultural Ecology* (in press).

Pimentel, D., S. A. Levin, and A. B. Soans. (1975). On the evolution of energy balance in exploiter-victim systems, *Ecology*, **56**, 381-390.

Pimentel, D., E. C. Terhune, R. Dyson-Hudson, S. Rochereau, R. Samis, E. Smith, D. Denman, D. Reifschneider, and M. Shepard. (1976). Land degradation: effects on food and energy resources, *Science*, **194**, 149-155.

Pimentel, D., C. Shoemaker, E. L. LaDue, R. B. Rovinsky, and N. P. Russell. (1977). Alternatives for reducing insecticides on cotton and corn: economic and environmental impact, Environ. Res. Lab., Off. Res. Develop., EPA, Athens, GA (issued in 1979), 145 pp.

Pimentel, D., D. Nafus, W. Vergara, D. Papaj, L. Jaconetta, M. Wulfe, L. Olsvig, K. Frech, M.

Loye, and E. Mendoza. (1978a). Biological solar energy conversion and U.S. energy policy, *BioScience*, **28**, 376–382.

Pimentel, D., J. Krummel, D. Gallahan, J. Hough, A. Merrill, I. Schreiner, P. Vittum, F. Koziol, E. Back, D. Yen, and S. Fiance. (1978b). Benefits and costs of pesticide use in U.S. food production, *BioScience*, **28**, 772, 778–784.

Pimentel, D., J. Krummel, D. Gallahan, J. Hough, A. Merrill, I. Schreiner, P. Vittum, F. Koziol, E. Back, D. Yen, and S. Fiance. (1979). "A cost-benefit analysis of pesticide use in U.S. food production," in T. J. Sheets and D. Pimentel, Eds., *Pesticides: Their Contemporary Roles in Agriculture, Health and the Environment*, Humana, Clifton, NJ, 186 pp.

Pimentel, D., E. Garnick, A. Berkowitz, S. Jacobson, S. Napolitano, P. Black, S. Valdes-Cogliano, B. Vinzant, E. Hudes, and S. Littman. (1980a). Environmental quality and natural biota, *BioScience*, **30**, 750–755.

Pimentel, D., D. Andow, R. Dyson-Hudson, D. Gallahan, S. Jacobson, M. Irish, S. Kroop, A. Moss, I. Schreiner, M. Shepard, T. Thompson, and B. Vinzant. (1980b). Environmental and social costs of pesticides: a preliminary assessment, *Oikos*, **34**, 127–140.

Pimentel, D., S. Fast, W. L. Chao, E. Stuart, J. Dintzis, G. Einbender, W. Schlappi, D. Andow, and K. Broderick. (1982a). Water resources in food and energy production, *BioScience*, **32**, 861–867.

Pimentel, D., C. Glenister, S. Fast, and D. Gallahan. (1982b). Environmental Risks Associated with the Use of Biological and Cultural Pest Controls, Final Report, NSF Grant PRA 80-00803, 165 pp.

Pimentel, D., C. Glenister, S. Fast, and D. Gallahan. (1984). Environmental risks of biological pest controls, *Oikos*, **42**, 283–290.

PSAC. (1965). *Restoring the Quality of Our Environment*, Report of Environmental Pollution Panel, Pres. Sci. Adv. Comm., The White House, Washington, D.C., 317 pp.

Radomski, J. L., W. B. Deichmann, E. E. Clizer, and A. Rey. (1968). Pesticide concentrations in the liver, brain, and adipose tissue of terminal hospital patients, *Food Cosmet. Toxicol.*, **6**, 209–220.

Risser, J. (1981). A renewed threat of soil erosion: It's worse than the dust bowl, *Smithsonian*, **11**, 121–131.

Root, R. B. (1973). Organization of a plain-arthropod association in simple and diverse habitats: The fauna of collards (*B. oleracea*), *Ecol. Monogr.*, **43**, 95–124.

Root, R. B. (1975). "Some consequences of ecosystem texture," in S. A. Levin, Ed., *Ecosystem Analysis and Prediction*, Soc. Ind. Appl. Math., Philadelphia, pp. 83–97.

Schalk, J. M., and R. H. Ratcliffe. (1976). Evaluation of ARS program on alternative methods of insect control: Host plant resistance to insects, *Bull. Entomol. Soc. Am.*, **22**, 7–10.

SCS. (1977). *Cropland Erosion*, Soil Conservation Service, USDA, Washington, D.C., 50 pp.

Silbert, S. (1984). Panhandle spray disaster cripples farmers, Texas Pesticide Watch, Issue 8, Texas Center for Rural Studies, Austin, Texas, 2 pp.

Simmonds, F. J., and D. J. Greathead. (1977). "Introductions and pest and weed problems," in J. M. Cherrett and G. R. Sagar, Eds., *Origins of Pest, Parasite, Disease and Weed Problems*, Blackwell, Oxford, pp. 109–124.

Spurr, S. H., and B. V. Barnes. (1980). *Forest Ecology*, 3rd ed., Wiley, New York, 687 pp.

Sterling, W. (1984). Action and Inaction Levels in Pest Management, Texas Agr. Expt. Stn., B-1480, College Station, Texas, 20 pp.

Stern, V. M. (1969). Interplanting alfalfa in cotton to control lygus bugs and other insect pests, *Proc. Tall Timbers Conf. Ecol. Anim. Contr. Habit. Mgmt.*, **1**, 55–69.

Stevens, N. E., and W. O. Scott. (1950). How long will present spring oat varieties last in the central corn belt? *Agron. J.*, **42**, 307–309.

Swanson, E. R., and C. E. Harshbarger. (1964). An economic analysis of effects of soil loss on crop yields, *J. Soil Water Conserv.*, **19**, 183–186.

Tahvanainen, J. O., and R. B. Root. (1972). The influence of vegetational diversity on the population ecology of a specialized herbivore, *Phyllotreta cruciferae* (Coleoptera: Chrysomelidae), *Oecologia*, **10**, 321–346.

Timmons, J. F. (1981). Protecting agriculture's natural resource base, *J. Soil Water Conserv.*, **38**, 5–11.

USBC. (1983). *Statistical Abstract of the United States*, 1982–1983, 103rd ed., U.S. Bureau of the Census, U.S. Government Printing Office, Washington, D.C., 1008 pp.

USDA. (1965). *Losses in Agriculture*, U.S. Department of Agriculture, Agr. Handbook No. 291, Agr. Res. Service, Washington, D.C., 120 pp.

USDA. (1973). *Final Environmental Statement on the Cooperative 1973 Gypsy Moth Suppression and Regulatory Program*, USDA Forest Service and Animal and Plant Health Inspection Service, 258 pp.

USDA. (1975). Farmers' use of pesticides in 1971—extent of crop use, Econ. Res. Serv., Agr. Econ. Rep. No. 268, 25 pp.

USDA. (1983). *Agricultural Statistics 1982*, U.S. Department of Agriculture, Washington, D.C., 566 pp.

USWRC. (1979). *The Nation's Water Resources. 1975-2000*, Vols. 1-4, Second National Water Assessment, United States Water Resources Council, Washington, D.C., 86, 376, 1342, 724 pp.

Van der Plank, J. E. (1968). *Disease Resistance in Plants*, Academic, New York, 206 pp.

Walker, J. C., R. H. Larson, and A. L. Taylor. (1958). Diseases of cabbage and related plants, USDA, Agr. Handbook No. 144, Agr. Res. Service, Washington, D.C., 41 pp.

Ware, G. W., W. P. Cahill, P. D. Gerhardt, and M. J. Witt. (1970). Pesticides drift IV. On-target deposits from aerial application of insecticides, *J. Econ. Entomol.*, **63**, 1982–1983.

Wassermann, M., D. P. Nogueira, L. Tomatis, A. P. Mirra, H. Shibata, G. Arie, S. Cucos, and D. Wassermann. (1976). Organochlorine compounds in neoplastic and adjacent apparently normal breast tissue, *Bull. Environ. Contam. Toxicol.*, **15**, 478–483.

Wassermann, M., D. P. Nogueira, S. Cucos, A. P. Mirra, H. Shibata, G. Arie, H. Miller, and D. Wassermann. (1978). Organochlorine compounds in neoplastic and adjacent apparently normal gastric mucosa, *Bull. Environ. Contam. Toxicol.*, **20**, 544–553.

Whalon, M. E., and B. A. Croft. (1983). "Implementation of apple IPM," in B. A. Croft and S. C. Hoyt, Eds., *Integrated Management of Insect Pests of Pome and Stone Fruits*, Wiley, New York, pp. 411–447.

White, G. B., and P. Thompson. (1982). The economic feasibility of tree fruit integrated pest management programs in the Northeast, *J. Northeast Agr. Econ. Council.*, **11**, 39–46.

# 12

# AGROECOSYSTEMS— STRUCTURE, ANALYSIS, AND MODELING

*Paul G. Risser*

## 12.1. INTRODUCTION

The presumption is that natural ecosystems are composed of abiotic and biotic components, which over long time periods have developed into functioning systems with some order among the interacting components. This order, however, is not a constant set of behavioral characteristics, but rather reacts via understandable principles to an ever-changing array of external and internal conditions. By studing natural ecosystems, it has been possible to describe some of these general principles of ecosystem behavior, especially in relation to productivity, nutrient cycling, and stability of these processes under various environmental conditions.

Agroecosystems are ecosystems but differ from natural ecosystems in several fundamental ways. For example, agroecosystems are usually driven by human-controlled external subsidies of energy, water, and nutrients and are modified by chemical and mechanical technology and by introduced genetics, and are socio-economic systems because of the conscious human control exerted in response to sociological and economic goals. The primary goal in the development and management of agroecosystems is usually cost-effective primary production from a few selected species. Indeed, 66% of the world food supply comes from cereals, and 80% from just 11 species (Loomis, 1983).

Despite these distinctions, agroecosystems involve biological processes characteristic of natural ecosystems. That is, energy transfer among the components adheres to the laws of thermodynamics with the resulting trophic structure; energy cycles of agroecosystems also fundamentally behave as they do in natural ecosystems though the various amounts of storage and the rates of transfer among components are modified by cultural and agricultural practices. This

321

concept is depicted in Fig. 12.1, in which three systems are contrasted in terms of human inputs and outputs on the abscissa and natural feedbacks on the ordinate axis. Natural ecosystems are characterized by minimal amounts of human inputs and outputs but are characterized by high amounts of natural feedback processing. Obversely, urban-industrial complexes are largely controlled by human inputs and outputs. Agroecosystems, however, are intermediate on both axes, thus indicating their dependence upon human inputs and outputs, but also noting that agroecosystems retain some portion of the natural feedback processes. This demanding situation means that understanding and managing agroecosystems requires a simultaneous treatment of both human and natural processes. Thus, the thesis of this chapter is that partial but essential understanding of agroecosystems can be gained by comprehending natural ecosystems.

Understanding natural ecosystems is not yet complete and, unfortunately, this incomplete understanding occurs in ecosystem attributes crucial to understanding agroecosystems. Considerable analysis of natural ecosystems has been focused on watersheds and the ways in which various management techniques affect the primary productivity and the nutrient dynamics of the watershed. The early emphasis on studying watersheds was a powerful approach made possible because the system boundaries could be defined precisely. Furthermore, the movement of water, and with it the transport of minerals and organic matter,

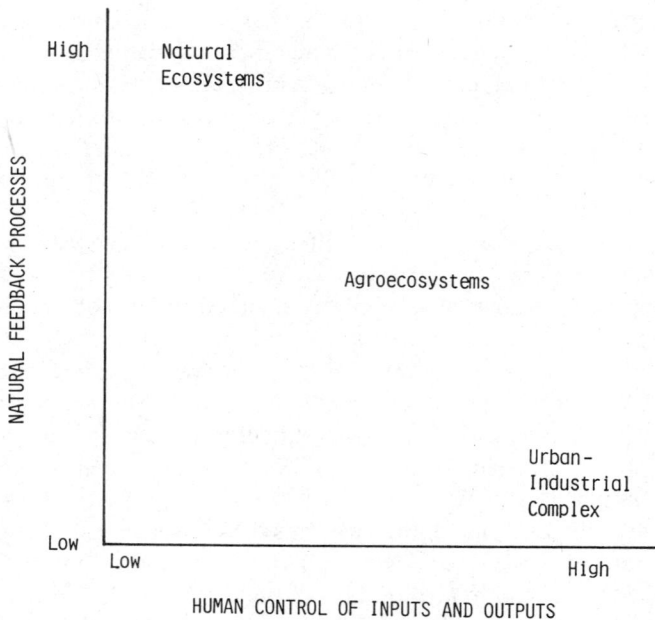

Fig. 12.1 Relation of agroecosystems to natural ecosystems and urban-industrial complexes with respect to natural feedback processes and human inputs and outputs.

was contained within the system and could be treated as a black box, that is, the inputs and outputs of the entire system could be measured and compared under different management treatments applied to the watershed. These analyses led to early principles about the retention of nutrients and some predictions about the important pathways of nutrient storage and transfer (Bormann and Likens, 1981). Far less is known about the influence of spatial pattern of vegetation on these watershed dynamics, though there is a clear recognition of the importance of these spatial characteristics (Peterjohn and Correll, 1984). Similarly, there are gaps in what is known about the obviously significant role of microbial processes on the retention and release of nutrients (Coleman et al., 1984). Also, the behavior and population dynamics of the most abundant vertebrate organisms are frequently well described, but little is known about the myriad of less conspicuous species that nevertheless may be important in understanding the ecosystem characteristics and population dynamics of the entire system (Davidson et al., 1984; Seastedt and Crossley, 1984; Simberloff, 1981). Such phenomena as the spatial pattern of vegetation in relation to nutrient patterns, the interactions of microbial populations and soil characteristics as they relate to nutrient storage and availability processes, and insect–food resource relationships are important constituents in understanding agroecosystems. Although these processes are not completely known for natural ecosystems, many of the general principles are developing and can be applied to agroecosystems.

Thus, the task at hand is to consider first the ways in which natural ecosystems are structured and have been analyzed and modeled. This process must recognize the incompleteness of our understanding of natural ecosystems, but not permit these deficiencies to restrict unnecessarily the extension of ideas about ecosystem structure and function to agroecosystems. Second, it is necessary to recognize the distinctions between natural ecosystems and agroecosystems, and then to develop appropriate analytical and modeling techniques. Clearly, this is a cybernetic process and the development of powerful ways to analyze agroecosystems should lead also to a better understanding of natural ecosystems. The approach used in this paper is to discuss first natural ecosystems and then agroecosystems. Then the two types of ecosystems are compared to demonstrate the appropriate analytic techniques.

Natural ecosystems and agroecosystems are subjected to widespread natural and anthropogenic influences, such as air pollutants, acid precipitation, changes in atmospheric carbon dioxide and methane, and to short- and long-term climatic changes (Burgess, 1984; Golitsyn, 1985). Although the responses of natural ecosystems and agroecosystems may be different, both ecosystem types are subjected to these pervasive external influences. However, agroecosystems are uniquely the recipients of a variety of human influences and contrivances designed to maximize harvestable yield of certain food and fiber products, and in this endeavor natural ecosystem processes are frequently altered. Some human influences have already changed the fundamental processes in these human-driven ecosystems (Loomis, 1984), and the future trajectory of other influences is already clear. Therefore, the last section of this paper discusses ways in which agroecosystems are likely to be managed in the future and how analytic and modeling procedures could be designed to acccomodate these future management strategies.

## 12.2. ANALYSIS OF NATURAL ECOSYSTEMS

Analysis of natural ecosystems grew largely from the energy pyramid and food chain concepts of forest and grassland biological systems (Murphy, 1977; French, 1979; Edwards et al., 1980; O'Neill and DeAngelis, 1980). Energy or primary production was the commodity under consideration, and the analysis framework was the movement of energy either from one trophic component to another or from prey to predator. At first, the analytic procedure was to identify compartments such as aboveground biomass or soil organic matter. Then the contents of each compartment were measured through a time period, and the flow of materials between compartments was inferred. Later analyses have focused on the flows between and among compartments and the biological and chemical processes that cause these flows. For example, the energy and material transfers in a stylized ecosystem are shown diagrammatically in Fig. 12.2. This diagram or model can be converted to a functional model by writing difference or differential equations to approximate the movement of materials or energy among the components, or to simulate the processes that control the flows.

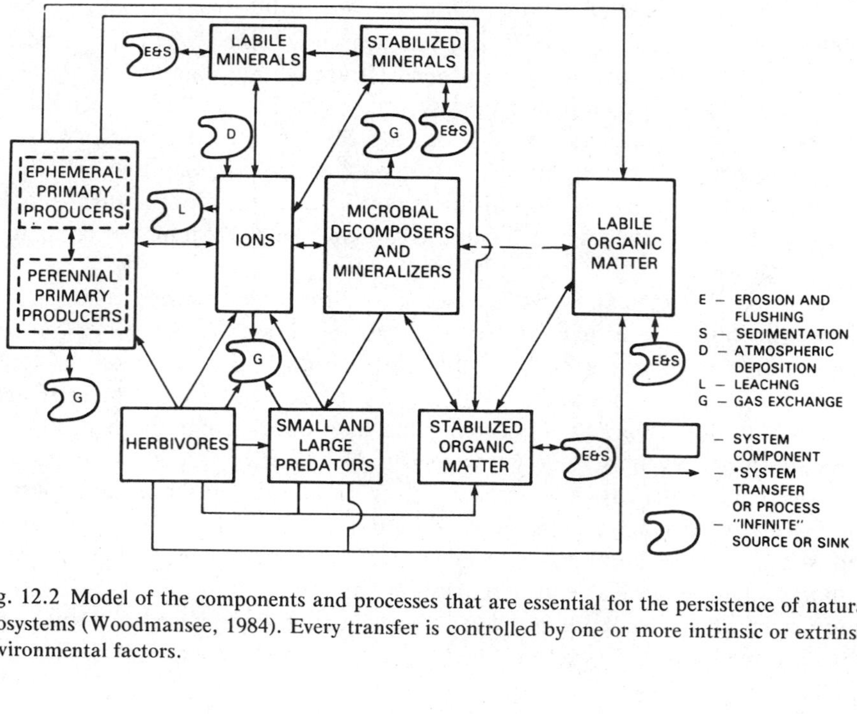

Fig. 12.2 Model of the components and processes that are essential for the persistence of natural ecosystems (Woodmansee, 1984). Every transfer is controlled by one or more intrinsic or extrinsic environmental factors.

If portions of the ecosystems are exposed to more detailed study, then the relevant components and flows can be further subdivided. Figure 12.3 shows such as analysis for the nitrogen cycle of the tallgrass prairie. Here, the components are more finely defined and the depicted processes are more explicit. Large ecosystem models (Innis, 1978) include the detail of Fig. 12.3 in models as large as the scope of Fig. 12.2. These complex models have been used to evaluate the flow of materials and energy in seminatural ecosystems subjected to various management techniques (Risser et al., 1981). Constructing large ecosystem simulation models has been of great heuristic value since researchers were required to study and mathematically describe the simultaneous interactions of many ecosystem components. Such modeling efforts have forced a more logical and

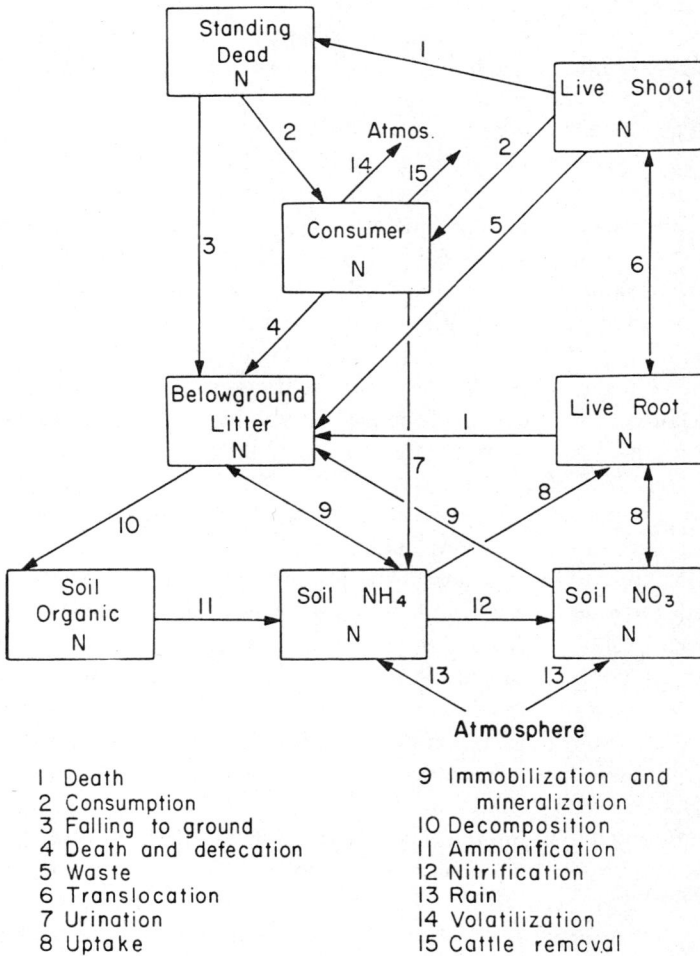

| 1 Death | 9 Immobilization and |
| 2 Consumption | mineralization |
| 3 Falling to ground | 10 Decomposition |
| 4 Death and defecation | 11 Ammonification |
| 5 Waste | 12 Nitrification |
| 6 Translocation | 13 Rain |
| 7 Urination | 14 Volatilization |
| 8 Uptake | 15 Cattle removal |

Fig. 12.3 Nitrogen cycle of the tallgrass prairie (Risser and Parton, 1982).

thorough understanding of natural ecosystems. These models now are routinely used for anticipating the consequences of changes in the driving variables or for predicting the transport and fate of materials in ecosystems.

A few very large ecosystem simulation models have been built (Innis, 1978) consisting of hundreds of equations. This approach arose from the assumption that to simulate ecosystem behavior, most of the known biological and physical–chemical processes should be included. These large models are expensive to develop and operate, and frequently become so complex that interpretation of the results is difficult because of the myriad interactions. Also, the data required to build and test such a model are difficult to obtain, especially in the comprehensive manner implied in the model. As a result, most natural ecosystem models now either address parts of the ecosystem (Detling et al., 1979) or are designed to include the minimum complexity that allows a reasonable approximation of system behavior (Wight et al., 1984).

Current ecosystem analysis of natural systems has now begun to focus on the processes that control ecosystem behavior. For example, the complexity of the relationship between microorganisms and microinvertebrates as both types of organisms contribute to the decomposition process (Coleman, et al., 1984) is particularly amenable to analysis with a model. Rates of decomposition of soil organic matter depend on variables such as the amount of organic material, its chemical constituents, carbon-to-nitrogen ratio, soil water content and temperature, and the population densities of bacteria, fungi, and invertebrate species. All of these variables are interrelated and change over time. Furthermore, some of the processes appear to be quite random. For instance, in decomposing logs, Fager (1968) found that the population structure of arthropods was most clearly approximated if the model assumed that the frequency of occurrence of the decomposer fauna was a measure of the probability that a species would invade the log. With more study, it may become clear that the invasion and establishment processes are not random, although at present including this element as a stochastic process provides the best estimate of arthropod populations. But the point is that natural ecosystem models include some processes in which the mechanisms are quite well known, whereas other processes appear to be either driven by or are a function of random or stochastic events. Thus, the experience with analyzing natural ecosystems does not permit a definitive description of all processes that will be important in understanding agroecosystems.

Increased attention is also being paid to the phenology of plant and animal growth as these growth and mineral scavenging characteristics relate to nutrient dynamics of both individual vegetation types and entire watersheds. As an example, Van Hook et al. (1980) measured the biomass and density of a *Liriodendron* aphid population and found that maximum aphid populations coincided in June with maximum nitrogen in the phloem and favorable canopy temperature for the aphids. There was a second aphid density peak in September just prior to leaf abscision when nitrogenous compounds were again mobile in the phloem. The impact of this consumption on the productivity of the ecosystem was not precisely known, but the aphids consumed about 1% of the annual photosynthate production and an amount equivalent to 17% of the annual standing crop of foliar nitrogen. Here again, however, it is important to note that the analysis of natural ecosystems has moved from the study of entire watersheds to a current emphasis on the actual biological processes that determine the nutrient dynamics of the system as a whole. As will be discussed subsequently, it is ultimately important to analyze the behavior of ecosystems at both levels of resolution.

In summary, analysis and modeling of natural ecosystems has resulted in an understanding of ecosystem structure, a reasonably complete catalog of the standing crop of major components, and some comparative estimates of flows between and among components. Current analysis is now directed toward understanding the processes that control the behavior of ecosystems over both short and long time periods and at various spatial scales.

## 12.3. ANALYSIS OF AGROECOSYSTEMS

Early recognition that agroecosystems were amenable to ecosystem analysis led to evaluations of the energetic efficiency of crop systems (Lemon et al., 1971) and to the development of simulation models describing individual crops. Subsequently, it has become clear that simply extrapolating the physiological behavior of a crop plant to the level of the ecosystem (or even the field) failed to capture the necessary understanding of the system as a whole. As will be shown below, this deficiency represents a major challenge and one that occupies a central place in agroecosystem research today.

Agricultural ecosystems have been evaluated using systems analyses (Spedding, 1975; Brockington, 1979; Loomis, 1984), but, in general, the emphasis has been on management strategies rather than a fundamental behavior of the ecosystem, for example, nutrient cycling (Loucks, 1977). These management strategies have included both marketing decisions and approaches to cropping schemes. Simulation models have been developed for crops grown under different environmental conditions, and these models have been used to suggest optimum strategies for irrigation, fertilization, pest control, and crop harvest (Dayan et al., 1981; Conway, 1984). After examining agricultural models, Van Dyne and Abramsky (1975) concluded that most agricultural simulation models were deterministic optimization models, primarily devoted to energy or production and using agricultural economics, and that there were few that incorporated the simulation of ecological processes. This analysis of most agricultural models is still largely true, though there is now a beginning convergence of the analytical techniques used for optimizing production and for capturing the essence of the ecological processes (Klopatek and Risser, 1982).

In considering agroecosystem analysis, it is helpful to emphasize that in addition to the ecological system behavior at the field level, there are system behavior processes at the farm or ranch level. Similarly, the farm systems are part of the suprasystem of the agriculture business of the region, state, nation, and, indeed, the world. Hart (1984) has identified five information processes that affect agroecosystems: (1) ecological environment, (2) agricultural resources, (3) household, (4) other agroecosystems, and (5) the state of the agroecosystems. The "ecological environment" consists of some relatively static factors such as soil properties and solar radiation as well as more dynamic and less predictable factors as rainfall and insect pest invasion patterns. In this scheme, "agricultural resources" refers to the direct factors such as land, labor, capital, chemicals, and machine energy. But the term also includes the factors that affect the resource availability, for example, socioeconomic environment, household, and the state of other agroecosystems on the farm. The "household" determines the objectives, for example, minimizing the economic risk, producing a specific product, or selling a product at a particular time. The "other agroecosystem" category includes considerations such as those related to the movement of materials from one field on the farm to another. An example might be the availability of alfalfa hay from one field for feeding livestock on a winter pasture elsewhere on the farm. The "state of the agroecosystem" is a general knowledge of the whole system and is used for making decisions based on comparisons of the real and predicted state of the agroecosystem. Thus, the cumulative experience of the farmer with the performance of the agroecosystem is important.

Figure 12.4 is a conceptual model of the management of such an agroecosystem (Hart, 1984). Here, the socioeconomic objectives are assumed to be an output from the household, and the farmer makes a number of decisions throughout the year. Some of these decisions are made over longer time intervals, such as the type of crop to plant each year, and other decisions are made at short intervals, such as whether to use a rescue insecticide. In managing an agroecosystem, humans must contend with socioeconomic information, but also the same biological processes as in the natural ecosystem. However, even the more comprehensive analyses of agricultural systems have continued to focus on the management objectives and less on the precision with which the physical and biological processes are represented (Loucks, 1977).

Throughout the world, the analysis of agroecosystems has been directed toward the evaluation of productivity from different cropping systems (Potts and Vickerman, 1974; Cox and Atkins, 1979; Edens and Koenig, 1981; Toky and Ramakrishnan, 1981; Edens and Haynes, 1982; Altieri, 1983; Loomis, 1984). In parts of the world, agricultural systems are frequently characterized by a large reliance on human and animal labor, as well as biomass fuels and organic fertilizers. Over the past 10 years (Pimentel et al., 1980; Mitsch et al., 1982; Costanza, 1980; Loomis, 1984), energy analyses have been applied to a large range of agricultural systems to determine the energy inputs and outputs. These energy analytical techniques involve calculating the energy inputs in terms of human and animal labor, raw materials, machinery, fossil fuels, electricity, and chemicals. These inputs are then traced through the system using a diagrammatic model to show the flows of energy (Spedding et al., 1983). In most cases to address the issue of different market values, the nonenergy components are converted to some constant value of energy equivalents (Odum, 1972; DeLeage et

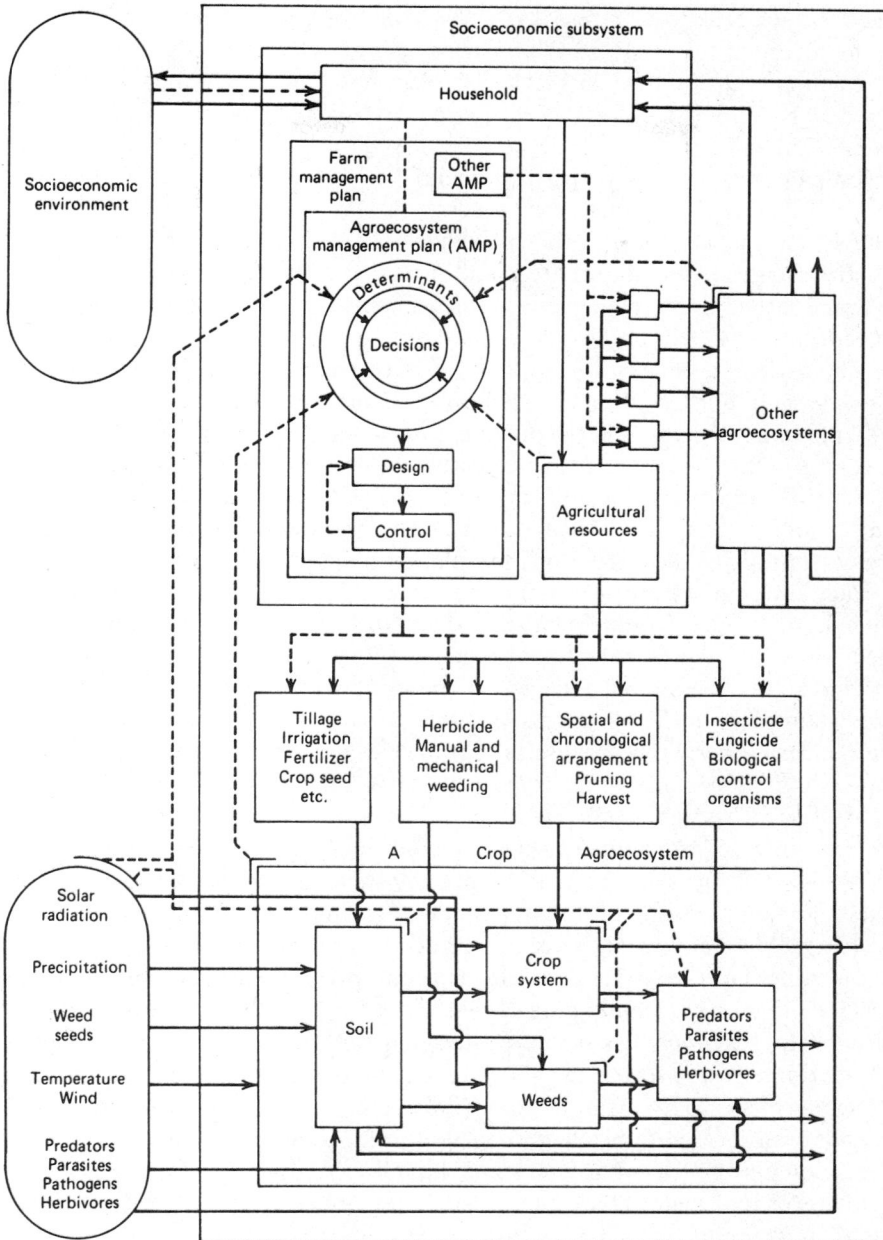

Fig. 12.4 Flow of energy and materials and information in a farm system (Hart, 1984).

329

al., 1979; Zucchetto and Jansson, 1979). Then, the outputs can be compared with the inputs and the relative efficiencies can be calculated (Rappaport, 1971; Uhl and Murphy, 1981). This technique does not account for differences in nutritional requirements, nor does it separate essential energy transfers from such events as luxury consumption. Smil (1981) used this broad analysis technique to investigate the agroecosystems of China and found that the total annual net primary productivity of China's land ecosystems was about 0.13% of solar insolation which is lower than the global average of 0.3% because of the large unproductive mountain and desert areas and extensive deforestation. Energy costs of external inputs into farming were approximately 80% of the energy value produced, but only 20% of the total annual crop produced.

The topic of cycling of minerals in agroecosystems has been exhaustively reviewed and published as a special issue of the journal, *Agro-Ecosystems* (Frissel, 1977a). Some 65 ecosystems from around the world were compared according to a common flow model that depicted the major transfers of nutrients among plant, livestock, and soil compartments (Fig. 12.5). Subsequently, these individual compartments were compared and some general conclusions were drawn from synthesis of the data. For example, on a world scale only 8–14% of the annual additions of nitrogen to vegetation was from fertilizer application, although, of course, there were large local and regional differences. Also, the efficiency with which grass takes up nitrogen fertilizers is about 60%, but the transfer from grass to animals is low, so the overall efficiency between fertilizer application and animal products on a world basis is only about 15%. Conclusions of this general type are useful, but the authors of the symposium plead for more simulation models that contain the necessary detail to make them adequate descriptors and predictors of the behavior of nutrients in agroecosystems.

Nutrient budgets have been calculated for some agricultural systems (Todd et al., 1984), and one such example is a grazing system in Denmark (Bulow-Olsen, 1980). In comparison with natural systems, grazing was shown to increase the exchangeable amounts of calcium, magnesium, manganese, and phosphorous in the upper layers of the soil profile. This interpretation of the system behavior was facilitated by examining the system as a whole. Indeed, such an evaluation of natural ecosystems has demonstrated that the detrital–saprohytic pathways are extremely important in both natural and human-dominated systems (Coleman et al., 1976). From 20 to 60% of the total photosynthate in crop plants may be translocated to below-ground root production and the same or similar allocation may be true of at least some natural grassland systems. Considerable amounts of root material are consumed by underground consumers, so the actual heterotrophic costs may be higher than previously recognized. If so, the photosynthesis/respiration ratios of highly productive crops, such as alfalfa, might be about 1.9 and much closer to the 1.3 ratio of managed natural systems.

In summary, various analytical techniques have been used to evaluate agroecosystems, but much of the effort has been directed toward either farm management systems emphasizing economics, or energy and nutrient budgets for comparative purposes. Despite ecosystem analyses that demonstrate important system-level events, such as the increase of nutrients in the upper soil layers

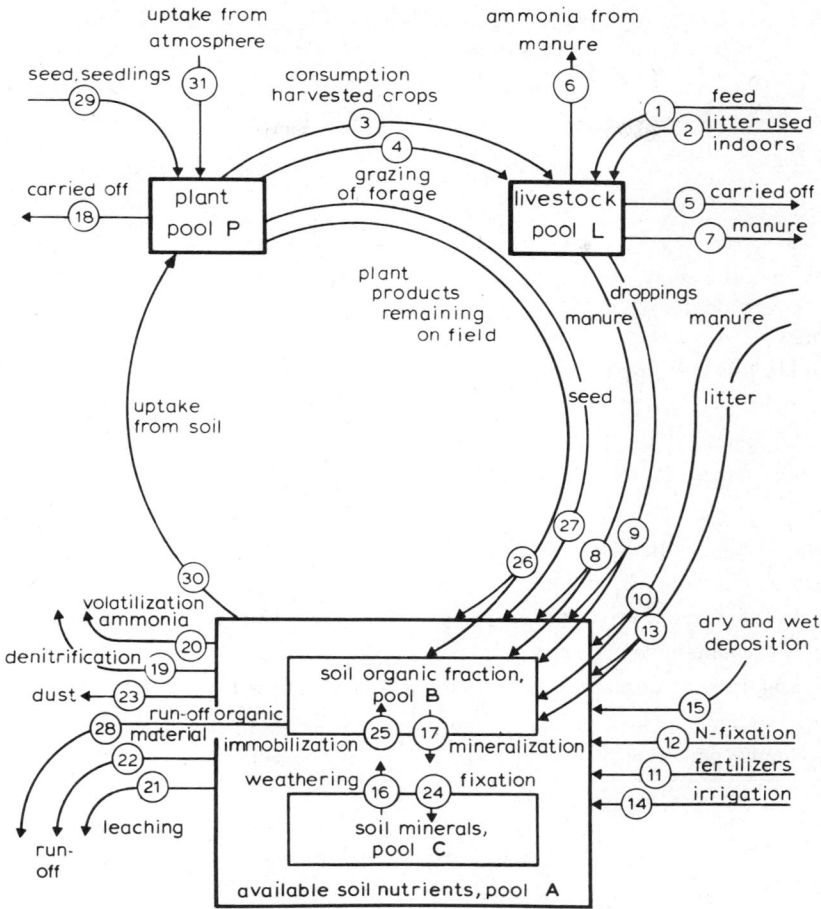

Fig. 12.5 Flow chart of the nutrient transfers for an agroecosystem (Frissel, 1977b).

under grazed conditions or the unexpectedly large magnitude of belowground herbivory, agroecosystem models have remained at a general level and, as such, have not incorporated significant attention to these important ecological processes. This omission is surprising since the analytical techniques applied to natural ecosystems have demonstrated the potential significance of explicitly including these or other ecological mechanisms related to energy efficiency and nutrient cycling.

## 12.4. COMPARATIVE ANALYSIS OF NATURAL ECOSYSTEMS AND AGROECOSYSTEMS

Woodmansee (1984) has contrasted the characteristics of natural and cultivated ecosystems (Table 12.1). Some of the contrasts are obvious from rather casual

**Table 12.1 Characteristics of Natural and Cultivated Ecosystems that Directly Influence Their Propensity to Accumulate Nutrients and Persist through Time[a]**

|                                              | Natural | Cultivated |
|----------------------------------------------|---------|------------|
| **Abiotic**                                  |         |            |
| Infiltration rates                           | High    | Low        |
| Runoff                                       | Low     | High       |
| Erosion                                      | Low     | High       |
|   Presence of canopy                         | High    | Low        |
|   Litter and debris                          | High    | Low        |
|   Rocks                                       | High    | Low        |
| Soil water loss to transpiration             | High    | Low        |
| Soil colloids                                | High    | Low        |
| Leaching losses                              | Low     | High       |
| Soil temperature                             | Low     | High       |
| **Biotic**                                   |         |            |
| Internal cycling by plants                   | High    | Low        |
| Synchrony of plant–microorganism activity    | High    | Low        |
| Temporal diversity of organism activity      | High    | Low        |
| Balance of plant–microorganism activity      | 1       | <1         |
| Structural diversity of plants               | High    | Low        |
| Genetic diversity                            | High    | Low        |
| Reproductive potential                       | High    | Low        |

[a]Data from Woodmansee (1984).

inspection, for example, erosion potential, genetic diversity, and structural diversity of the plant canopy. However, the more interesting comparisons are those of ecosystem characteristics that could not have been discovered without the analysis of entire ecosystems. Synchrony of plant–microorganism activity became known only as the result of studying nutrient cycles of ecosystems and a careful analysis of how the nutrient-conserving mechanisms operated in these systems. In natural ecosystems, most plant growth, microbial decomposition, and mineralization processes occur during the growing season. Thus, when nutrients are freed by the decomposition process, actively growing plants absorb nutrients from the labile pool. In agricultural systems of single crops, organic matter breakdown occurs primarily during the growing season but also throughout the year when conditions are favorable. When the growing crop is not present, released nutrients from microbial decomposition of organic matter may be more vulnerable to loss by leaching. Also, many agricultural systems are designed to remove portions of the plant canopy, a part of the system that not only contains nutrients but also contributes organic matter to the soil. This soil organic material, in turn, contributes to the nutrient retention capability of the ecosystem. Continuous removal of plant canopy thus removes a portion of the source of soil organic material. As a result, the fundamental ecosystem processes operate, but because of human intervention into the phenology of the system, agroecosystems do not retain all the natural feedback loops and are thus more likely to permit the escape of nutrients.

The management techniques used on a grassland or rangeland ecosystem usually do not change the fundamental nature of the system. That is, excessive grazing may cause a change in the species composition of the vegetation, or may contribute to increased soil erosion which then leads to decreased primary productivity and diminished resistance to climatic conditions such as drought. However, the natural feedback processes predominate. In agroecosystems with more intensive management, nutrition for the primary consumers (livestock) may be supplemented, certain land areas may be converted to monospecific pastures of introduced species, and the system may be subsidized with water or nutrients. Here, the natural feedback processes are reduced further. With intensive agriculture, some of the general feedback mechanisms are controlled by human intervention. Thus, the separation between a natural ecosystem and an agroecosystem is not distinct, but is frequently a gradation even on a single farm (Spedding, 1975). This blurring of the distinctions between agroecosystems and natural ecosystems contributed to the dearth of early comparisons of natural and agricultural ecosystems.

In one such comparison, Ovington and Lawrence (1967) compared the amounts of energy and chlorophyll during 1959 in a Minnesota prairie, savanna, oak wood, and maize field. At that time, the conclusion regarded as important was the realization that only a very small amount of energy was captured by these ecosystems in comparison to the total available from the sun. This small portion was true even for the natural ecosystems which had a longer growing season than the maize crop. Field methodology was not advanced so there was a concerted attempt to find convenient field measurements of energy transfers. On the basis of comparisons between natural and crop systems, the authors concluded that chlorophyll content was not a useful indicator of energy content of either type of ecosystem.

Other early ecosystem analyses built on the theoretical work of Lindeman (1942) and developed the analytical framework for considering the energetics of quasi-natural ecosystems. Golley (1965) developed a compartment model of the energetics of a successional old-field, tracing the energy inputs and outputs through the trophic structure. Although subsequent studies have contributed to our understanding of food web energetics (Jager et al., 1984), this was one of the earliest attempts to combine the energetics of the producers with those of at least the "conspicuous" consumers. Again, the system was found to be quite inefficient relative to the total incoming radiation, but more energy was captured by the system than was lost. This concept of energy accretion is part of the existing but still developing theory concerning successional ecosystems and contributed to concepts about developing natural ecosystems (Odum, 1969).

Several studies have been designed to compare the production efficiencies between natural ecosystems and agroecosystems. As an example, in the Amazon Basin, Uhl and Murphy (1981) used energy analysis to compare the productivity of a slash-and-burn agricultural system with successional vegetation. During the first year after burning, the agricultural field was more productive than the adjacent successional field; however, by the second year the successional vegetation was more than twice as productive as the field. In Oklahoma, Klopatek and Risser (1982) used energy analysis to compare the productivity efficiencies of rangeland ecosystems with improved pasture ecosystems where the latter consisted of introduced vegetation as well as irrigation and fertilizer subsidies. Energy requirements to maintain improved pasture systems ranged from 10 to 100 times that to maintain native range ecosystems. The model used indicated that fuel-based subsidies enhanced the grazing system's capacity to make use of solar energy, but there was an upper limit to the ratio of beef that can be obtained for fuel-based inputs. In this study, the maximum possible efficiency of beef production to fossil field subsidies was 14%. Loomis (1984) discussed the energetics of agroecosystems and provided some comparisons on a regional basis, but the important point is that these models, which involve some deficiencies noted earlier, provide general comparisons on system behavior but do not include detailed evaluation of the ecological processes.

There is currently considerable interest in the short- and long-term implications of various tillage practices, especially types of minimal tillage for conventionally heavily tilled land under intensive agricultural management conditions (Hadas et al., 1980). In most, but not all, situations there may be greater pesticide requirements with certain minimum tillage methods (House and Stinner, 1983). Clearly, however, these minimum cultivation practices reduce soil erosion and require fewer tractor trips across the field (Mueller et al., 1981). In Georgia, Stinner et al. (1984) collected data on nutrient budgets for conventional-tillage, no-tillage, and old-field systems. Leaching losses of nitrogen and calcium were greatest in conventional tillage, followed by no-tillage, and least in the old field. Decomposition of litter and mineralization was more rapid in conventional tillage than in either the no-tillage or old-field ecosystems. These studies of conservation tillage practices demonstrate the need for analytical techniques that comprehend the natural system properties in such a manner that the consequences of human intervention can be quantified on both the long- and short-term basis.

Existing ecological concepts would predict that soil invertebrate species density and diversity would be greatest under untilled conditions, followed in decreasing order by minimum tillage and coventional tillage. However, a 1-year study in Georgia (Blumberg and Crossley, 1983) showed that soil surface arthropod density was higher in the no-till than either the old-field or conventional tillage. Furthermore, the arthropod assemblage in the no-till had a greater proportion of predators and parasites. Since the sorghum yields were not different under the two tillage practices, the authors concluded that insecticide application might actually disrupt the normal diversity of the arthropod complex, harming the predators and parasites, and ultimately leading to a less stable system. Just how the invertebrate faunas react to different tillage practices and how faunas in agricultural ecosystems compare with those of natural ecosystems is not yet clear (Stinner et al., 1984).

The previous paragraphs have described analysis and modeling techniques at the level of the field or farm, but such individually managed areas are a functional part of the landscape. Different nutrient retention characteristics of natural ecosystems and agroecosystems must ultimately be considered at the landscape level of interaction (Lowrance, et al., 1984). That is, the spatial configuration of vegetation types along a catena may affect the nutrients and organic matter that are carried from the land into the receiving stream; and these processes, though at a larger spatial scale than normally considered, are important. Peterjohn and Correll (1984) examined the nutrient dynamics in an agricultural watershed near Chesapeake Bay. By examining inputs and outputs, it was possible to show that an approximately 50-m-wide riparian forest trapped large amounts of particulates, organic and nitrate nitrogen, ammonium, and particulate phosphorous. Of the total nitrogen removed from the cropland, 64% was in the harvested crop, 9.2% in surface runoff, and 26% in groundwater flow. Groundwater appeared to be the major flux between the cropland and the riparian forest. On the other hand, phosphorous losses were 84% in the harvested crop and 16% in the surface runoff. By knowing routes taken by various nutrients and by understanding how each vegetation type captures and releases nutrients, it should be possible to design agricultural landscapes to maximize the retention of soil and nutrients. Indeed, the analysis of agroecosystems cannot ignore the landscape level interactions with respect to the distribution and redistribution of energy and materials (Risser et al., 1984).

## 12.5. FUTURE EMPHASIS IN STRUCTURE, ANALYSIS, AND MODELING

The previous sections of this paper have described the ways in which natural ecosystems and agroecosystems have been analyzed, and examples have been provided of analytical comparisons between the two types of systems. It is now appropriate to anticipate the future of analytical methods for agroecosystems. Several topics in agricultural systems will need to be analyzed in future models of agroecosystems (Wittwer, 1980; Loucks, 1984): soil erosion and loss of organic matter, chemical residues in soil and soil water, air pollution, climate variations, agricultural technology, pest control strategies, genetic engineering, and economic conditions. Ecosystem analysis has demonstrated the power of considering ecosystems as systems, that is, the interactions among the various components. Once these systems are understood, then it will be possible to design ecosystems to maximize certain characteristics such as productivity and resistance to pests or weather, or to minimize the necessary energy and chemical subsidies.

The spatial attributes of agroecosystems will become increasingly a part of analytical procedures. An example can be described using a modular system that has been proposed for the rural lowland tropical region of southwestern Mexico (Gliessman et al. 1981). The farming unit is constructed to include basic components to recycle energy and nutrients. In the center or elsewhere, depending upon topographic conditions, a tank is located to catch dissolved nutrients and particles of soil and organic matter. Fish, ducks, and other aquatic organisms are produced with aquatic plants, and enriched sediments are used for fertilizer on other parts of the module. Adjacent to the tank are raised plots or platforms used for intensive vegetable production. The soils of these chinampas are enriched from the tank, and excess material from the chinampas is fed to corralled animals nearby. The surrounding croplands are planted to the regionally acceptable crop combinations. Finally, the outer belt of forest serves as a windbreak, a source of natural predators and parasites for biological control, a source of firewood and building materials, and as biological reserves for the diversity of plants and animals normally present in these tropical ecosystems. This system is designed to internalize the nutrient cycles, minimize the introduction of chemicals and external fuels, and maintain the biological diversity of natural ecosystems—all active feedback components of natural ecosystems.

Most entomological models have been focused at the population level and there are a number of detailed pest management models (Conway, 1984). However, because many species of insects require resources from two or more landscape units to complete their life cycle, analysis of agroecosystems may eventually incorporate the spatial pattern of host and pest populations as well as the habitat characteristics required for both types of population. Mayse and Price (1978) studied the spatial distribution of arthropod species in three central Illinois soybean fields. The number of species found at the edge of the fields was greater than at the center of the field. These differences, however, were greater for herbivore arthropod species than for predator/parasitoid species. The authors also measured "habitat space" as an estimate of the actual leaf area of the soybean plant. A relatively constant number of arthropod species per habitat space existed throughout most of the season, but the number of species per habitat space was greater at the edge of the field as compared with the center for both herbivores and predators/parasitoids. By modifying the habitat space, and perhaps by adding plant species that would provide pollen and nectar sources for predators and parasitoids, significant effects could be made on the arthropod populations.

Currently, there are a few simulation models that have been constructed to consider such spatial and temporal dynamics (Stinner et al., 1974), but because of the lack of experience and refinement, the use of these models in managing pests and their potential predators is infrequent (Stenseth and Hansson, 1981; Risser et al., 1984). By taking account of the spatial patterns of habitats and habitat requirements of insects, and then developing models to consider the trade-offs in terms of production and harvest costs, eventually it should be possible to design agricultural fields and borders to enhance the biological control of pest insects (Price, 1976; Stinner et al., 1977; Altieri and Letourneau 1982; Altieri, 1983, 1984a, 1984b; Barney et al., 1984; Stephens, 1984). Many potential management practices are known (Altieri and Letourneau, 1982), but the systems have not been subjected to sufficient ecological and economic analyses, nor is there enough practical experience to encourage widespread adoption of these techniques.

The issue of species diversity and ecosystems stability has been a discussion topic for 30 or more years (Zaret, 1982; Kimmerer, 1984). Ecologists have intuitively believed that more diverse ecosystems, with more pathways for the movement of energy and materials, were more stable because of a greater number of feedback mechanisms. In attempting to understand these concepts theoretically, Holling (1985) noted the importance of recognizing and analyzing certain specific ecosystem characteristics, namely, that ecosystems may not be restricted to just one stage of stability, but rather can reach more than one position or domain of stability (Holling, 1973); ecosystems involve coupling biotic and abiotic subsystems that have cycles with different temporal lengths (Levin, 1978); and ecosystem hierarchical organization relates to resilience and stability of that system (Allen and Starr, 1982). Thus, the suggestion that diversity generally led to decreased stability in randomly connected networks, was confusing (May, 1974). But in Holling's (1985) valuable discussion, he notes that this simply means that ecosystems are not randomly connected, and, furthermore, the meanings of stability and resilience have not always been well or consistently distinguished. The key characteristic is connectedness; that is, the higher the connectedness, the more complexity and the lower the probability of stability. Succession in natural ecosystems leads to connectedness which leads to increased likelihood of instability. This instability leads to a discontinuous change where the interconnectedness is then reduced, which eventually leads to reorganization and renewal. This destabilizing effect caused by connectedness leads to variability which leads to resilience. In other words, long-term stability of a system is enhanced by short-term disturbances. These disturbances permit the simultaneous occurrence of various stages of ecosystem development. Obviously, high-intensity agriculture, with all parts of the system in essentially the same stage, does not maintain a matrix of connectedness with these short-term disturbances.

Systems that develop under a uniform pattern of external control demonstrate high stability, but low resilience (e.g., the constant tropics.) Temperate systems that develop under high climatic variability show low stability and high resilience. Thus, the latter are more likely to withstand human disturbance, but are less stable than the susceptible tropical systems. In a practical way, this theory indicates that in the future, analysis of agroecosystems must deal specifically with connectedness—not the superficial surrogate of species diversity. Future analyses must identify the important connecting pathways, whether these pathways involve, for example, predators of pest insects or nutrient retention capabilities of root systems.

Most agroecosystem models are designed to provide the farmer with a prediction, or at least a description, of the likely consequences of certain management or marketing decisions. Simply providing a decision framework is valuable because it provides an organization for even intuitive decisions. However, future models will need to incorporate several new characteristics. First, many decisions require a continuing stream of information that must be up-to-date if the best decision is to be made. For example, whether to apply a rescue insecticide depends upon knowing the intensity of the pest problem and the presence of parasites and predators, the weather, and the relative costs of various options. The best decision depends upon rapidly changing events. Therefore, pest simulation models need to be coupled with plant production models and run with current economic and weather information. Second, future analyses must address the vulnerability of agroecosystems to the interactions of climate, pests, and economic forces which shape the demand for food and the utilization patterns of different types of agricultural land (Clark and Holling, 1985). Meaningful scenarios need to be evaluated that anticipate how, for example, drought will affect food supplies in various parts of the world. Then, a risk assessment of agroecosystems can be designed to minimize the risk, or at least quantify the expected risks. Third, present farm management models do not account for the ecological processes known to exist in agroecosystems, for example, nutrient retention characteristics of the soil and the landscape pattern of land use. Given our experience with models of natural ecosystems, it is clear that not all ecological processes need to be included. However, there must be a systematic evaluation of which processes are significant under the conditions in question, and these processes need to be incorporated into the models. Fourth, future agroecosystem models that display economic trade-offs must also account for the long-term ecological costs and benefits. The costs of certain conditions must be made explicit and entered into the model, for example, increased or continued soil erosion, loss of soil organic matter, effects of air pollutants, accumulation of chemical residues in the soil, and increased resistance to pesticides. Similarly, ecological benefits must be included in the analyses, such as the retention of nutrients in the field or watershed. Only when the ecologic and economical factors are considered simultaneously will the true costs and benefits be known for use in the management and design of agroecosystems.

This chapter is intended to prompt a convergence between what is known about the analytical techniques of natural ecosystems and agroecosystems. The fundamental processes in these two ecological systems are the same, but in agroecosystems, some of these processes are altered by human intervention. As was described, analysis of natural ecosystems had led to a general understanding of ecosystem behavior. Current emphasis is on understanding the mechanisms of the processes that control ecosystem dynamics. Analysis of agroecosystems must incorporate this information as needed. But just as importantly, the development of analytical techniques for agroecosystems will be successful only by focusing on the future and designing models to anticipate the needs of the agriculturist at local, regional, and global scales. The concluding section of the chapter has sought to identify the demands that will be made on the techniques for analyzing and modeling agroecosystems.

## ACKNOWLEDGMENTS

The author gratefully recognizes the reviews provided by Marcos Kogan, David Pimentel, Bill Ruesink, and Daniel Simberloff.

## REFERENCES

Allen, T. F. H., and T. B. Starr. (1982). *Hierarchy: Perspectives for Ecological Complexity*, University of Chicago Press. Chicago.

Altieri, M. A. (1983). Agroecology. *The Scientific Basis of Alternative Agriculture. Division of Biological Control*, University of California, Berkeley.

Altieri, M. A. (1984a). Patterns of insect diversity in monocultures and polycultures of brussels sprouts, *Prot. Ecol.*, **6**, 227-232.

Altieri, M. A. (1984b). Pest management technologies for pheasants: a farming system approach, *Crop. Prot.*, **3**, 87-94.

Altieri, M. A., and D. K. Letourneau. (1982). Vegetation management and biological control in agroecosystems, *Crop Prot.*, **1**, 405-430.

Barney, R. J., W. O. Lamp, E. J. Armburst, and G. Kapusta. (1984). Insect predator community and its response to weed management in spring-planted alfalfa, *Prot. Ecol.*, **6**, 25-3.

Bormann, F. H., and G. E. Likens. (1981). *Pattern and Process in a Forest Ecosystem*, Springer-Verlag, New York, 253 pp.

Blumberg, A. Y., and D. A. Crossley, Jr. (1983). Comparison of soil surface arthropod populations in conventional tillage, no-tillage, and old-field systems, *Agro-ecosystems*, **8**, 247-253.

Brockington, N. R. (1979). *Computer Modeling in Agriculture*, Oxford University Press, Oxford, 156 pp.

Bulow-Olsen, A. (1980). Nutrient cycling in grassland dominated by *Deschampsia flexuosa* (L.) Trin. and grazed by nursing cows, *Agro-ecosystems*, **6**, 209-220.

Burgess, R. L. Ed. (1984). *Effects of Acidic Deposition on Forest Ecosystems in the Northeastern United States: An Evaluation of Current Evidence*, ESF 84-016, College of Environmental Science and Forestry, Institute of Environmental Program Affairs, State University of New York, Syracuse, 108 pp.

Clark, W. C., and C. S. Holling. (1985). "Sustainable development of the biosphere: Human activities and global change," in T. F. Malone and J. G. Roederer, Eds., *Global Change*, Cambridge University Press, London, England, pp. 474-490.

Coleman, D. C., R. Andrews, J. E. Ellis, and J. S. Singh. (1976). Energy flow and partitioning in selected man-managed and natural ecosystems, *Agro-ecosystems*, **3**, 45-54.

Coleman, D. C., C. V. Cole, and E. T. Elliott. (1984). "Decomposition, organic matter turnover, and nutrient dynamics in agroecosystems," in R. Lowrance, B. R. Stinner, and G. J. House, Eds., *Agricultural Ecosystems. Unifying Concepts*, Wiley-Interscience, New York, pp. 83-104.

Conway, G. R., Ed. (1984). *Pest and Pathogen Control, Strategic, Tactical and Policy Models. International Series on Applied System Analysis*, Vol. 13, Wiley-Interscience, New York, 488 pp.

Costanza, R. (1980). Embodied energy and economic evaluation, *Science* **210**, 1219-1224.

Cox, G. W., and M. D. Atkins. (1979). *Agricultural Ecology*, Freeman, New York.

Davidson, D. W., R. S. Inouye, and J. H. Brown. (1984). Granivory in a desert ecosystem: experimental evidence for indirect facilitation of ants by rodents, *Ecology*, **654**, 1780-1786.

Dayan, E., H. Van Keulen, and A. Dovrat. (1981). Experimental evaluation of a crop growth simulation model. A case study with Rhodes grass, *Agro-ecosystems*, **7**, 113–126.

DeLeage, J. P., J. M. Julien, N. Sauget-Naudin, and C. Souchon. (1979). Eco-energetics analysis of an agricultural system: The French case in 1970, *Agro-ecosystems*, **5**, 345–365.

Detling, J. K., W. J. Parton, and H. W. Hunt. (1979). A simulation model of *Bouteloua gracilis* dynamics on the North American shortgrass praire *Oecologia*, **38**, 167–191.

Edens, T. C., and D. L. Haynes. (1982). Closed system agriculture: Resource constraints, management options, and design alternatives, *Annu. Rev. Phytopathol.*, **20**, 363–395.

Edens, T. C., and H. E. Koenig. (1981). Agroecosystem management in a resource-limited world, *BioScience*, **30**, 697–701.

Edwards, N. T., H. H. Shugart, Jr., S. B. McLaughlin, W. F. Harris, and D. E. Reichle. (1980). "Carbon metabolism in terrestrial ecosystems," in D. E. Reichle, Ed., *Dynamic Properties of Forest Ecosystems. International Biological Programme 23*, Cambridge University Press, England, pp. 499–536.

Fager, E. W. (1968). The community of invertebrates in decaying wood, *J. Anim. Ecol.*, **37**, 121–142.

French, N. R. (1979). "Principal subsystem interactions in grasslands," in N. R. French, Ed., *Perspectives in Grassland Ecology*, Springer-Verlag, New York, pp. 173–190.

Frissel, J. J., Ed. (1977a). Cycling of mineral nutrients in agricultural ecosystems, *Agro-ecosystems*, **4**, 1–354.

Frissel, J. J. (1977b). Method of data presentation, *Agro-ecosystems*, **4**, 27–32.

Gliessman, S. R., R. Garcia, and M. Amador. (1981). The ecological basis for the application of traditional agricultural technology in the management of tropical agro-ecosystems, *Agro-ecosystems*, **7**, 173–185.

Golitsyn, G. S. (1985). "The changing atmosphere," in T. F. Malone and J. G. Roederer, Eds., *Global Change*, Cambridge University Press, London, England, pp. 114–126.

Golley, F. B. (1965). Structure and function of an old field broomsedge community, *Ecolog. Monogr.*, **35**, 113–137.

Hadas, A., D. Wolf, and E. Stibbe. (1980). Tillage practices and crop response–Analyses of agroecosystems, *Agro-ecosystems*, **6**, 235–248.

Hart, R. D. (1984). "Agroecosystem determinants," in R. Lowrance, B. R. Stinner, and G. J. House, Eds., *Agricultural Ecosystems. Unifying Concepts*, Wiley-Interscience, New York, pp. 105–119.

Holling, C. S. (1973). Resilience and stability of ecological systems, *Annu. Rev. Ecol. Syst.*, **4**, 1–23.

Holling, C. S. (1985). "Resilience of ecosystems: Local surprise and global change," in T. F. Malone and J. G. Roederer, Eds., *Global Change*, Cambridge University Press, London, England, 512 pp.

House, G. J., and B. R. Stinner. (1983). Arthropods in no-tillage soybean agroecosystems: Community composition and ecosystem interactions, *Environ. Manag.*, **7**, 23–28.

Innis, G. S., Ed. (1978). *Grassland Simulation Model. Ecol. Studies*, Vol. 26, Springer-Verlag, New York, 298 pp.

Jager, H. I., R. H. Gardner, D. L. DeAngelils, and W. M. Post. (1984). *A Simulation Approach to Understanding the Processes That Structure Food Webs*, ORNL/TM-8904, Oak Ridge National Laboratory, Environmental Sciences Division Publication No. 2227, Oak Ridge, TN, 174 pp.

Karlovsky, J. (1981). Cycling of nutrients and their utilization by plants in agricultural ecosystems, *Agro-ecosystems*, **7**, 127–144.

Karlson, R. H., and L. W. Buss. (1984). Competition, disturbance and local diversity patterns of substratum-bound clonal organisms: A simulation, *Ecol. Model.*, **23**, 243–255.

Kimmerer, W. J. (1984). Diversity/stability: A criticism, *Ecology*, **65**, 1936–1938.

Klopatek, J. M., and P. G. Risser. (1982). Energy analysis of Oklahoma rangelands and improved pastures, *J. Range Manag.*, **35**, 637–643.

Lemon, E., D. W. Stewart, and R. W. Shawcroft. (1971). The sun's work in a cornfield, *Science*, **174**, 371–378.

Levin, S. A. (1978). "Pattern formation in ecological communities," in J. H. Steele, Ed., *Spatial Pattern in Plankton Communities*, Plenum, New York, pp. 433–465.

Lindeman, R. L. (1942). The trophic-dynamic aspect of ecology, *Ecology*, **23**, 399–418.

Loomis, R. S. (1983). "Productivity of agricultural systems," in O. L. Lange, P. S. Nobel, C. B. Osmond, and H. Ziegler, Eds., *Physiological Plant Ecology IV, Ecosystem Processes, Encyclopeadia of Plant Physiology*, N. S., Springer-Verlag, Berlin/Heidelberg, pp. 151–172.

Loomis, R. S. (1984). Traditional agriculture in America, *Annu. Rev. Ecol. Syst.*, **15**, 449–478.

Loucks, O. L. (1977). Emergence of research on agro-ecosystems, *Annu. Rev. Ecol. Syst.*, **8**, 173–192.

Loucks, O. L. (1984). Role of basic ecological knowledge in the mitigation of impacts from complex technological systems: Agriculture, transportation, and urban, Conference on Long-Term Environmental Research and Development, October 9–10, 1984, Council on Environmental Quality, National Science Foundation, Washington, D.C., 28 pp.

Lowrance, R. R., R. L. Todd, and L. E. Asmussen. (1984). Nutrient cycling in an agricultural watershed: I. Phreatic movement, *J. Environ. Quality*, **13**, 22–27.

May, R. M. (1974). *Stability and Complexity in Model Ecosystems*, Princeton University Press, Princeton, NJ, 265 pp.

Mayse, M. A., and P. W. Price. (1978). Seasonal development of soybean arthropod communities in east central Illinois, *Agro-ecosystems*, **4**, 387–405.

Mitsch, W. J., R. K. Ragade, R. W. Bosserman, and J. A. Dillon, Jr. (1982). *Energetics and Systems*, Ann Arbor Sciences, Ann Arbor, Michigan, 132 pp.

Mueller, D. H., T. D. Daniel, and R. C. Wendt. (1981). Conservation tillage: Best management practice for nonprofit runoff, *Environ. Manag.*, **5**, 35–53.

Murphy, P. G. (1977). Rates of primary productivity in tropical grassland, savanna and forest, *Geo-Eco-Trop.* **2**, 95–102.

Odum, E. P. (1969). The strategy of ecosystem development, *Science*, **164**, 262–270.

Odum, H. T. (1972). "An energy currency language for ecological and social systems. Its physical basis," in B. C. Patten, Ed., *Systems Analysis and Simulation Ecology*, Vol. 2, Academic, New York, pp. 129–211.

O'Neill, R. V., and D. L. DeAngelis. (1980). "Comparative productivity and biomass relations of forest ecosystems," in D. E. Reichle, Ed., *Dynamic Properties of Forest Ecosystems, International Biological Programme 23*, Cambridge University Press, England, pp. 411–449.

Ovington, J. D., and D. B. Lawrence. (1967). Comparative chlorophyll and energy studies of prairie, savanna, oakwood, and maize field ecosystems, *Ecology*, **48**, 515–524.

Peterjohn, W. T., and D. L. Correll. (1984). Nutrient dynamics in an agricultural watershed: observations on the role of a riparian forest, *Ecology*, **65**, 1466–1475.

Pimentel, D., L. E. Hurd, A. C. Bellotti, J. J. Forster, I. N. Oka, O. D. Sholes, and R. J. Williams. (1973). Food production and the energy crises, *Science*, **182**, 443–449.

Pimentel, D., P. A. Oltenacu, M. C. Nesheim, J. Krummel, M. S. Allen, and S. Chick. (1980). The potential for grass-fed livestock: resource constraints, *Science*, **207**, 843–848.

Potts, G. R., and G. P. Vickerman. (1974). Studies on the ceral ecosystem, *Adv. Ecol. Res.*, **8**, 107–147.

Price, P. W. (1976). Colonization of crops by arthropods: Non-equilibrium communities in soybean fields, *Environ. Entomol.*, **5**, 605–611.

Rappaport, R. A. (1971). The flow of energy in an agricultural society, *Sci. Am.*, **225**, 116–132.

Risser, P. G., and W. J. Parton. (1982). Ecosystem analysis of the tall-grass prairie: Nitrogen cycle, *Ecology*, **63**, 1342–1351.

Risser, P. G., E. C. Birney, H. D. Blocker, S. W. May, W. J. Parton, and J. A. Wiens. (1981). *The True Prairie Ecosystem*, Hutchinson and Ross Publ. Co., Stroudsburg, PA, 557 pp.

Risser, P. G., J. R. Karr, and R. T. T. Forman. (1984). *Landscape Ecology. Directions and Approaches*, Illinois Natural History Survey Special Publication Number 2, Champaign, IL, 18 pp.

Seastedt, T. R., and D. A. Crossley, Jr. (1984). The influence of arthropods on ecosystems, *BioScience*, **34**, 157–161.

Simberloff, D. (1981). "Community effects of introduced species," in M. Nitecki, Ed., *Biotic Crises in Ecological and Evolutionary Times*, Academic, New York, pp. 53–81.

Smil, V. (1981). China's agro-ecosystem, *Agro-ecosystems*, **7**, 27–46.

Spedding, C. R. W. (1975). "The study of agricultural systems," in G. E. Dalton, Ed., *Study of Agricultural Systems*, Applied Science, London, pp. 3–19.

Spedding, C. R. W., A. M. M. Thompson, and M. R. Jones. (1983). Energy and economics of intensive animal production, *Agro-ecosystems*, **8**, 169–181.

Stenseth, N. C., and L. Hansson. (1981). The importance of population dynamics in heterogeneous landscapes: Management of vertebrate pests and some other animals, *Agro-ecosystems*, **7**, 187–211.

Stephens, C. S. (1984). Ecological upset and recuperation of natural control of insect pests in some Costa Rican banana plantations, *Turrialba*, **34**, 101–105.

Stinner, B. R., D. A. Crossley, Jr., E. P. Odum, and R. L. Todd. (1984). Nutrient budgets and internal cycling of N, P, K, Ca, and Mg in conventional tillage, no tillage, and old-field ecosystems on the Georgia Piedmont, *Ecology*, **65**, 354–369.

Stinner, R. E., R. L. Rabb, and J. R. Bradley, Jr. (1974). Population dynamics of *Heliothis zea* F. in North Carolina: A simulation model, *Environ. Entomol.*, **3**, 163–168.

Stinner, R. E., R. L. Rabb, and J. R. Bradley, Jr. (1977). "Natural factors operating in the population dynamics of *Heliothis zea* in North Carolina," in *Proceedings of the XV International Congress on Entomology*, August 19–27, 1976, Washington, D.C., pp. 622–642.

Todd, R. L., R. Leonard, and L. Asmussed, Eds. (1984). *Nutrient Cycling in Agricultural Ecosystems*, Ann Arbor Scientific Publications, Ann Arbor, MI, 602 pp.

Toky, O. P., and P. S. Ramakrishman. (1981). Cropping and yields in agricultural systems of the northeastern hill region of India, *Agro-ecosystems*, **7**, 11–25.

Uhl, C., and P. Murphy. (1981). A comparison of productivities and energy values between slash and burn agriculture and secondary succession in the upper Rio Negro region of the Amazon Basin, *Agro-ecosystems*, **7**, 63–83.

Van Dyne, G. M., and Z. Abramsky. (1975). "Agricultural systems models and modeling: An overview," in G. E. Dalton, Ed., *Study of Agricultural Systems*, Applied Science, London, pp. 23–106.

Van Hook, R. I., M. G. Nielsen, and H. S. Shugart. (1980). Energy and nitrogen relations for a *Macrosiphum iriodendri* (Homoptera:Aphididae) population in an east Tennessee *Liriodendron tulipifera* stand, *Ecology*, **61**, 960–975.

Wight, J. R., C. L. Hanon, and D. Whitmer. (1984). Using weather records with a forage production model to forecast range forage production, *J. Range Manag.*, **37**, 3–6.

Wittwer, S. H. (1980). "Future trends in agriculture technology and management," in *Long-Range Environmental Outlook*, Proceedings of a Workshop, November 14–16, 1979, National Academy of Sciences, Washington, D.C., pp. 64–107.

Woodmansee, R. G. (1984). "Comparative nutrient cycles of natural and agricultural ecosystems: A step toward principles," in R. Lowrance, B. R. Stinner, and G. J. House, Eds., *Agricultural Ecosystems. Unifying Concepts*, Wiley-Interscience, New York, pp. 145-156.

Zaret, T. M. (1982). The stability/diversity controversy: A test of hypotheses, *Ecology*, **63**, 721-731.

Zucchetto, J., and A. Jansson. (1979). Total energy analysis of Gotland's agriculture: a northern temperate zone case study, *Agro-ecosystems*, **5**, 329-344.

# INDEX

**345**